Water Wells

Monitoring, Maintenance, Rehabilitation

Other books available from E. & F.N. Spon

Microbiology in Civil Engineering
Edited by P. Howsam

Organized by the Federation of European Microbiological Societies, this Symposium highlights those areas of civil engineering where microbiological activity can have a significant impact during design, construction and operation phases of projects. The book is in eight parts: introduction; overviews of the main microbiological processes relevant to civil engineering; water supply engineering; engineering materials; groundwater engineering; land drainage and reclamation and waste disposal; geotechnical engineering, diagnosis, monitoring and control.

Hardback, ISBN 0 419 16730 7, 400 pages

Geomembranes – Identification and Performance Testing
Edited by A.L. Rollin and J.M. Rigo

Geomembranes are increasingly being used in transportation, environmental and geotechnical applications to control gas and liquid movements. This book provides authoritative guidance on testing of geomembranes. It has been prepared by an international committee of experts under the auspices of RILEM, the International Union of Testing and Research Laboratories for Materials and Structures.

Hardback, ISBN 0 412 38530 9, 368 pages

Vegetable Plants and their Fibres as Building Materials
Edited by H.S. Sobral

This book examines the state-of-the-art on plants and fibres as building materials, emphasizing their use, properties, fabrication, new procedures and future developments. It makes available research results on new techniques for fibre reinforcement of concrete, stabilized clay and other matrices and analyses procedures for making fibre-based and wood-based building materials for use in developing countries.

Hardback, ISBN 0 412 39250 X, 400 pages

Publisher's note
This book has been produced from camera ready copy provided by the individual contributors. This method of production has allowed us to supply finished copies to the delegates at the Conference.

Water Wells

Monitoring, Maintenance, Rehabilitation

Proceedings of the International
Groundwater Engineering Conference held at
Cranfield Institute of Technology, UK.

6–8 September, 1990

Edited by

P. HOWSAM
Silsoe College
Cranfield Institute of Technology
UK

Routledge
Taylor & Francis Group
LONDON AND NEW YORK

First published 1990 by Taylor & Francis

2 Park Square, Milton Park, Abingdon, Oxfordshire OX14 4RN
52 Vanderbilt Avenue, New York, NY 10017

Routledge is an imprint of the Taylor & Francis Group, an informa business

First issued in paperback 2019

ISBN 978-0-419-16750-1 (hbk)
ISBN 978-0-367-86350-0 (pbk)
ISBN 978-0-442-31286-2 (USA)

British Library Cataloguing in Publication Data

International Groundwater Engineering Conference (*1990: Cranfield Institute of Technology*)
Water wells.
1. Water supply. Wells. Maintenance & repair
I. Title II. Howsam, P. (Peter)
628.114

Library of Congress Cataloging-in-Publication Data

Available

Contents

Preface

Groundwater abstraction plays a very important part in water supplies throughout the world. Yet too many wells and boreholes operate inefficiently or are abandoned.

It really is quite surprising that until relatively recently, very many water well operators had valuable assets in the form of groundwater abstraction works, of which they really had little idea of condition or operating efficiency. Inadequate monitoring and little if any maintenance occurred. As long as some water was gained little else seemed to matter. If a well failed for some reason, then it was often abandoned and water obtained quickly via an engineering solution varying from drilling a new well, to manipulation of other parts of the supply/distribution system. Rehabilitation, if attempted, was on a 'suck-it-and-see' basis and frequently failed.

Such approaches prevail in many parts of the world. Yet it must be regarded as particularly serious in developing countries where weak economies could well do without the added strain of supporting inefficient systems or where failure can simply lead to going without.

Now however attitudes are changing. Whereas maintenance and rehabilitation used to lack glamour, and were often tainted with the embarrassment of something gone wrong, engineers are now keen to spot inefficiencies which if corrected can yield economic advantages. New knowledge and acquired experiences about treatment processes now exist, which should enable more confident application and better results.

Awareness and understanding of a particular problem are however, as important as its cure. Yet proper diagnosis requires a range of hydrogeological and operational information, which experience shows is all too often not available because appropriate monitoring has not taken place. Guide-lines on cost effective monitoring need to be established.

It is to be hoped this publication will go some way to improving knowledge in many aspects of the monitoring, maintenance and rehabilitation of water wells. With contributions from, or relating to, many parts of the world (Argentina, Australia, Belgium, Canada, Ethiopia, France, Germany, India, Netherlands, Oman, Pakistan, Peru,

Uganda, UK, USA, Yugoslavia, Zambia) it forms a significant contribution to the aims of the conference, namely to provide a forum for the exchange of practical experience and scientific knowledge, to the benefit of both industrial and developing nations.

Peter Howsam
May 1990

Acknowledgements

This publication owes its existence to all the contributors.

The conference and therefore this publication would not have happened without the sponsorship and support for some overseas participants from:

Christian Aid
Danish International Development Agency (DANIDA)
Overseas Development Administration, UK (ODA)
United Nations Children's Fund (UNICEF)
United Nations Development Programme (UNDP/World Bank)
World Bank (UNDP/IBRD/IDA)

and the support in a variety of ways provided by the following:

American Water Works Association
Anglian Water, UK
Association for Geoscientists in Development (AGID)
Association of Groundwater Scientists and Engineers (AGWSE) (USA)
British Geological Survey, Overseas Hydrogeology (BGS), UK
Centre for Groundwater Management & Hydrogeology, Australia
Mott MacDonald Group Ltd, UK
Regina Water Research Institute (RWRI), Canada
Rofe Kennard & Lapworth, UK
Scott Wilson Kirkpatrick & Partners, UK
S.A. Smith Consulting Services, USA
Southern Water, UK
University of Kebangsaan, Malaysia
Water Research Centre (WRC), UK
Waterworks Testing & Research Institute (KIWA), Netherlands
Wessex Water, UK
World Water, Thomas Telford, UK

Finally members of the conference technical steering committee helped to make it all happen by providing much useful guidance and by performing numerous tasks with great care and enthusiasm.

The technical steering committee:

Kees van Beek, KIWA, Netherlands
Richard Carter, Silsoe College, UK
Lewis Clark, WRC, UK
Roy Cullimore, RWRI, Canada
Ken Edworthy, Consultant, UK
Robin Herbert, Overseas Hydrogeology, BGS, UK
Carolyn King, Silsoe College, UK (Conference Secretary)
Bruce Misstear, Mott MacDonald, UK (Committee Chairman)
Shamy Puri, Scott Wilson Kirkpatrick & Partners, UK
Ed Smith, Anglian Water, UK
Stu Smith, Consultant, USA
Sean Tyrrel, Silsoe College, UK

My many thanks to all of them.

Peter Howsam
Silsoe College

PART ONE

OVERVIEWS

1 BOREHOLES AND WELLS: THEIR MONITORING, MAINTENANCE AND REHABILITATION

L. CLARK
Water Research Centre, Medmenham, UK

The idea that a borehole or well requires regular maintenance just like any other engineering structure is a relatively modern concept in UK and, I suspect, in most of the rest of the world. In UK there still appears to be resistance to setting up regular monitoring and maintenance programmes but attitudes are changing and I trust that this Conference will provide the impetus to maintain those changes in our new water companies.

The three aspects of well management: monitoring, maintenance and rehabilitation, are integral parts of a cohesive programme rather than separate functions. Monitoring is required to show the need for active maintenance measures, maintenance (servicing of both the pump works and well structure) is needed to maintain well performance and rehabilitation is required to repair damaged pump or well structures. These three phases of a maintenance programme are familiar to anyone owning a car. Few would expect a car to run for years without maintenance yet this has been expected of wells and pumps; this makes neither engineering nor common sense.

The monitoring of groundwater levels through observation boreholes is established practice in groundwater resource management. The monitoring of water levels in production wells however, is not common. Many pumping stations were equipped with water level measuring devices when pumps were first installed but the devices are now rarely used and very commonly have been neglected or deliberately vandalised. The only instrumentation deemed necessary by many organisations are cut-off electrodes to protect the pumps and discharge meters to fulfill licence demands. This purely 'protective' philosophy of well maintenance is changing in UK, particularly with the introduction of remote telemetry, but one could argue that the change is not deep enough, wide enough nor fast enough.

The need for monitoring production wells is, in my view, self-evident; the viability of a water resource as a whole can be monitored through the water levels in observation boreholes but the condition of a well can only be monitored through the changes in the water level/discharge rate relationship in the pumped well. Assuming a constant discharge rate, then a slight but continuous drop in the water level may be due to depletion of the overall resource or to the onset of deterioration of the pump or well structure, which could be halted or reversed by proper routine maintenance. A marked continuous fall in water level that is not seen in nearby observation wells is most likely due to deterioration of the well structure. A reduction in the discharge rate may be caused by well deterioration or pump wear. Either case is a clear sign that action is needed.

The constant monitoring of the discharge/water level relationship is valuable for showing the onset of problems in a well. A more quantitive assessment of well behaviour is provided by the step drawdown test. A step test run on completion of a borehole provides a basis for comparison with well performances at future dates. Serial step tests at regular intervals provide the best quantitive measure of changes in well performance (Clark et al, 1988) and, together with the constant monitoring of pumped water levels, should be the basis of all monitoring programmes. Severn Trent Water in 1985 (Skinner, 1988) began to set up the first regional borehole performance evaluation programme in UK. Anglian Water are now in the late stages of issuing a Maintenance Manual for their groundwater sources with the aim to regularly review the status of their boreholes and pumps. Their monitoring of boreholes comprises CCTV inspections each time a pump is withdrawn, together with constant monitoring of the drawdown/discharge relationship by telemetry (pers. comm.). The two organisations, Severn Trent and Anglian Water, have set the UK water industry an example to follow.

A monitoring programme should also include the monitoring of the water quality. An increase in suspended solids, for example, may be the first indication of the perforation of well screens as a borehole structure deteriorates.

Maintenance of wells is commonly equated with well rehabilitation; this is not so. Maintenance involves a programme of routine actions taken to prevent borehole deterioration while rehabilitation is the action needed to repair a well that has failed through inadequate monitoring and maintenance. Maintenance work covers two fields: the maintenance of the pump-works and maintenance of the well structure. In UK and industrialised countries in general, the need for maintenance of the pump-works is recognised and is undertaken on a routine basis. The main lack of pump-work maintenance is met in the developing countries. The source of the problem is less a lack of engineering skills than one of logistics; the lack of national or regional maintenance programmes, lack of funds to maintain such programmes and, in several countries, the impossibility of maintaining such programmes in times of civil war or revolution. Whatever the cause, the end results are large 'aid' projects for regional well-rehabilitation schemes.

The maintenance regime for well structures will depend on the local conditions in the particular aquifer. Biofouling of screens is most common in areas of sand aquifers and relatively iron-rich groundwater while encrustation is important in areas of hard water but not in soft water. The periodicity of maintenance and the action needed to maintain well performance will both vary but in all cases should be based on data from the monitoring programme. No logical maintenance schedule can be set up without that monitoring. I do not know of a UK organisation where such a maintenance schedule is fully implemented though the Severn Trent and Anglian Water actions show a beginning.

Rehabilitation or restoration implies a total breakdown of a pump or well structure. The breakdown of pump-works through lack of maintenance is not common in developed countries but is probably the most common reason for borehole failure in developing countries (Figure 1). Borehole 'restoration' programmes in such countries are usually pump servicing or replacement programmes. It should be stressed that the need for restoration implies a total failure of the borehole management programme.

Fig.1 Broken windmill pump in Nigeria

Pump out of action for several months because of simple broken con-rod on windmill. People had to revert to buckets to get their water supply.

Thirty years ago in Uganda there was a well established regional borehole maintenance organisation run by the Geological Survey. This has lapsed and WATER AID is undertaking a rehabilitation programme in Busoga Region where, of 430 boreholes, about 75% are not operational. This is in part because of silting up but all pumps need refurbishing. WATER AID is undertaking a similar exercise in Tanzania around Dodoma where most of the deep boreholes have failed. In this case almost all have failed because of the pumps and power sources. So far, about 60 Lister diesels and associated pumps have been refurbished (WATER AID, Pers. Comm.).

The results of pump failure can be catastrophic where there are no alternative sources of water; villages or stock holdings can be destroyed. The need in such cases for maintenance programmes and local repair facilities is absolute. Aid schemes should devote an element of their budgets to setting up the long-term maintenance organisations needed to avoid the need for large scale rehabilitation schemes.

The rehabilitation of boreholes themselves has tended to be a neglected technology in the UK. The techniques used in rehabilitation are extensions of those used in well development:- acidisation, jetting or the use of chemical dispersants. The need for rehabilitation implies that a well has failed, possibly through such processes as encrustation or biofouling, and implies that methods are needed to restore the well performance. Improved techniques have become available over the past decade, such as high-pressure jetting equipment with a rotating head. A great difficulty met in UK has been in obtaining proof that a technique is efficient and whether a well's performance has been restored. The first use of high-pressure jetting in UK was with the East Surrey Water Company in a Greensand borehole in 1979 when an improvement in well performance was demonstrated by repetitive step tests (Clark, 1980). After that, for several years, although this technique was used in many boreholes, clients were reluctant to pay for the step tests necessary to show that the rehabilitation had actually worked. This meant that case histories showing the efficiency of individual techniques, or allowing a comparison of different techniques, were quite rare. The record of one such case history from Italy is given in Figure 2 to demonstrate the value of such repetitive testing.

The case histories presented at this conference will be extremely valuable to those people responsible for well maintenance programmes throughout the world.

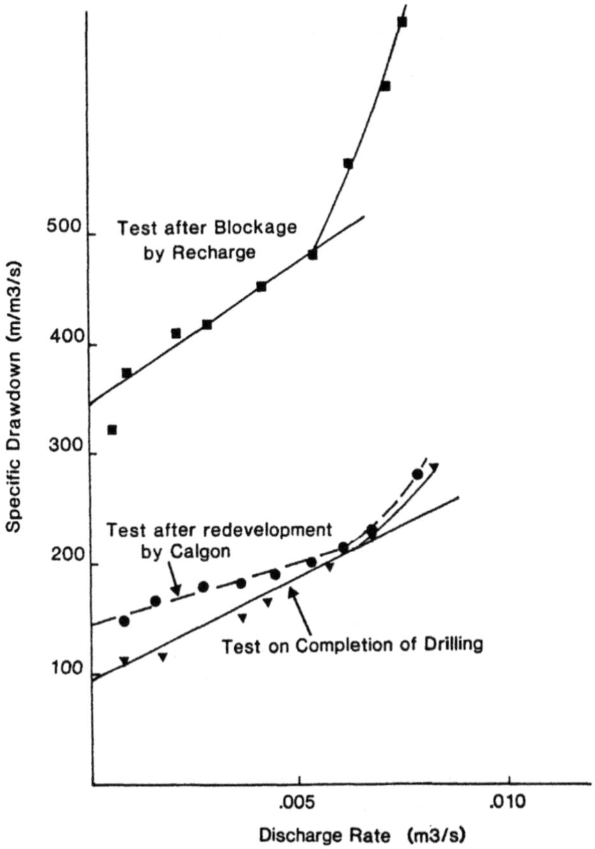

Fig.2 Step-drawdown tests on recharge borehole

References

Clark, L., Radini, M. and Bison, P.L. (1988) Borehole restoration methods and their evaluation by step-drawdown tests: the case history of a detailed study in Northern Italy. **Quart. J. Engng. Geol.**, 21, 315-328.

Clark, L. (1980) The evaluation of high pressure water jetting as a redevelopment technique in boreholes: Flower Lane 'A' test site. Report ER 742. Water Research Centre, Medmenham.

Skinner, A.C. (1988) Practical experience of borehole performance evaluation. J. **IWEM**, 2, 332-340.

2 WELL MAINTENANCE AND REHABILITATION IN NORTH AMERICA: AN OVERVIEW

S.A. SMITH
Ada, Ohio, USA

Abstract
Well rehabilitation in the United States and Canada is a relatively old practice which is now undergoing a renewal after a period of apparent stasis and neglect that masked significant actual advances. The economics of well restoration were studied in detail in the 1950s, but not extensively since. Well restoration is increasing favored due to problems with new public water supply well siting, and increasingly effective due to recent technological developments.

Well maintenance has largely been confined to mechanical and electrical maintenance of well pumps. Early warning detection and prevention of incrustation, biofouling and corrosion are sporadically practiced.

The extent and cost of well maintenance problems are difficult to ascertain due to lack of statistics and recent studies. In one study of utilities, engineers, and water well contractors, all three groups reported widespread well maintenance problems, predominantly corrosion, sand pumping, and iron-related biofouling.

There is a significant need for communication and education to bring current preventative maintenance information and methods to the well designing and operating community.
Keywords: Well Rehabilitation, Well Maintenance, History, Costs

1 Introduction

A reawakening of interest in well rehabilitation or restoration and maintenance methods is underway in the water supply utility industry in North America. A result of this interest is a desire by the utility industry to better understand its well service needs and also a desire by many companies to provide well maintenance and restoration services.

This paper presents available information and commentary on the general state of well maintenance and rehabilitation practices in North America, results of one specific study, and expectations for the near future. Well maintenance generally covers those practices that prevent the deterioration of well structures and equipment. Well rehabilitation or restoration are terms that refer to the practice of repairing, cleaning, and redeveloping wells that have declined in performance or product water quality.

2 Historical perspective

Attempting to provide a definitive history of well maintenance and rehabilitation practice and innovation in the USA and Canada is well beyond the scope of a paper such as this. However, it is useful to know where we have been in judging the overall costs of well deterioration, its impact on water supply, and the potential market for services and products in that field.

Systematic well rehabilitation in North America can be traced at least to the early 20th Century. By the 1920s and 30s, USA and Canadian water well contractors had applied steam treatments to clear iron-related bacteria from wells and even chlorine flooding of wellfields (without much success in the latter case). Wedge wire-wound and louvered shutter screen designs were available, allowing for better well maintenance. The Johnson's Drillers Journal in the 1930s already discussed fairly sophisticated surging and jetting tools. Essentially, by 1940, most modern water well technology was already available in some form.

By the end of World War II, inexpensive hypochlorites were widely available and used in well disinfection. Other chemicals such as glassy phosphates were also available and promoted for well cleaning by the early 1950s (with varying degrees of success). Steam and hot water faded as alternatives as steam engines left the well construction inventory. The late 1940s brought the very modest beginnings of what is now the largest groundwater association in the world, the National Water Well Association (NWWA) and the second significant groundwater industry publication, the Water Well Journal.

The postwar period also began nearly 30 years of apparent hard times for well restoration with booming new home construction and the abandonment of old village wellfields and private house wells for the surface-water community supplies the engineers preferred. As with the rest of the North American economy, there was a fascination with the "new", combined with a firm belief in the efficacy of existing methods (which persists to this day).

Still, the 1950s and 60s were an era of innovation in well rehabilitation and maintenance. The patented vibratory incrustation-cleaning method, Sonar-Jet, was developed in California. Downhole still and video cameras were developed and brought into commercial use. The first chemical specifically developed for use against iron bacteria, LBA, came onto the market. At the end of the 1950s, Koenig (1960a,b,c, 1961) conducted the only definitive published studies of well restoration effectiveness and cost-utility. Like Koenig's papers, many of the North American papers of significance in well maintenance and rehabilitation would be published in the Journal of the American Water Works Association.

The 1970s began a modern era of well restoration. In-well sand-separators were developed (again in California) for all those poorly designed wells. Roy Cullimore heard about iron bacteria, and he and collaborators at the Regina Water Research Institute in Canada brought studies of bacterial well biofouling and treatment back to life (e.g., Cullimore and McCann, 1977, Cullimore, 1981a,b). The end results of some of these studies will be presented elsewhere in this meeting and in Cullimore (1990).

Elsewhere on the information front, Campbell and Lehr (1973) summarized useful knowledge of well rehabilitation derived from oilfield and water well industry sources. The experience of the petroleum industry with well corrosion had been previously neglected in the water well industry.

The 1980s brought rehabilitation back into the economic mainstream as new wells and wellfields became more difficult and expensive to site and bring on line. The few professional water well firms who carried out rehabilitation regularly and professionally benefited from the increased business and included new innovations.

By the end of the 1980s, we have sophisticated and cost-effective treatment programs for biofouled wells (much of the details shrouded in industrial secrecy), and widespread use of hydraulic fracturing in some parts of the USA. Hydraulic fracturing, as it is known (high-pressure water flooding: often it is just cleaning out fractures), is revolutionizing well construction in low-yield hard-rock areas. Well owners at long last have some hope of obtaining a reasonable well yield where it was not possible before.

Among the very few really new items of water well technology were suction flow control devices to modify near-well hydraulics, being applied in the USA by Aquastream Inc. and Layne-Western (see Pelzer and Smith elsewhere in this proceedings). Practical in-well centrifugal sand separators, mounted on the pump intake, are also gaining acceptance.

On the information front, this author, Cullimore, and some others also worked during this period to bring information about well deterioration and restoration to a broader groundwater market (e.g., Smith, 1980, Smith and Tuovinen, 1985, Cullimore, 1981a,b). The NWWA published an EPA-financed manual on well rehabilitation technology (Gass et al., 1981). A new and very heavy edition of the standard industry reference, Groundwater and Wells, with lots of rehabilitation information was assembled by Driscoll (1986). A first-ever manual on hydraulic fracturing was also commissioned by the NWWA (Smith, 1989). We had our first ever (?) "think tank" and symposium on aquifer biofouling (re: Cullimore (ed.), 1987a,b). As we enter the 1990s, a comprehensive manual of well restoration methods is being developed by the NWWA (authors M.J. Hendry, S.A. Smith, and M.A. Borch) for the American Water Works Association (AWWA) Research Foundation (AWWARF), tentatively scheduled for publication in 1991.

3 North America as a groundwater market and what we know about it

3.1 Market as a whole

North America is a vast groundwater market, both in terms of the amount of groundwater used and the need for services. The total number of operating wells on the continent is still a matter of conjecture. The last NWWA estimate seen was about 13 million, while some U.S. Environmental Protection Agency (EPA) people say there are 40 million active wells in the USA. The following table provides some useful water well data to help judge the scale of the market:

Table 1. Miscellaneous North American Water Well Statistics

Annual USA well construction:	approx. 500,000 (1)
Annual USA well pump shipments:	approx. 500,000 (2)
Abandoned wells, Kansas, USA:	approx. 250,000 (3)
Irrigation wells, USA (1984):	353,243 (4)
USA community water supply systems (EPA):	72,000 (5)
Public systems serving <500 persons):	38,000 (6)
USA population using groundwater:	51 % (7)
Small systems using groundwater:	97 % (8)
1981 USA average well rehabilitation cost (NWWA):	$1,428.
Annual cost of well deterioration in Saskatchewan, Canada ($ Can.)	$2 million (9)

Sources of data:
(1) Water Well Journal surveys, 1950s-1980s.
(2) Water Systems Council press releases, 1980s.
(3) NWWA
(4) U.S. Bureau of the Census (1984).
(5) Longtin (1988)
(6) Rozelle (1987).
(7) U.S. Geological Survey (1985).
(8) U.S. EPA information.
(9) Cullimore and McCann (1977)

3.2 Rehabilitation market

What do we know about the rehabilitation and maintenance aspects of this market? Unfortunately, we know relatively little. Specific useful maintenance-related information on well pumping and equipment characteristics and well problems for well-using water purveyors has not been compiled for public use. For example, while the AWWA maintains data on overall groundwater usage, it does not have any information on the numbers of wells, types of pump or pump riser, etc., operated by its utility members.

Similarly, the U.S. Federal Reporting Data System has an estimate for the number of community groundwater supplies and the NWWA WellFax database has similar data on private well owners, but neither has more specific well system information useful for assessing well restoration or maintenance needs in the USA. Parallel Canadian data apparently are not compiled.

A similar information gap exists for the types and extent of well maintenance problems faced by well operators in North America. These situations are highly site-specific, and this characteristic is reflected in the literature. Individual well problems and solutions have been described elsewhere (e.g., Cullimore, 1983; Smith 1987) and the overall problem of well maintenance and rehabilitation discussed (e.g., Gass et al., 1981; Hackett and Lehr, 1986), but none of these adequately detail the numbers, types, and costs of well deterioration. Similarly, reports on regional "iron bacterial" occurrences (e.g., Lueschow and Mackenthun, 1962; Valkenburg et al., 1976) do not generally tackle the economic significance of these problems. The major exception is Cullimore and McCann (1977).

Methods of estimating costs and cost-effectiveness of individual well restoration work are available for the North American economic situation (e.g., Tavangar, 1987). Combined with the derivative data tables in Gass et al. (1981), these are good tools for individual problems. Koenig's (1960b,c) calculations and percentages remain the North American standard, but desperately need updating. The problem remains of determining the national scope of well deterioration costs in the USA and Canada to better focus attention on research needs, capital, and energy costs.

The closest recent approximation to a large scale estimation of well deterioration costs was provided by Alford et al. (1984), which provided specific cost data for well deterioration case histories. A survey taken at the 1986 American Water Resources Association meetings on aquifer and well biofouling (and yet to be published) promises to provide more specific information.

4 Well Deterioration Survey

To partially lift this veil for a manufacturer client (Angus Fire Armour USA), the author conducted a search of information sources to provide information on well and pump products usage, performance, and the occurrence of water conditions that unfavorably affect well equipment. A summary of the results of these studies is available from the author. (A summary of just the water quality data from this preliminary study was adapted for Smith, 1988.)

As part of the Angus study, a survey was conducted of USA water well contractors and hydrogeologists/engineers (nationwide) and AWWA-member water utilities in the southeast and south-central regions of the USA. Water utilities were surveyed to determine the dimensions and types of well equipment used, and known maintenance problems. Water well contractors, engineers, and hydrogeologists serving utility customers were surveyed to ascertain perceptions and knowledge of well characteristics and problems. This information was compared to existing published and otherwise available information.

The south central and south Atlantic regions were selected to provide a broad range of sizes and groundwater conditions in a manageable list size. Contractors known or likely to install significant numbers of municipal or commercial wells and pumps were polled. Consultants in the USA were selected from a list of firms with a significant groundwater interest. The survey provided a lot of useful information, some of which is summarized here. Among the big disappointments were the vague and contradictory responses from consultants.

Among the most useful information from the survey is the number of respondents reporting known well deterioration problems. "Conventional wisdom" in the groundwater industry is that well deterioration is widespread. Survey data (Tables 2 and 3) would seem to bear this out.

Table 2. Water Utility Well Deterioration Problems

Problem	Yes responses	Percent of total (154)*
Sand pumping	61	39.6
Hydrogen sulfide	51	33.1
Corrosion	43	27.9
Iron bacteria	39	25.3
H_2S + Sand pumping	22	14.3
H_2S + Corrosion	20	12.9
Iron/manganese clogging	12	7.7
Mineral clogging	7	4.5
Fe/Mn + H_2S	4	2.5

* Total groundwater-using water utilities responding. (Percentage total greater than 100 indicates multiple answers per response.)

Table 3. Contractor-Cited Well Deterioration Problems

Problem	Yes responses	% of Total (15)
Iron bacteria	3	20
Corrosion	3	20
Hydrogen sulfide	3	20
Iron/manganese	2	13
Sand pumping	2	13
Mineral clogging	2	13

About one-half of the utility respondents reported no discernible problems. Most of the utility respondents with well deterioration problems reported multiple well symptoms. Putting these responses in the context of over 300,000 active high-capacity irrigation wells in the USA and a certainly conservative figure of 200,000 active municipal wells (see following discussion), there are a large number of wells with noticeable problems.

Other information from the survey can be compared to published data to gain some sense of the magnitude of the potential for well restoration services in North America and the cost to well operators (Table 4).

Table 4. Comparison of Published and Survey Data

	Published	Utilities	Contractors
Av. Pump setting (ft)	145 (1)	293.5	100-399
Av. discharge (gpm)	830 (1)	846	--
Wells per utility	5 (2)	16.33	--
Charges for pump work	$29-50+/hr (3)	--	$30-$145/hr

Sources of data:
(1) Irrigation wells (U.S. Bureau of Census, 1984)
(2) Dixon and Lee (1988)
(3) Ground Water Age (1988)

The above figures help to give some dimension to the costs of this problem. Each time a pump is pulled and reset, that average 145 feet (published) or 294 feet (survey) of column pipe is moved twice. The average round trip for column pipe was around 4 hours, with 17 hours devoted to pump work. The labor cost does not included well cleaning, the cost of the pump (typically in the thousands of dollars for a high capacity turbine), the cost of down time and loss of water, or the secondary costs of sand, bacteria, clogging, and corrosion exported out into the treatment plant and distribution lines.

What then is a realistic figure for the annual cost of well restoration, including pump repair and replacement, in North America? Using some rough calculations based on data we have, and using a rule of thumb derived from our data and life spans presented by Gass et al. (1981), an average of 10 % of wells are undergoing service in any one year.

This author would propose that the direct cost of this work is conservatively between 200 and 285 million U.S. dollars per annum for utilities and irrigators, and probably above $1,000 million per annum when private water supply wells are included. We can only guess at the indirect cost in terms of pipeline and treatment plant maintenance and loss of confidence among customers due to poor water quality.

5 State-of-the-Art?

New developments in well restoration and maintenance from North America will be presented elsewhere in this meeting, but it is useful to summarize some of our conclusions in this plenary presentation.

At the present time, we see sweeping improvements underway in the technical state-of-the-art, especially in rehabilitating biofouled wells. The problem is known and effective methods are applied by some. Unfortunately, most of this innovation is tied up in a very few well rehabilitation firms, one of which is the largest water well contracting firm in the USA. Out in the "countryside", very little has changed in the well treatment methods employed.

The vast majority of well deterioration problems are discovered due to noticeable performance decline, failure of key components, or pumpage of sandy or otherwise poor quality water. Very little "early warning" maintenance monitoring is being practiced. In other words, crisis management rather than preventative medicine is the routine practice. Consequently, the typical rehabilitated well is an impaired patient suffering many of the complications on a chronic basis that caused its loss of performance or failure in the first place.

New maintenance monitoring methods to provide early warning for well biofouling are being tested by a number of North American workers for routine use and should eventually gain acceptance in the market if (1) they are proven to be practical, and (2) proper publicity and education are applied. The AWWARF well restoration guidance manual should be a useful tool in making maintenance more effective and accepted in routine operations.

6 Need for More Information and Applying it Better

In order to better serve the reawakening interest in well restoration and maintenance in the water utility industry, much more information on the costs, causes, and "cures" for well deterioration are needed. Among the leading information gaps are in the areas of well riser and pump deterioration, life-cycle costs of well deterioration, the cost-effectiveness of restoration vs. abandonment or maintaining the status quo, and the sheer dimensions of the market.

The other half of the information problem is applying it better in the mainstream water supply industry. Much of the establishment water supply industry and regulatory communities are just now rediscovering groundwater, let alone restoration of wells. The AWWA is doing an admirable job of addressing this education need. Still, spreading the "gospel" of appropriate well restoration and maintenance from the diverse group of visionaries that provide the research and development to the conventional thinking engineers and operators that design and operate water supply wells is going to take some effort.

7 References

Alford, G., et al. (1984) **Contamination of Water Wells by Organisms**. Arkansas Water Resources Research Center, Fayetteville, AR, USA.

Campbell, M., and Lehr, J.H. (1973) **Water Well Technology**. McGraw-Hill, New York.

Cullimore, D.R. (1981a) The Bulyea experiment. **Can. Water Well**, 7(3), 18-21.

Cullimore, D.R. (1981b) Physical ways to control build-up of "rust" in wells. **Johnson Drillers J.**, 53(1), 8, 9, 23, 25.

Cullimore, D.R. (1983) The study of a pseudomonad infestation in a well at Shilo, Manitoba. **Ground Water**, 21, 558-563.

Cullimore, D.R., ed. (1987a) **IPSCO 1986 Think Tank on Biofilms and Biofouling.** Regina Water Research Institute, Regina, Saskatchewan, Canada, 119 pp.

Cullimore, D.R., ed. (1987b) **Proc. International Symposium on Biofouled Aquifers: Prevention and Restoration.** American Water Resources Assn., Bethesda, MD, USA, 183 pp.

Cullimore, D.R. (1990) Evaluation of the available technology to conveniently monitor and control biological fouling of water wells. **Proc. Int'l Conf. Microbiology in Civil Engineering**, Cranfield, Beds., UK.

Cullimore, D.R., and McCann, A.E. (1977) The identification, cultivation and control of iron bacteria in ground water. **Aquatic Microbiology**, F.A. Skinner & M.M. Shewan, eds. Academic Press, London, pp. 219-260.

Dixon, K.L., and Lee, R.G. (1988) Occurrence of radon in well supplies. **Jour. AWWA**, 80, 7, 65-70.

Driscoll, F. (1986) **Groundwater and Wells**. Johnson Division, St. Paul, MN,

USA.
Gass, T.E., et al. (1981) **Manual of Water Well Maintenance and Rehabilitation Technology**. National Water Well Association, Worthington, OH, USA.
Ground Water Age (1988) 1988 water well industry survey. **G.W. Age**, 23(1), 21-26.
Hackett, G., and Lehr, J.H. (1986) **Iron Bacteria Occurrence, Problems and Control Methods in Water Wells**. National Water Well Association, Worthington, OH, USA.
Koenig, L. (1960a) Economic aspects of water well stimulation. **Jour. AWWA**, 52, 631-637.
Koenig, L. (1960b) Survey and analysis of well stimulation performance. **Jour. AWWA**, 52, 333-350.
Koenig, L. (1960c) Effects of stimulation on well operating costs and its performance on old and new wells. **Jour. AWWA**, 52, 1499-1512.
Koenig, L. (1961) Relation between aquifer permeability and improvement achieved by well stimulation. **Jour. AWWA**, 53, 652-670.
Longtin, J.P. (1988) Occurrence of radon, radium, and uranium in groundwater. **Jour. AWWA**, 80, 7, 84-93.
Lueschow, L.A. and Mackenthun, K.M. (1962) Detection and enumeration of iron bacteria in municipal water supplies. **Jour. AWWA**, 54, 751-756.
Pelzer, R. and Smith, S.A. (1990) Eucastream suction flow control device: an additional element for optimization of flow conditions in wells. **Proc. Int'l. Conf. Monit. Maint. Rehab. of Wells**, Cranfield, Beds, UK.
Rozelle, L.T. (1987) Point-of-use and point-of-entry drinking water treatment. **Jour. AWWA**, 79(10), 53-59.
Smith, S. (1980) A layman's guide to iron bacteria problems in wells. **Water Well J.**, 34(6), 40-41.
Smith, S. (1987) Case histories of iron bacteria and other biofouling. **Water Well J.**, 41(2), 30-32.
Smith, S. (1988) A geographical look at water quality in the US. **Water Techn.**, 11(9), 33-36.
Smith, S.A. (1989) **Manual of Hydraulic Fracturing for Well Stimulation and Geologic Studies**. National Water Well Assn., Dublin, OH, USA, 66 pp.
Smith, S.A., and Tuovinen, O.H. (1985) Environmental analysis of iron-precipitating bacteria in ground water and wells. **G.W. Mon. Rev.**, 5(4), 45-52.
Tavangar, J. (1987) Evaluation of ground water supply system rehabilitation. **Proc. NWWA FOCUS Conference on Midwestern Ground Water Issues**. National Water Well Assn., Dublin, OH, USA, pp. 457-470.
U.S. Bureau of the Census (1986) **1984 Farm and Ranch Irrigation Survey**, AG84-SR-1. U.S. Dept. of Commerce, Washington, DC.
U.S. Geological Survey (1985) **National Water Summary 1984**. Water Supply Paper 2275, USGS Water Supply Division, Reston, VA, USA.
Valkenburg, N. et al. (1976) **Occurrence of Iron Bacteria in Ground Water Supplies of Alabama**. Circular 96, Geological Survey of Alabama, University, AL, USA.

PART TWO
CAUSES

3 WELL PERFORMANCE DETERIORATION: AN INTRODUCTION TO CAUSE PROCESSES

P. HOWSAM
Department of Agricultural Water Management,
Silsoe College, CIT, UK

Abstract
The practical significance of reduced well performance is dependent on the ease with which this can be avoided or rectified. Changes in drilling, design and/or operation practice can obviate many problems, whilst various treatments can be used to rehabilitate the wells. None of these measures can be implemented in a rational and optimised manner without first both identifying and understanding, the processes which give rise to the problem. This paper outlines the main types, nature and extent of processes involved.
Keywords: Biofouling, Clogging, Corrosion, Drilling damage, Economics, Encrustations, Particle migration/mixing, Permeability impairment, Rehabilitation,

1 Introduction

Performance of a groundwater abstraction system is measured in terms of how much water is produced, either absolutely or per unit input of effort. Various factors can play a part in adversely affecting this performance, with the net result that less water is produced.
The primary processes involved may be categorised as:

Physical
Chemical
Microbial

With important secondary factors being:

Operational
Structural

Rarely do these factors occur in isolation. They can affect various parts of the groundwater abstraction system:

Aquifer
Well
Pump
Distribution

The aquifer is defined here as including the face of the drilled hole as well as that part of the water bearing formation influenced by the presence and operation of the well

Well will be used as a general term meaning borehole, tubewell, etc, and is defined here as including the well casing and screen string together with such associated material(s) as a gravel-pack, formation stabilizer, filter, seals

The pump is defined simply as the down-hole pumping components, including the pump body, diffuser(s),impeller(s), inlet screen, strainer, shroud, and in the case of submersible pumps the motor.

The distribution is defined as all pipework and fittings associated with the groundwater abstraction/supply system, including the rising main, well-head works and subsequent supply mains and/or distribution network.

2 Cause processes

2.1 Physical

(P1) Clogging (or reduction in material permeability) by the redistribution of particulate matter, which can affect the aquifer and the well. The processes involved include:

- Drilling fluid invasion damage at the time of construction;
- Inter-mixing of aquifer horizons as a result of wash-out/caving during drilling and development;
- Inter-mixing of aquifer and gravel-pack material due to over aggressive development;
- Migration of fines from the aquifer towards the well and into the gravel-pack material;
- Migration of aquifer material into the well causing it to be infilled;

It is not always appreciated that the mixing of two materials, ie the gravel-pack and the formation, which on their own may have a good permeability, can lead to severe permeability impairment. Better understanding and quantification of development processes such as surging, and more carefully selection and execution of drilling methods, would reduce such problems.

There are also some misconceptions about drilling fluid damage. Bentonite has a bad reputation for causing severe skin damage on the drilled well face, but modern drilling polymers have the capacity to cause internal damage within the formation, which may be equally difficult remove because it is difficult to get at. All fluids, which are used in part to stabilse a hole during drilling, have the potential to cause formation damage, by virtue of the fact that they are all particle laden (drilling fines) and with a positive gradient from the hole into the formation, will flow into the formation carrying these particles into and clog the pores of the formation. Some polymers change their properties with time and so lose the ability to carry back out the particles they carried in.

Also drilling with 'clean' water does not mean no formation damage.... how clean is 'clean' water?

(P2) Abrasion by the flow of particle laden water, which can affect the well, the pump and pipework.

The process involves relatively high velocity particle laden water causing abrasion damage. This is usually encountered in well casing /screens adjacent to high flow fissures and in parts of the pump and distribution. Pump impeller wear is commonly reported but there have been cases of holes being 'sand-blasted' through thick steel well linings.

2.2 Chemical

(C1) Clogging by processes of chemical precipitation, which can affect the aquifer, the well, the pump and pipework.

The most commonly reported encrustations are those of iron oxyhydroxides (sometimes associated with manganese deposits) and calcium carbonate. The former occurs when ferrous bearing anaerobic groundwater becomes oxygenated causing the ferrous to ferric conversion and the precipitation of insoluble ferric oxyhydroxides. In groundwater environments the ferrous/ferric balance is influenced by pH and Eh.

The precipitation of calcium carbonate is often quoted but apparently less commonly observed. The process is generally explained by saying that as groundwater approaches/enters a well there is a drop in pressure, which causes carbon dioxide to be released, which in turn affects the carbonate/bicarbonate equilibrium, with the result that calcium carbonate is precipitated. For many circumstances in and around a well however, the pressure/flow/temperature/time conditions as generally perceived, are not adequate to cause the deposits sometimes observed or predicted. This may imply that our understanding of flow conditions around a well is not as good as we think; or that other factors play a significant part.

(C2) Electro-chemical corrosion, which can affect the well, the pump and pipework, where these have components made of corrodible material such as iron and steel.

The process is well described in the literature, as are the consequences, such as corrosion of casing/screen joints, leading to sand ingress and possible complete structural failure of the well; corrosion and enlargement of screen openings, leading to ingress of formation/filter material into the well; corrosion and failure of pump components and rising main.

2.3 Microbial

(M1) Clogging by the processes of biofouling, which can affect the aquifer, the well, the pump and pipework.

There is now an increasing interest in biofouling problems in wells. Improved awareness and understanding amongst hydrogeologists and groundwater engineers, of microbial process is however hampered by various factors:

Groundwaters and aquifers, which are not obviously polluted, are widely perceived as being bacteriologically 'pure'. Although groundwater does contain far fewer microorganisms than surface water it is by no means sterile. The soil through which much groundwater passes on its way to the aquifer is absolutely teeming with microorganisms. Some of these can adapt to conditions in the aquifer and become residents, some can be introduced into the aquifer during drilling (migrants) and some described as ultra-microcells, with extremely slow rates of metabolism and by shedding surplus cellular material, become itinerants able to travel long distances through the aquifer

The consequences of biofouling will in some cases be visually obvious, ie a slimy material can be observed clogging the strainer of a retrieved pump or soft filamentous material can be observed with a down-hole CCTV camera, covering the slots of a well screen. In other cases however the appearance of the clogging material, eg brittle encrustation, clay-like sludge, will give no clue that microbial activity is, or has been, involved. There is growing evidence that encrustations perceived as chemical in orogin , probably in fact involved microbial processes. There are many bacteria (eg Gallionella) which can initiate/enhance the formation of commonly encountered iron deposits and others which are associated with precipitation of calcium carbonate in natural environments.

(M2) Microbially enhanced corrosion, which can affect the well, the pump and pipework.

Many engineers will now be aware of the existence of a group of anaerobic microorganisms called sulphate-reducing bacteria which are associated with the corrosion of steel and concrete. These microbes, which attach to and create biofilms on surfaces, are particularly hardy having been found to exist in environments of extreme temperature, pressure and chemical condition. What is of particular interest is the fact that examination of typical corrosion-encrustation cell material (eg rust tubercles on a ferrous pipe surface) has revealed the presence of sulphate-reducing bacteria. Another significant factor is that biocorrosion can occur in macro-environments which would not normally be considered as corrosive. This is because micro-environments can exist within a biofilm. A typical situation might be: biofilm develops on surface in aerobic, apparently non-corrosive, conditions; oxygen is used up by bacteria in upper surface layers of biofilm, creating anaerobic conditions, suitable for growth of corrosion-enhancing, sulphate-reducing bacteria; this can all happen within a biofilm only a few millimetres thick !

2.4 Operational

(O1) Inappropriate operating schedules, which can enhance many of the primary processes affecting the well, the pump and pipework.

Examples of this are intermittent pumping which may lead to increased particle redistribution (sand pumping and fines migration) as a result of the higher velocities generated at start up. Intermittent pumping may also increase the oxygenation of groundwaters in and around the well. Pumping wells at too high a rate or pumping a group of wells together with high levels of interference, may cause pumping water levels to fall below the top of the screen. This again enhances oxygenation and therefore the potential for iron fouling. It also necessitates placing the pump within the screen , thereby increasing the potential for sand pumping and fines migration where there is intermittent pumping.

(O2) Abstraction from aquifer exceeds recharge, which can affect well yield

This should be obvious and the first factor to check when there is a reduction in well yield. Yet it is all too often the case that appropriate monitoring of regional/pumping/static water levels and abstraction, which would provide the basis on which to assess this factor, is not taking place. Furthermore when monitoring does occur the data is sometimes just filed away and forgotten!, thus precluding any chance of predicting problems and allowing avoidance measures to be planned.

2.5 Structural

(S1) Poor design and construction, which can compound problems caused by the primary processes affecting the well, the pump and pipework.

Poor design problems should (we hope) largely relate to old wells which were designed in the absence of current knowledge and experience. Cases where groundwater mining has occurred or where abstraction has exceeded estimated recharge, over long periods of time, may mean that pumping water levels now fall within screened sections of the wells. Such conditions, as mentioned above, enhance the potential for problems such as iron fouling.

Poor construction may include such cases as insufficient care taken when installing/joining the casing/screen string which may lead to parting at joints and or particle ingress to the well; inappropriate use and improper placement of grouting materials, which may lead to ingress of polluted or oxygenated waters into the well; poor selection and installation of gravel-pack, which may lead to movement of material through the screen causing damage to the screen itself and the pump, or to clogging of the gravel pack by formation material; poor selection and implementation of drilling method, leading to hole instability, caving and collapse of the well.

(S2) Use of inappropriate well casing/screen materials/types, which can influence other processes.

The obvious case here is the use of ferrous materials where there is a risk of corrosion. Corrosion can directly contribute to causing sand pumping by enlarging screen slots and creating gaps in casing joints; pump failure by corrosion of mechanical parts; clogging by increasing iron concentrations and hence iron fouling. This problem, with improved water quality monitoring, together with the range of suitable non-corrodible materials (particularly the plastics, including glass-reinforced plastics) now available, is very largely a problem of the past. Except where perhaps economic restraints force people to reduce water quality testing and/or to take chances with inappropriate materials.

3 Discussion

This introductory paper serves to highlight, not only the main factors involved in well performance deterioration, but also to stress the principle that to be effective in both prevention and rehabilitation it is necessary to fully understand the nature of the cause processes involved. Poor design and construction practices and the SIS ('Suck-It-and-See') or AR (Abandon and Replace) options should now be less justifiable, with the better awareness and improved understanding of cause processes, which should now exist.

In many cases the various cause processes have inter-related elements of which we all should have been aware for a long time. Should we really need to be reminded?:

Don't over-pump relative to well design and aquifer capacity
Don't allow pumping water levels to fall below top of screen
Don't employ intermittent pumping, if possible
Do take care over screen/gravel pack design
Do use appropriate construction materials
Do employ good drilling practice in all cases

Finally, let us put this subject into context. World-wide groundwater abstraction costs (including operation and maintenance) must be of the order of several billion pounds per annum. Whilst a vast number of wells throughout the world are operating relatively efficiently, without any problems, there is too large a proportion (I would estimate at least 25%) which are not. The cost of not being aware, of not avoiding problems and of not correcting them is too high. Trends in world economics suggest that in the future prestige will be gained from the efficient running of existing systems rather than, as in the past, the creation of brand new ones.

4 AN EVALUATION OF THE RISK OF MICROBIAL CLOGGING AND CORROSION IN BOREHOLES

D.R. CULLIMORE
Regina Water Research Institute, University of Regina, Regina, Canada

Abstract
Microbial clogging and corrosion of the well is difficult to predict due to the insidious nature of the sessile growths around a water well. A number of strategies are proposed based around an active sampling program after a period of quiescence. One example given is the triple three scenario where samples are taken three minutes, three hours and three days after the initiation of pumping. It should be noted that the well should be quiescent for at least twelve hours prior to pumping. A number of microbiological techniques are discussed which allow a monitoring of this event and subsequent interpretation.
Keywords: Clogging, Water Wells, Corrosion, Pump Tests, Iron Bacteria, Biodetection Systems.

1 Introduction

It is now generally recognized that groundwater systems are as prone to microbial infestations as are surface-waters (Cullimore, 1987). A major difference between a groundwater event and a parallel occurrence in a surface-water is that the surface area to volume ratio is much greater in the groundwater system. One product of this very high surface area exposure is that the majority of the microbial colonizers will, in fact, attach to these available surfaces and form extensive biofilm (Caldwell, 1986), often slow growing, which will then interact with hydraulic flows through the aquifer media. The primary sequence of events occurring during an infestation is: (1) attachment; (2) primary colonization; (3) primary biofilm formation; (4) stratified biofilm formation; (5) corrosivity initiating at the surface area interface with the biofilm; and (6) occlusion of the biofilm due to excessive swelling and confluence of the biofilm into an integrated matrix. A major difference between a groundwater infestation and a similar event occurring in the surface-water is that the former cannot be directly observed whereas, in many cases, a surface-water event would be clearly observable.

By nature, therefore, the occurrence of a severe microbial infestation within a water well installation and in the aquifer immediately surrounding that environ often goes unrecognized and undiagnosed. It has been recognized in the last ten years there is the need to develop a management-response cycle which would be able to predict, monitor and evaluate the control strategies applied to such infested installations (Cullimore, 1989). This paper approaches a sequential management strategy to monitor the level of occlusivity and corrosivity within such an installation.

2 Routine monitoring for microbial infestations in groundwater systems

Three approaches can be routinely employed in the monitoring of potentially serious nuisance microbial events in water wells. These relate to: (1) degenerative observables; (2) routine evaluation of particulates; and (3) routine analysis of the postdiluvial planktonic microbial populations.

Routine observation of the water well for the qualitative and quantitative aspects of the postdiluvial flow can be observed and recorded as an index of potential degeneration within the installation (Smith, 1984).

Degenerative observables which can be used as an index of biofouling (e.g., corrosivity, plugging) can include a direct observation of the water for clarity, colour, taste, turbidity and odor; increases in drawdown during the pumping procedure; extension of pumping times to achieve target production; and a greater occurrence of equipment failure:

Direct observation of the postdiluvial effluent water can provide an early indication of an aggressive infestation. One early symptoms can be the loss in optical clarity of the water as a result of the increasing particulate loading in the water. It cannot be presumed that an optically clear water (turbidity of less than 1 NTU) is free from a significant microbial population. Such "crystal clear" waters have been found (Cullimore, Alford and Morrell, 1990) to contain up to 400,000 colony forming units (cfu/ml). While this occurs, the highest probability is that the incumbent micro-organisms are in a planktonic (suspended) phase or are occurring in relatively small particulate masses with a high incumbency of cells within the polymeric structures. When there are significant releases of particulate materials into the water, a cloudiness may be imparted to the water (Characklis, Cunningham, Escher & Crawford, 1987). This cloudiness may generate a grey colour to the water if no pigmentation is involved. However, where there is a significant uptake of pigmented ions such as the ferric and manganic forms of oxides and hydroxides, the suspended particles may be pigmented (i.e., pigmented particulates, pp). Generally, where iron and manganese are involved being carried in either dissolved or particulate forms in the postdiluvial water, the water will generate a colour ranging from yellow through orange to brown and occasionally black. These colour changes in water are often diagnosed as being related to iron bacterial infestations of the well. It should be noted that these colours may also be reflected in any slime deposits building up around leaking gaskets and valves. These are caused by a biofilm growing in the trickling water film moving away from the fractured site and supporting, very commonly, an aerobically dominated biofilm which is bioaccumulating the metals being released through the fracture site.

Of the taste and odours observed in water, the most common is the "bad egg" smell associated with the presence of hydrogen sulphide. This hydrogen sulphide is created during the anaerobic growth of sulphate and sulphur-reducing bacteria which reduce the sulphate or sulphur to hydrogen sulphide during the normal growth cycle. Additional or alternative sources of hydrogen sulphide are generated by anaerobic bacteria degrading the sulphur-containing amino acids by proteolysis. Hydrogen sulphide is particularly of concern in the operation of the well because of its intense corrosivity potential through the initiation of the electro-chemical corrosion process.

Some micro-organisms as they grow and metabolize will impart taste and odours to the water. Two major odours occasionally observed in waters are a fishy and a faecal odour. In general, the fishy odour is most commonly associated with the presence of a significant population of pseudomonads in the water while a faecal odour is relatable in many cases to the occurrence of a significant population of enteric bacteria (coliforms).

As an occlusive type of biofilm forms around the water well, an increasing resistance occurs to hydraulic flow into the well. As this increases, so there will be a greater drawdown in the head of the water column to compensate for this increasing resistance. In practice, an increase in drawdown can also be related to a reduction in the height of the saturated zone within the aquifer as the groundwater capacity of the aquifer falls as a result of the pumping. Radical increases in drawdown are, however, more likely to be associated with a biologically more chemically induced occlusion of the media in the vicinity of the well. A common practice is to refer to these types of production losses as relating to a "silting" or encrustation process which has traditionally been linked to the admission of silt-like particles into the well and a degenerations resulting from "salt-like" encrustations in and around the well screen. Both of these events can involve a biological component in terms of the generation of larger particulate masses which may then resemble silt or biologically generated encrustations which could appear to be similar to a chemically generated crystalline salt deposit.

Concurrent with increased drawdowns in a water well can be the extension of pump times in order to achieve the desired production requirements set for the well. As the resistance to flow increases around the well as a result of biofouling, so prolonged periods would be required in order to pump the same quantity of water to ground level and into the water transmission system. Corrosivity can concurrently be a major factor in biofouling and increases in the occurrence of equipment failure may also be observed as the levels of corrosion increases.

3 Routine evaluation of suspended particulates in water

As a biofilm matures, there are periodic shearing events which cause parts of the biofilm to detach and slough into the water. Erratic occurrences of significant increases in the total particulate volumes in the postdiluvial water may therefore reflect the occurrence of sloughing biofouling event within the media associated with the water well. While very significant particulates may be directly recordable as degenerative observables, frequently this increase in particulates may not be sufficiently significant to be directly observed. A variety of technologies exist to monitor these particulate releases including the entrapment of particulates by filtration on or within a filter media where changes in texture and flow rates may be observed (e.g., moncell system) or by measuring the suspended particles using a lazer driven particle counter whereby the size, shape and total volume of the particulates can be subsequently computed. Of the three technologies, the lazer driven particle counter is probably the superior since it does not rely upon the particulates being pigmented to easily observable. Primarily, the suspended particulates in well water can be considered to be pigmented particulates (pp) or achromogenic pigments (ap) which will not impart colour but will cause cloudiness when in high concentrations in water. Pigmented particulates are relatively convenient to determine since these can be readily concentrated from the water by the process of membrane filtration through a pore size of .45 microns. These pp aggregates will collect on the surface of the membrane filter where their density can be measured after drying the filter medium. Various methods of measurement can be used ranging from a direct comparison with colour scales to determination of the absorption of the pp as the inverse of the reflectometric characteristics of the filter media itself. The greater the density of these particulates, the less the reflected light is emitted from the filter and the greater the absorptive density of the pp. The reflectometric potential (irp) for a given sample can be measured and used to estimate the amount of suspended pp in the water.

The precise determinations of the total particulate mass suspended in the water can be achieved using a lazer driven particle counter which will record the total volume commonly as parts per million of total suspended solids (ppm TSS). Very often when a well is pumped for the first time after an extended period of quiescence (e.g., greater than 12 hours) there will be an accumulation of both sedimentary and sloughing particulates within the water phase which will be removed preferentially by the initiation of pumping. An indication of the degree of biofouling can therefore be achieved by the comparison of the total suspended solids measured by this technique at various time intervals after the initiation of pumping. The greater the factorial increase in total suspended solids during the early phases of pumping over the concentration of TSS recorded after the particulate discharges have stabilized, generally, at very much lower levels can be used to indicate the level of biofouling.

From field experience, this factorial can vary from as low as x3 where little biofouling is occurring to as much as x100 or higher where there is an extreme biofouling using the equation:

$$F = TSS_i \;/\; TSS_v$$

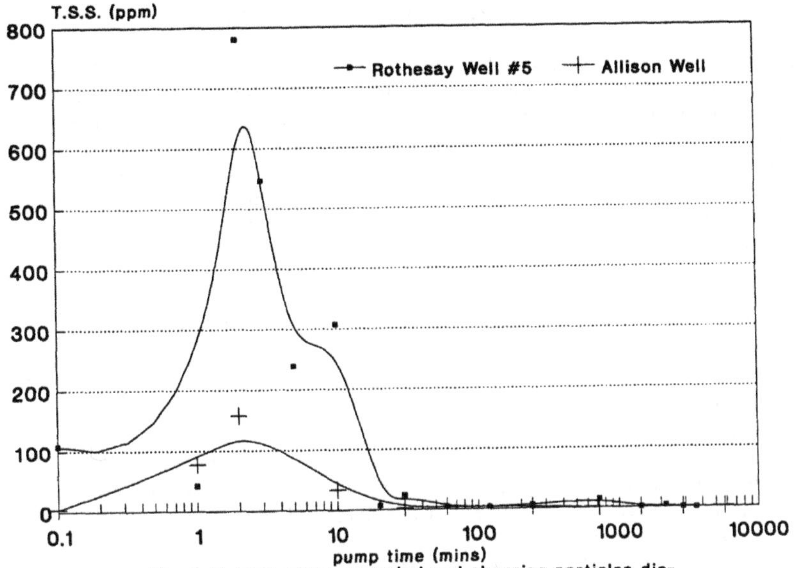

Fig. 1. Surging of resuspended and sheering particles discharged from two wells following a period of quiescence.

where F is the factorial, TSS_i is the initial highest total suspended solids recorded as ppm by the lazer driven particle counter and TSS_v is the background total suspended solids recorded after prolonged pumping when the particulate content has stabilized generally at a lower level than the TSS_i. Examples of these initial surges in TSS are shown in Figure 1.

Particulates can also be conveniently monitored by interceding a filter media into the flow down a bypass line. Where the particulate entraps into the filter media where it will cause increases in resistance to hydraulic flow, differential shifts in the water pressure upstream and downstream around the device and also cause visible changes to the filter media itself where observable. One such device which has been successfully applied to monitor these occlusive particulates (op) can be readily estimated. The moncell forms one such system which was designed (Howsam & Tyrrel, 1989) to facilitate a more rapid rate of plugging than would be expected to occur in the borehole itself. These particulates may be made up of inorganic materials along with such viable and inert organic materials as may be found within polymeric matrices. The incumbency of viable units in the volume of particulates may be used to give an indication as to the status of the biofouling event. Where there is a high incumbency within the sheared particulates, it may be projected that there is a very actively growing biofilm formation which is sheared to produce this event.

4 Routine microbiological evaluation of biofouling

Traditionally, the dominant microbial presence observed in boreholes is bacterial. These are generally subdivided into a number of interconnected groupings. Included as identifiable groups are: gross aerobic, heterotrophic, anaerobic saprophytic, sulphate-reducing, sufphur-reducing, coliform, enteric, iron-related, iron oxidizing, iron reducing, slime forming, streptomycete, and sulphur oxidizing bacteria. Additionally, the bacteria present in the system may be referenced by the environmental conditions within which they have been observed to grow. Temperature, for example, is relatively stable in groundwater systems when compared to surface-water so that the incumbent bacteria within a biofouling event within an aquifer medium may be expected to grow within a relatively narrow tolerance limit for temperature changes (e.g., 3°C temperature range rather than 20°C temperature range for a temperate surface-water). Where the ground water is permanently below 15°C, the dominant bacteria will be psychro- trophic, that is, able to grow at temperatures below 15°C. Where the groundwater temperature may vacillate below and above 15°C, the organisms would be facultatively psychrotrophic and able also to grow at higher temperatures. Where the ground water is consistently above 15°C, the facultative psychrotrophs could still grow but an additional group of bacteria able to grow only within the range of 15° to 45°C would now be able to function. These bacteria are referred to as mesotrophs. In preparing to conduct a microbiological examination of ground waters, it is therefore important to utilize temperatures which reflect the natural temperature expected in the ground water and borehole rather than the artificial temperatures which are often generated in the laboratory for culturing bacteria. For example, where water has been obtained from a ground water which is consistently below 15°C, the ideal temperature to culture the incumbent bacteria must be at less than 15°C (e.g., 10°C) in order to ensure that the strictly psychrotrophic bacteria are able to grow and be recorded during the testing sequence. In practice, much of the microbiological testing is conducted either at room temperatures (i.e, 18° to 25°C) or using an incubator set at temperatures such as 26°, 30° and 37°C. In all of these latter cases, the strictly psychrotrophic bacteria would be inhibited from growing and therefore would not be recorded in any subsequent testing.

A variety of cultural techniques are normally applied for the observation of microbial activities in ground water. Where there are a variety of rapid test methods relying upon the concentration of enzymes within the water, the number of cells and level of activity of metabolic activity within the water. More commonly, an examination of ground water involves the utilization of cultural techniques involving the use of specific agar or broth media. Of these, the most commonly used enumeration technique for bacteria is the application of a serial dilution of the ground water to spread agar plates in which the medium composition and the temperature of incubation are key selective factors. In general, most spread plate enumerations are performed aerobically recognizing an assumption that the most severe biofouling is likely to occur in the presence of oxygen. Incubation time and temperatures are two further constraints applied to the spread plate technique. The most common temperature used is room temperature or 25-26°C. On occasion, growth is thought to be stimulated by raising the temperature to 28 or 30°C. Where there is a direct interest in the potential for contamination of the ground water with sewage, the temperature may be elevated further to 37 and/or 45°C. These latter conditions are specifically applied when testing for faecal contamination by the presence of coliform bacteria.

Historically, spread plate evaluations have been performed on clinical, industrial and food products where there is a very active microbial population and relatively rapid consequent growth. As a result of this, incubation periods are generally short (i.e., one, two, five and/or seven days). The need to deliver the data as quickly as possible to the user has initiated a shortening of the incubation period to that most convenient to the laboratory and the user. For groundwater samples, the changes in environment which occurs during the sampling and enumeration procedure can frequently cause the micro-organisms to enter a prolonged induction phase prior to growth. The use of short incubation periods may eliminate such traumatized organisms from growing to form colonies and, therefore, from being enumerated. Under ideal conditions, the incubation temperature should be close to that of the original water sample and ideally between one degree below and two degrees Celcius above that value. Incubation times should be minimally seven days with an option to extend to fourteen and twenty-one days if there is no significant colonial formation on a particular agar plate. From direct field experiences, it has been noted that a series of "pulses" of colonial growth can occur on a spread agar plates over periods as long as forty-two days.

The agar media used in the spread plate analysis is by its very nature selective to the types of organisms that can be recovered by colonial growth. Primarily, the microbial cell must be able to produce an observable colony formation in order to be directly observed. Agar media that have been recommended vary with the target bacteria to be observed. Common media used in the spread plate evaluation include (target bacterial group in brackets): R2A (heterotrophic bacteria); Winogradsky's Regina (iron-related bacteria); M-Endo Les (coliforms); Brain Heart Infusion Agar at quarter strength (gross aerobes); Abiola's agar (fluorescing section 4 bacteria); and Sabaro's agar (yeast and molds).

While the spread plate does give a quantitative evaluation of targeted bacterial groups, some of the viable components within a biofouling event may not, in fact, be observable. To encourage the growth of bacteria which may be inhibited in, or on, an agar matrix and be stimulated by growing within a liquid medium, an alternative cultural evaluation can be performed using the groundwater sample dissolved within the selective media into a liquid phase. One example of such a technology is the patented biological activity reaction test system (BART™) which stratifies the type of microbial growth that would occur within the liquid medium (Cullimore & Alford, 1990). A floating intercedent device (FID) is floated on the surface of the culture medium so that sessile aerobic microbial growth can occur at the interfaces between the air, the FID and the liquid. Such growths, not only produce significant "rings" of cloudiness and polymeric material around the FID, but also prevent the diffusion of oxygen into the deeper parts of the medium. Here, the intrinsic oxygen is rapidly utilized by the incumbent aerobic bacteria so that this deeper zone becomes anaerobic and supports a different type of microbial growth. BART^{TM} tests have been developed for the iron-related (IRB-BART^{TM}), slime forming (SLYM-BART^{TM}), sulphate reducing (SRB-BART^{TM}) bacteria and the fluorescing pseudomonads (FLOR-BART^{TM}). These tests (manufactures by Droycon Bioconcepts Inc., Regina, SK., Canada) are initiated by the addition of 15 ml of groundwater sample to each specific test. The dried culture media dissolves and diffuses into the water to create a selective function which will allow the target groups of organisms to grow. When incubated, different reaction patterns are observed. For the tests listed above, the number of reactions which can be observed are 9, 6, 4 and 3 respectively. By noting the reaction pattern that has occurred, it is possible to begin to evaluate the types of organisms present in the water sample. The length of the sampling period before the noted reaction occurred (days of delay, dd) inversely relates to the size of the population of the target organisms. For example, a groundwater sample with a population of 1,000,000 cfu/ml or iron-related bacteria may react very quickly in the IRB-BART^{TM} causing a reaction with a dd of 1 or 2. On the other hand, a sample with a population of only 10 cfu/ml of IRB may not trigger a reaction until a typical dd of fifteen days. From the field data generated to date, a dd of less than 6 generally relates to a microbial population in excess of 500 cfu/ml while a dd of 6 or greater tends to have fewer than 500 cfu/ml of the target organism. For convenience, it is easier to consider the number of days of positivity (dp) for the particular test on the assumption that the test is negative if no reaction has occurred on day 15. The equation for calculating the dp is:

$$dp = 15 - dd$$

where dd is the number of days of delay, and dp is the days of positivity and the assumption is made that the test is negative if there is no reaction on day 15. The BART^{TM} test therefore allows a semi-qualitative and semi-quanitative determination of the presence of specific bacterial groups within ground water and thus, allows a monitoring tool to be conveniently applied to determine whether a treatment scenario (e.g., shock chlorination) has successfully suppressed the amount of microbial activity associated with the water and possible biofouling events.

5 Sampling times to evaluate biofouling

It has already been noted that a biofouling event may be present within a borehole, but not easily observed when water is being continuously pumped from the well. If biofouling is to be observed, the well should be held in a quiescent state for a minimum of twelve hours before testing. During this period, there would not be the turbulence within the aquifer and intimate contact with the borehole which would cause shearing. Consequently, some additional biofilm generation may occur which would be vulnerable to shearing once the turbulence was initiated by pumping. Each individual well may be expected to have a slightly different form of biofouling event which makes a common sampling system difficult to apply. It is, however, essential that the sample being taken for testing be obtained aseptically from the water being pumped contaminated as little as possible by organisms which may have grown around the sampling port. Standard aseptic techniques can reduce this risk.

One method for determining the amount of biofouling in a well would be to sample at three points in time after the initiation of pumping. One such scenario (TRIPLE THREE, TT) would involve sampling at three minutes (after a stable water temperature has been obtained in the pumped water), three hours and three days into the pumping period. The factorial (F) factor calculated using either the TSS values obtained by lazer driven particle counting or by utilizing the dp or cfu/ml values obtained by bacteriologically testing systems would give an index of the likelihood of a biofouling event occurring in the well.

6 Summary

It is relatively impractical to obtain samples of biofouled material from around the borehole without the partial contamination of the sample and destruction of the environment by the acquisition process. Any monitoring procedure therefore has to rely upon indirect evidence relating to the particles shearing from the biofilm during pumping along with the releases of any inorganic or organic product which can be linked to the biofouling event. The types of organisms recorded during the monitoring will indicate the probably form of the biofouling. For example, very high populations of iron-related bacteria would indicate that a severe plugging is occurring with the bioaccumulation of iron and/or manganese. The presence of relatively large populations of sulphate reducing bacteria would indicate that the biofouling was extensively anaerobic, sulphates were present in significant concentration and that hydrogen sulphide was being produced which could stimulate electro-chemical corrosion. When remediation techniques are applied such as disinfection, application of heat and penetrants, an evaluation of the amount of particulates and the incumbency of different microbial groups within the particulate structures may be used as evidence for a successful or failed treatment. Subsequent monitoring of the boreholes on a routine basis (e.g., monthly) can then be used to determine the speed with which the biofouling is recurring within the well.

7 Acknowledgements

The author wishes to acknowledge the National Science and Engineering Research Council of Canada for a grant-in-aid, G. Alford, N. Mansuy,

A. Abiola, J. Reihl and M. Mnushkin for their considerable assistance, and N. Ostryzniuk for preparation of the manuscript.

8 References

Caldwell, D.E. (1987) Microbial colonization of surfaces, in International Symposium on Biofouled Aquifers (ed D.R. Cullimore), American Water Resources Association, Bethesda, pp. 7-9.

Characklis, W.G., Cunningham, A.B., Escher, A. and Crawford, D. (1987) Biofilms in porous media in International Symposium on Biofouled Aquifers (ed D.R. Cullimore), American Water Resources Association, Bethesda, pp. 57-78.

Cullimore, D.R. (1987) Think Tank on Biofilms and Biofouling in Wells and Groundwater Systems. University of Regina, Regina Water Research Institute, Regina, SK.

Cullimore, D.R. (1989) Looking for iron bacteria in water, Canadian Water Well, 15(3), 10-12.

Cullimore, D.R. and Alford, G. (1990) Method and apparatus for producing analytic culture. U.S. Patent 4,906,566.

Cullimore, D.R., Alford, G. and Morrell, R. (1990) Bacteriology of groundwater sources and implication for drinking water epidemiology, in National Environmental Health Association Conf. (in press).

Howsam, P. and Tyrrel, S. (1989) Diagnosis and monitoring of biofouling in enclosed flow systems--experience in groundwater systems. Biofouling, 1, pp. 343-351.

Smith, S. (1984) An Investigation of Tools and Field Techniques for the Detection of Iron Precipitating Bacteria in Groundwater and Wells. M.S. Thesis, The Ohio State University, Ohio.

5 INFLUENCE OF CHEMICAL REGENERATION ON IRON BACTERIAL POPULATION AND OCHRE DEPOSITS IN TUBEWELLS

F. BARBIČ
Institute for Technology of Nuclear and Other Mineral Raw Materials, Belgrade, Yugoslavia
I. SAVIĆ
Faculty of Science, University of Belgrade, Yugoslavia

Abstract
With the new substance "Skital B" the regeneration of a greater number of tube wells was done on the springs Obrenovac and Vreoci. Laboratory analyses proved that the substance is chemically effective with bactericidal properties. Through the combined regeneration method (chemical - mechanical) the instantaneous specific production rate of wells (before regeneration) was increased for 30-85% depending on the well and used technology. Besides the production rate increase a regeneration effect was observed through a change in the qualitative and quantitative composition of iron and manganese bacterial populations as well as through ochre sedimentation dynamics. Questions related to the regeneration process are discussed.
Keywords: Well. Chemical and Mechanical Regeneration. Regenerative Preparation. Specific Production Rate Decrease. Ochre Sedimentation. Iron and Manganese Bacteria.

1 Introduction

Chemical regeneration assumes a defined technology aimed towards prevention of further production rate decreases and establishing of a higher specific well production rate. This technology comprises certain chemical substances which soften and dissolve solid deposists (ochre sediments) on the screens of wells and destroys microorganisms at the same time with its bactericidal properties.

A specific production rate decrease of the Ranney wells on the Belgrade spring was observed a few years after the construction of the first wells. The severity of this problem popularly called aging of wells is best illustrated with the following facts. In 1963 ten Ranney wells on the Belgrade spring were under exploitation with the total annual water productivity rate of approximately 67 million m^3 and in 1978 fifty wells were under exploitation yet only approximately 169 million m^3 of water were produced. This suggests that the average specific production rate of 10 wells in 1963 was approximately 210 l/sec, and in 1978 it was approximately 108 l/sec for 50 wells. Having in mind that by 1978, a great number of new wells was included into exploitation. It comes out that production rate decrease of some wells was over 70%. (Barbič, 1983).

Such rapid production rate decrease has initiated a more complete investigation of this phenomenon and of making decisions on regeneration measures. The first intervention in this sense was the direct adding of a certain amount of sodium hypochlorate in the hole of a well, after the well had stopped abstracting and water level increased through adding of an extra water from the distribution system. In this way 12 wells were regenerated during 1964 - 1971. The amount of sodium hypochlorate for each regeneration was 1000 - 1700 l per well. At the same time, an attempt to regenerate with this preparation through adjacent piezometers was made, but this method was abandoned as it did not give expected results.

In the begining of seventies a mechanical-chemical regeneration of the Ranney wells on this spring was undertaken. The technology used is similar to that used today. The regeneration technology in brief comprises the following acts: the well is taken out of supply, pipe work openings closed water is pumped out, and the complete installation is removed. The necessary regeneration equipment is then placed on a platform on the well and sediment from the bottom of the well is pumped out. Each drain of Ranney well is washed with water under high pressure and then the chemical preparation is added. This is followed by repeated mechanical cleaning with water-air blowers. After desinfection, the well can be put back into supply (Technical Report, 1985). Today such regenerations are being regularly carried out on the wells of the Belgrade spring by a specialized skilled team. In this way four to six Ranney wells are being regenerated annually. Two months on average are necessary for regeneration of one Ranney well.

In respect to the production rate decrease, by apllying chemical and mechanical regneration, we can infuence the following areas of clogging: a) of the area around the well; b) of the area around the drainage galleries: and c) of the filtering part of the well. However, in the case of the clogging towards the river-groundwater interface of the recharge area of the well, and when the production rate decrease was due to interference and declined spring capacity, these methods were not effective. In the first case the production rate decrease curve has a constant slope with values always greater than zero (equation 1):

$$\frac{\Delta q}{\Delta t} > 0 \qquad (1)$$

The possibility of restoration of part of the lost specific production rate depends on the value of the terms and their interrelationship in equation 2, which applies to clogging in the cases a, b and c.

$$\frac{\Delta q}{\Delta t} = f \sum_{a}^{c} \qquad (2)$$

Besides the regeneration of the Ranney wells, which is continually being carried out on the Belgrade spring. In the last eight years regeneration of a greater number of tube wells has been carried out on the springs of Novi Sad and Obrenovac. In this case results are presented on the regeneration of four tube wells on the Obrenovac spring and of one well on the Vreoci spring. To explain some equations regarding regeneration we use the results and experience gained through the work on the regeneration of some other wells. With this paper we wish to draw attention to the usefulness of the regeneration and to the justification of this method, as well as to certain problems which accompany the regeneration (technical, professional, organizational) and have influence on the overall results of the performed regeneration.

2 Material and Methods

For regeneration of the tube wells, of which results are presented in this paper, we used the preparation defined by our Institute. It has been patented under the name "Skital B" and is manufactured in quantities ordered by customers. Chemical and bactericidal properties of the substance wil be described later.

Before starting with the regeneration process a plan is formalised for each chosen well, with a detailed description and sequence of operations, details of the necessary equipment, organization and work dynamics, etc. Technical accuracy of the wells, specific production rate, qualitative and quantitative coomposition of the iron and manganese bacterial population, ochre sedimentation dynamics and water quality of the wells are all determined. Cleaning of the bottom of a well with the air-lift system is done before the regeneration if necessary.

The regeneration of a well consists of the following. The well is put out of action and the underground water allowed to stabilize at the static rest level. Then the preparation is added into the well either directly of by pumping. The added preparation is allowed to stand in the well for a fixed time (for these wells 48 hours). The screens of the well are rinsed with the water-air high-pressure pumps. The sediment and dissolved ochre that result are removed trough the air-lift system due to action of the praparation and mechanical agitation. Afterwards follows the so-called "secondary chemical regeneration", i.e. repeated adding of the preparation inside the well. It is desirable that this time the preparation stands in the well for a longer time period (in this case two days). From time to time the decrease of the preparation concentration in the well was monitored. After 48 hours water with sediment is removed from the well with the silt pump or the air-lift. The well starts pumping into supply from the very moment when it is determined that all added preparation has been used up or the surplus has been remuved pumping.

After the regeneration, at defined time intervals (2, 10 and 30 days) the specific production rate is determined, as well as the condition of the bacterioflora and ochre sedimentation in each well. To satisfy nedds of comparasion we are now making efforts to perform corresponding measurements before and after the regeneration (e.g. observations of the underground water level, in connection with the specific production rate, are done by the same people whenever possible).

Of the large number of wells which were regenerated, significant parameters of the only one were observed for a long time, i.e. up to one year. Quality of water, concentration of iron and manganese and their compounds, chlorides and phosphates, carbone acid, possible surplus of hidrochloric acid and pH value were observed during the pumping of water, just before connecting the well to the water supply network. We observed these values in order to establish to reaction speed of our chemical and the moment when the water is potable.

The procedure described above respresents the "complete" regeneration, socalled chemical-mechanical-chemical regeneration. However, four years ago a well in the same aquifer was regenerated without the second chemical regeneration. Due to technical reasors, another well could not be regenerated mechanically such that only chemical treatment was applied. These two wells were used to make comparisons with the wells which were completely regenerated.

We observed the occurence of iron and manganese bacteria and dyinamics of ochre sedimentation before regeneration as well as their change due to the use of chemical (after regeneration) by use of the water samples and simulators in the way the authors described in their earlier papers (Bracilović et al., 1975; Barbič et al., 1987).

The locations of the Obrenovac aquifer and its wells and their traitswere given in detail in the previous paper (Barbič et al., 1990). The well of the Vreoci aquifer which was regenerated in 1989 is nine years old. The arrangement of lithological segments, construction and dimensions of the well built in 1985 are given in Figure 1. Quarternary sediments consist of coarse-granulated gravel about 3m thick, overlain by dark-yerllow clay. The other sediments are presumed of Tertiary age. From a hydrogeological point of view the aquifers have integranular permeability while aquicludes are clays. The well includes three water-bearing horizons adjacent to which the well screen was installed. Its dimensions and design depend on the permeability of the flow horizon.

The following parameters were obtained using data from a pumping test:

- Transmissibility, T=14,37 m^2 per day
- permeability, K=1, 28x10^{-3} cm/sec
- radius of influence, Ra= 245,88 m
- specific yield, q = 0,15 - 0,17 l/sec.m.
- well yield, Q = 10 l/sec

3 Results and discussion

3.1 Properties of "Skital B"

The chemical used is one of the basic factors which makes chemical regeneration successful or not. That is why "Skital B" was examined for its ability to dissolve ochre and its bactericidal effects on iron bacteria first in laboratory wells and then in experimental wells. "Skital B" is a complex chemical consisting of a number of organic and inorganic acids and other components in smaller concentrations (such as an inhibitor of corrosion, a surface active agent, a bactericidal component and a solution stabilizer). The efficiency of the chemical was firstly examined obsereving the effects of each component by itself (i.e. each acid by itself) and then observing the effects of the complete chemical.

We examined the bactericidal effect of each component observing the changes of the population of Thiobacillus ferrooxidans bacteria in the Leathen medium. Thus, we found that 0,12 mol/dm^3 concentration of hydrochloric acid is lethal for the population of these bacteria. Formic acid is more toxic and has lethal concentration of 0,01 mol/dm^3. The increase of Thiobacillus ferrooxidans population in the presence of different concentrations of "Skital B" is presented in Figure 2. The chemical even in the concentration of 12 ml/1 makes the oxidation of ferrosulphate slower in the medium. The increase of "Skital B" concentration in the medium prolongs the beginning of oxidation, i.e. lagphase becomes more dominant. The lethal concentration of the chemical is about 60 ml per litre of medium.

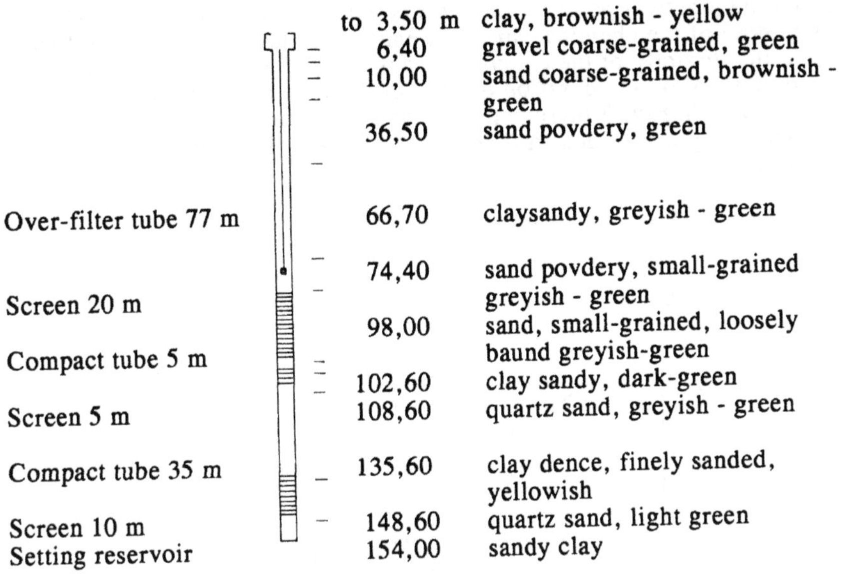

Fig. 1. Distribution of the filters and lithological segments of the tube well V-1

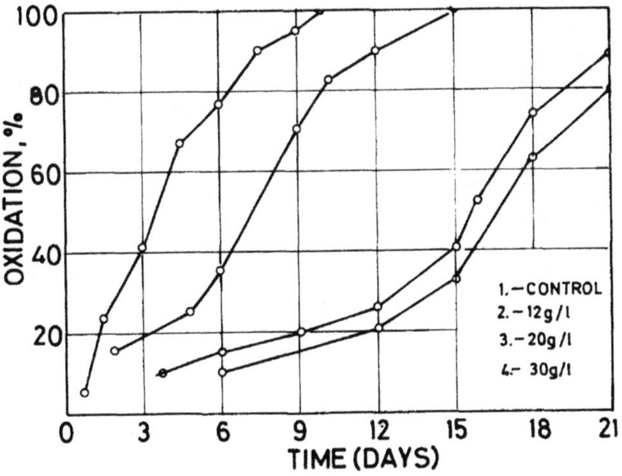

Fig. 2. Dynamics of the oxidation $FeSO_4$ by Thiobacillus ferrooxidans depending upon various concentrations of the preparation ("Škital B" in the medium)

From the practical point of view it was very important to determine the optimal dose of the chemical for the chemical regeneration of the well having in mind that toxicity of the chemical is not only a function of its concentration but also defined by the duration of contact between the toxic substance and the bacterioflora. In Fig. 3 there is presented the survial of Thiobacillus ferrooxidanns population as a function of time and the number of toxic doses of "Skital B".

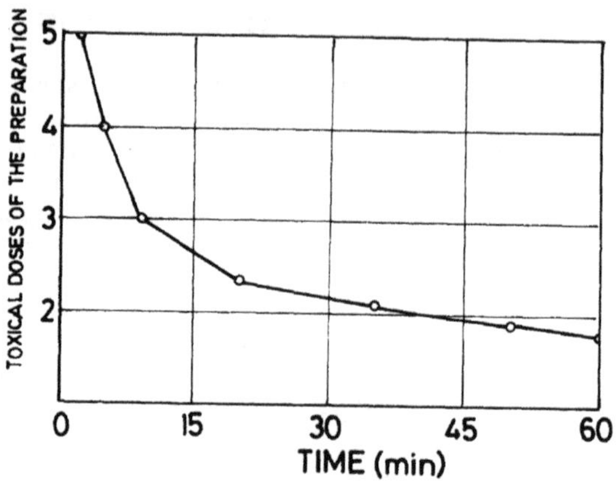

Fig. 3. Survial of the population Thiobacillus ferrooxidans in relation to the time and toxical doses of the preparation "Skital B"

The optimal dose of the chemical used for the regeneration must also be in accordance with the time in which the ochre sedimentation of the well will be dissolved in the greatest quantity. It is necessary to be aware of the cost as every reduction of the chemical activity time in the well makes it necessary to use greater quantity of the chemical (to provide greater concentration). To shorten the time to less than eight hours makes the regeneration very expensive and cannot be justified.

3.2 The effect of the chemical on the ochre sedimentation

How "Skital B" effects the dissolving and softening of ochre was investigated under laboratory conditions on a naturally occurring substance (ochre from the well) and on compounds which comprise the ochre. Using chemical analysis of ochre sedimentation we found that the most frequent compounds were due to transformation of iron and manganese, and consisted of: hydroxides, hydrated oxide, sulphide, carbonates of calcium and magnesium. The results indicated the following:

dissolution speed of iron compounds in hydrochloric acid depends on; when pH is higher than 2 hydroxide starts to separate from chloride compounds, ascorbic acid in acid medium has a reductive effect; the increase of pH makes soluble complexes giving rise to stable complexes with further pH increase, dissolution efficiency of iron compounds when the pH is lower (originating from the hydrochloric acid) depends on the concentration of ascorbic acid, dissolution of iron compounds in phosphoric acid is not significant when compared with hydrochloric acid, while ascorbic acid reacts in the same way as when hydrochloric acid is present, the reaction is slower and weaker,

calcium and magnesium sedimentation can be easily dissolved by use of acid components depending on their concentration.

The main components of the chemical "Skital B" (HCl, H_3PO_4, ascorbic acid) in the concentrations used in the regeneration corrode elemental iron.

Similar behaviour, though in a mild form, is exibited by the preparation "Skital B" without the addition of anti-corrosive compounds. By using a certain cation tenzid (we have used "Aks") in small concentrations, about 0,1 g/l the aggressive impact of the acids was eliminated. For these investigations steel tablets with various properties (filled or perforated) were used. Simulators were used to assess the dynamics of the ochre precipitation in the wells and on the original material (a brand of steel) of which the screens in Ranney wells were made.

On account of these investigations we can assert with certainty that the preparation for the chemical regeneration of wells fulfills all the necessary conditions. That is;

it does not leave harmful effects (instantaneous nor permanent) on the quality of water;

it is efficient in the sense of dissolution of the precipitated ochre;

it possesses the necessary bactericidal property (albeit the afterward addition of the bactericide component is lacking);

it is not aggressive towards metal, i.e. it does not hasten the corrosive process;

it is relatively easy to apply.

3.3 Results of the regenertion of the well

Table 1 contains some characteristics of the treated wells together with basic parameters which point to the effectiveness of the regeneration. Mark "V" signifies the well from the spring Vreoci, while mark "O" signifies wells from the spring Obrenovac. In three wells (V-1, 0-1 and 0-2) a complete chemical-mechanicl-chemical regeneration has been performed, in well 0-3 chemical-mechanical regeneration has been performed, while in well 0-4 only mechanical regeneration took place.

According to the established protocol, a month before the regeneration, the state of the populations of iron and manganese bacteria and the ochre precipitation in all wells that were planned for regeneration were obseerved by analysing the water and by inserting the simulators. This is summarized in Table 1 and is presented in greater detail in Table 2. Simulators, by which the qualitative and quantitative composition of bacterioflora was assessed, were also placed in the nearby piezometers.

The choice of wells suitable for regeneration is mainly governed by the magnitude of reduction of the specific productivity and secondly by the presence of iron bacteria and the amount of precipitated ochre. Wells presented here are characterized by intensive development of bacteria, that is, biooxidation of iron is very high, which is reflected in the ochre precipitation. Expect for the well V-1 (index of precipitation is 4), in all other wells the ochre precipitation and the bacterioflora are intensely present (index ranges from 6 to 8). As in other wells of this spring, genus Siderocapsa is dominant, and together with the genera Gallionella and Leptothrix they amount to over 90 percent of all the present species of iron and manganese bacteria.

Table 1. The main parameters of the well and their change after regeneration

Well symbol	V-1	0-1	0-2	0-3	0-4
Wel age (years)	9	11	8	13	9
Starting specific yield (l/sec/m)	10	10,5	8	11	10
Specific yield before regeneration (l/sec/m)	5,5	5,8	3	5,1	6
Index of ochre sedimentation and bacterio-flora before regeneration	4	8	7	7	6
Type of regeneration	HNH	HNH	HNH	HN	H
Specific yield 10 days after regeneration	8,8	9,4	5,4	7,4	8,2
Index of ochre sedimentation and bacterio-flora 25 days after regeneration	0	0	0	0-1	0
Index of ochre sedimentation and bacterio-flora 90 days after regeneration	2	4	2	4	3
Increase of specific yield (%)	60	62	81,5	45	37

HNH - chemico-mechanico-chemical regeneration
HM - chemico-mechanical regeneration
H - chemical regeneration

The increase of the specific yield of five treated wells (Table 1), compared to the pre-regeneration yield, ranges from 37 to 81%. The biggest increase of productivity was achieved in wells which were "completely" treated (chemical-mechanical-chemical treatment), and in such instances the increase is always above 60%. By such regeneration in some earlier cases the increace of the specific yield was more than 150% (compared to the productivity before regeneration). In the well 0-3, which has undergone chemical-mechanical treatment, the resulting increase was 45%, while in the well 0-4. Which has undergone only chemical treatment, the amount of increased productivity was 37%. The simplified conclusion drawn from these results would be that the "pure chemical regeneration" contibutes about 50% to the total effect of regeneration. Results of some previously committed regenerations involving many tube and Ranney wells, suggest that it is possible to accomplish an increase of over 250% by the chemical preparation only. Such a high percentage of production rate increase relative to production rate prior to regeneration was registered for the wells where the preceding production rate decrease was as high as 80%. Therefore, the percentage of increase should be considered together with the absolute values of production rate prior to regeneration (by utilizing sodium hypochlorate). The characteristic of the pure chemical regeneration is best reflected in the fact that after the regeneration the curve of decrease is very steep.

The age of the treated wells (Table 1) ranges from 9 to 12 years, and during this span their specific yields have decreased by about 55%, on average. According to our observations, this is the optimum period when the first regeneration should be carred out. Later on, when a production rate decrease exceeds 60%, this high percentage of regained specific yield does not give a proper ilustration of successfulness of a performed regeneration.

Table 2. Change of qualitative and quantitative composition of the populations of iron and manganese bacteria and of ochre caused by the regeneration of the well

Index		Well Symbol				
		V-1	0-1	0-2	0-3	0-4
	a	4	8	7	7	6
of ochre precipitation	b	0	0	0	0-1	0
and bacterioflora	c	0	1	0	2	1
	d	1	3	1	3	2
	e	2	4	2	4	3
	f			2		4-5
Siderocapsa	a	7	8	9	9	5
	b	0	0	0	0-1	0
	c	0	0-1	0	1	1
	d	1	4	1	4	2
	e	3	4	1	4	2
	f			3		5
Gallionella	a	3	7	6	7	5
	b	0	0	0	0	0
	c	0	0	0	0	0-1
	d	0-1	2	0	1	1
	e	0-1	2	1	3	1
	f			1		2
Leptothrix	a	3	4	7	7	6
	b	0	0	0	0	0
	c	0	0-1	0	0	0
	d	0	1	0	2	1
	e	1	1	1	2	1
	f			1		3

a - state before the regeneration
b - 25 days after the regeneration
c - 40 days after the regeneration
d - two months after the regeneration
e - three months after the regeneration
f - 6 month after the regeneration

Computer processing of the many parameters which relate to the regeneration process of the 15 tube wells, such as:

productivity,
characteristics of the spring,
age of well,
qualitative and quantitative composition of bacterioflora and changes within it,
previous interventions performed in the wells,
type of performed regeneration,
the amount of chemical preparation,
duration of chemical performance in the well,
duration of mechanical treatment of the filter parts of the well, and others,

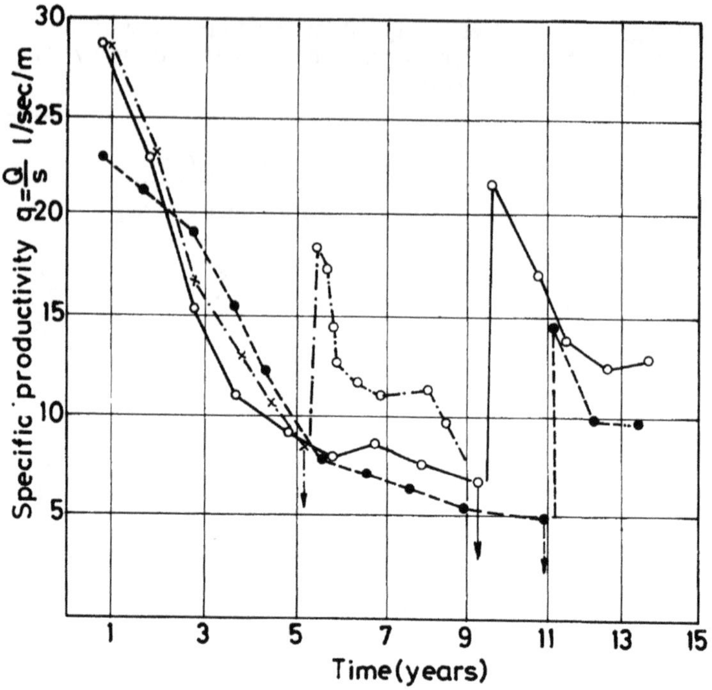

Fig. 4. Decrease of the specific productivity and its regain by the chemical regeneration

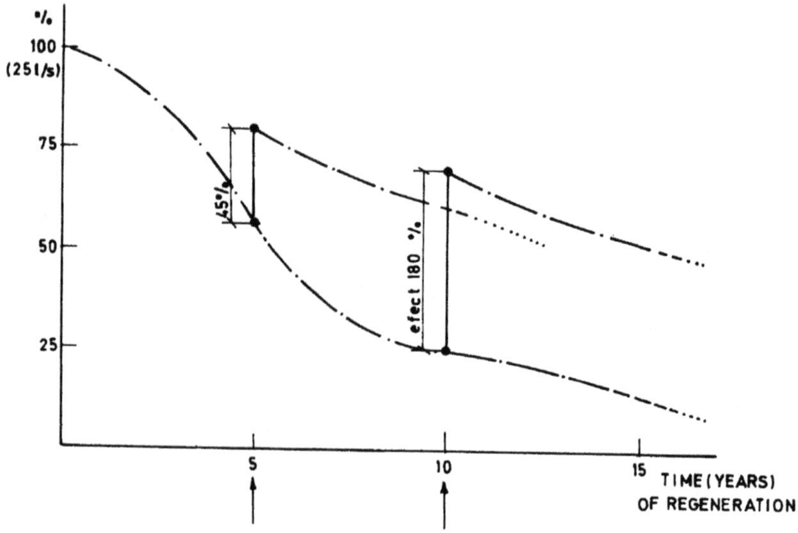

Fig. 5. Effect of the chemical and mechanical regeneration bz the preparation "Skital B" (computer data processing for 15 tube wells)

resulted in the relationship represented on Fig. 5. Noting the fact that a lot of wels have been regenerated twice, and interesting question occurred related to the time interval between regenerations. It is proposed that it is important to determine the critical time for the first regeneration, while the second one ought to be performed before the productivity falls bellow 50% of the initial productivity. On the average, with first regeneration an increase of about 45% (from 56% increased to 81%) is achieved, and with second regeneration the increase amounted to about 180% (from 25% to 67%). The possibility of regaining the lost specific productivity by the method of regeneration, as is represented in Fig. 5, holds true for the shallow tube wells, of a specific location (e.g. spring Obrenovac) and for the wells where the cause of the "clogging" was "blinding" of the filter parts of the well. Any generalization of the wells is untenable. Similarly, measures which are undertaken to hinder the decrease of productivity and to regain the productivity, exhibit variable success.

Regards of the type of the performed regeneration (chemical or combined) in all wells, the absence of iron and manganese bacteria as well as ochre precipitation has been observed (Table 2). Longer or shorter absence of the population of iron and manganese bacteria in the wells after regeneration depends, it is proposed, on the intensity of the chemical regeneration performed. In all the wells where the regeneration has been applied the detecable apperance of iron and manganese bacteria is observed only after 50 days has elapsed, and around the 90th day the presence of bacterioflora and of ochre precipitation on the simulators is only half the amount which was observed before the regeneration. In two wells where observations continued even after six months the difference in quantitative and qualitative structure of bacterioflora before and after the regeneration is still conspicuous. It is interesting to note that from the tree dominant genera of bacteria which occur in the wells when they are in normal exploitation, after the regeneration Siderocapsa appears first, while Leptothrix appears in some wells only after two months and even then in a very scarce number. Along with the reduction of bacterioflora after the regeneration, the quantity of the precipitated ochre in unit of time is several times lowered compared to the one prior to regeneration. The change in bacterioflora in the nearby piezometers, 20 and more meters away, was not significant. The only exeptions to this is the change recorded in a piezometer which was only 4,5 m away from one of the wells.

3.4 Cost effectiveness

Of all the important parameters which could be measured, the increase of the specific yield best indicates that with the performed regeneration the natural aging of wells is slowed down. Results presented here, as well as the best effect regarding the incrrease of the productivity of well is achieved when the chemical regeneration is combined with mechanical regeneration. The success of such a combination is understandable when bearing in mind the causes of the decrease in productivity, the impact of "Skital B" and the significance of the decisive factor of it's possible application. In assessing the total cost for one well, the following tree items participate equally:

specialistic of the job related to regeneration;
the price of mechanical regeneration;
the price of chemical regeneration.

According to the authors' calculations (taking the price of water as the basic for calculation) the regeneration in Yugoslavia is economically justifiable if at least 35% of increase of specific productivity in relation to the productivity prior to regeneration is achieved, while total expenses necessary for the regeneration do not exceed one fifth of the cost necessary for construction of a new well.

4 Conclusion

Years long investigation of springs at various localities of Serbia has shown an evident decrease of production rate of the wells constructed upon these springs. This decrease (so called "well aging") is always associated with the presence of iron and manganese bacteria and depositing of ochre upon filtering parts of the well.

By applyng the method of a combined regeneration (chemical and mechanical), the increase of specific yield of up to 80% (when compared to the one prior to regeneration) was achieved. For this purpose, our patented substance, "Skital B", of a proven physico-chemical and bactericidal properties, was used.

When carried out upon the well of the optimum age, when the possibility of "curing" still exist, the regeneration is economically justified since the overall costs of regeneration in Yugoslavia are stil lower then the value of water regained trough this intervention.

5 References

Barbič, F. (1983) Ph.D.Thesis: Iron and manganese bacteria population in groundwater of Belgrade spring and their ecological importance in water exploiting, University of Belgrade, Belgrade

Barbič, F. Savić, I. Krajčić, O. (1990) Complexity of causes of well yield decrease. Internat. Conference on Microbiology in Civil Engineering, Bedford, England (in press).

Barbič, F. Savić, I. Krajčić, O. Vuković-Pal, M. Babić, M. (1987)Benefits of upholding of iron bacteria for transformation of iron and manganese in groundwater. Mikrobiologija, 24/2, 129-138. Belgrade

Bracilović, D. Barbič, F. Djinjić, M. (1975) Chlorination for iron removal in well water. Water and Sewage Works. 1, 40-42.

Institute for Technology of Nuclear and Other Mineral Raw Materials (1985) Technical report on technological process of well regeneration. Belgrade.

6 TUBEWELL FOULING AT A SALINITY INTERCEPTION SCHEME, WAKOOL, NSW, AUSTRALIA

R.G. McLAUGHLAN and R.M. STUETZ
Centre for Groundwater Management & Hydrogeology
University of New South Wales, Kensington, Sydney, Australia

Abstract
Iron fouling of pumps and tubewells can significantly affect the economics and efficiency of a groundwater drainage scheme. Commonly collected information on pump performance is not always adequate to determine tubewell efficiency particularly at a site where recharge from irrigation may be significant. Hydrochemical monitoring of the water in the aquifer and after pumping, show increased oxygen levels which may affect fouling. Closed circuit television camera surveys indicated heavy iron fouling in tubewells, particularly around the pump inlet. Localized fouling over part of the tubewell inlet zone, indicates discrete hydrochemical zones, causing fouling. Iron related bacteria were found to be associated with these deposits.
Keywords: Fouling, Clogging, Tubewells, Bore, Incrustation, Iron hydroxide, Bacteria, Gallionella

1 Introduction

The Wakool-Tullakool Sub-Surface Drainage Scheme has been designed to reduce waterlogging and land salinisation problems in the surrounding irrigation areas.The scheme is located in the Riverine Plains of southern New South Wales (Figure 1). The $28 million scheme involves tubewells at 47 pump sites discharging saline water from shallow aquifers to a large evaporation basin. At some pump sites the tubewells and pumps have iron deposits which have affected pump performance and increased maintenance costs of the scheme.

The Wakool-Tullakool area is underlain by buried river channels and their associated floodplains. The pumpsites are located in and around the sands and fine gravels of these 'prior' streams.

Most of the pump sites have suction lift centrifugal pumps which will pump from a water level of about 7 metres. Several pump sites have a submersible pump due to deeper aquifers. The suction lift pumps have one or two tubewells connected to them. The tubewells are PVC and generally 200mm

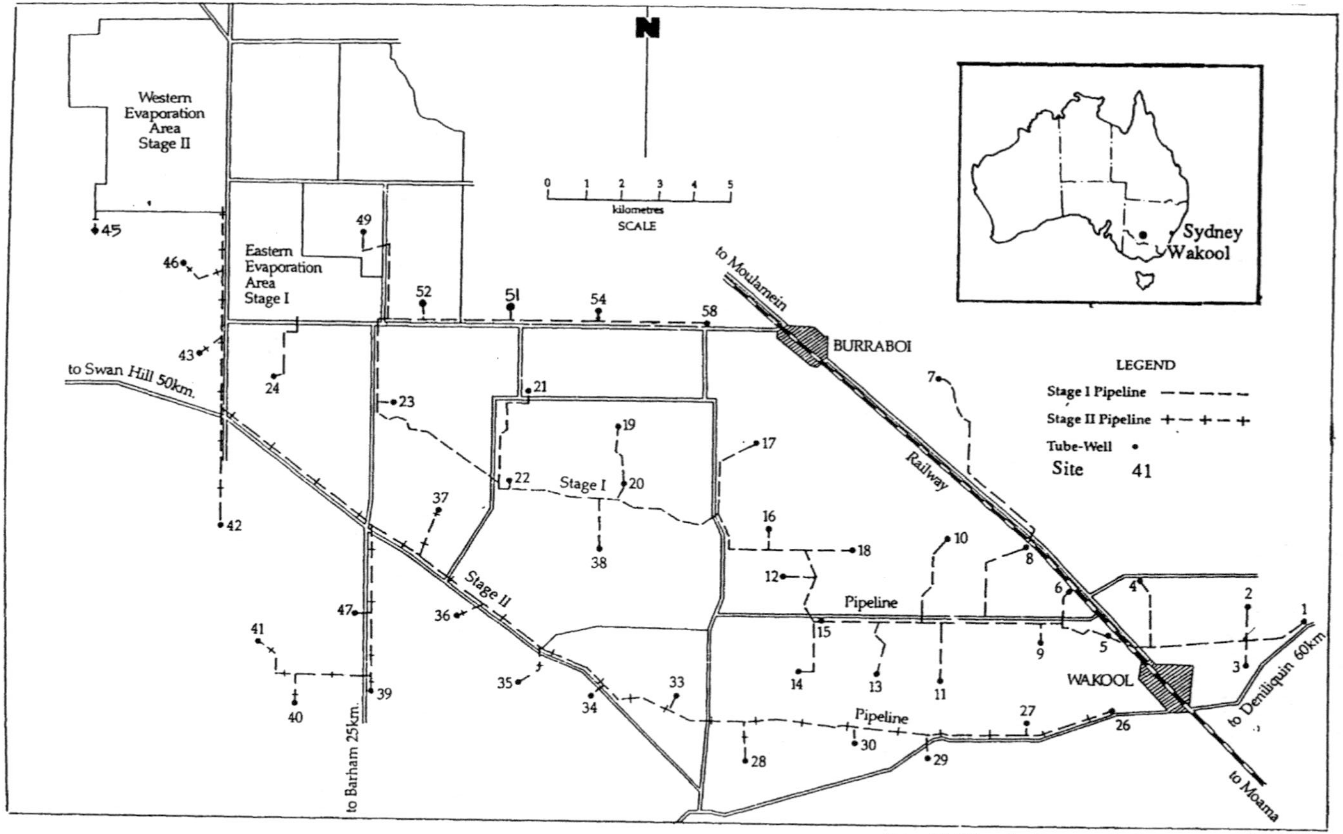

(Modified from Oxborrow 1987)

Fig. 1 The Wakool-Tullakool sub-surface drainage scheme and location map

diameter about 9 - 15m deep with a slotted interval of 2 - 8 metres. The wells have been gravel packed but the slotted PVC only has an open area of about 7%. Static water levels are 1 - 3 metres below ground level.

Ten pump sites were selected to evaluate biofouling problems. This study involved analysis of pump performance data, hydrochemical monitoring, chemical, microbial analysis of fouling deposits and CCTV camera surveys.

2 Materials and Methods

Two sets of water samples were collected from the tubewells. A Riha bladder sampling pump was used to indicate hydrochemical conditions in the aquifer. The other samples were pumped using either a centrifigal suction lift pump or a submersible pump under normal operating conditions. The tubewells were pumped for 20 to 60 minutes before sampling. Samples were pumped through a flow cell and measured by a YSI Water quality monitoring system (Model 3560) and a YSI dissolved oxygen meter (Model 58). Iron levels were determined in the field with a Merck Microquant iron kit (Model 14759).

Tubewell deposits were collected using a bailer type device. They were examined by phase-contrast microscopy using an Olympus microscope (Model BHC) fitted with an Olympus photomicrographic system (Model EMM-7) and scanning electron microscopy (SEM) using a Cambridge scanning electron microscope (Model Stereoscan 360) equipped with a Bio-Rad CryoTrans system (Model E7400/7450).

Samples were homogenised (1 g/10 mL of sterile distilled water) in a blender for 40 seconds and serially diluted for carbohydrate and microbiological analysis. Carbohydrate determinations were performed by the phenol-sulfuric acid assay (Dubois *et al.*, 1956). Microbiological analyses consisted of enumeration of Hetertrophic Iron bacteria (Kolbel-Boelke *et al*,. 1988) and Standard Plate Counts (AWWA, APHA, WPCF, 1981).

Dried samples were acid digested (perchloric:nitric acid /2:1) and then analysed using a Labtam inductively coupled plasma emission spectrography (ICP).

A closed circuit television (CCTV) system (Laval Underground Surveys) was used to examine fouling in the tubewells.

3 Tubewell performance

In order to determine the type and scale of the tubewell fouling problem it is necessary to quantify the performance of the tubewells. Tubewell performance can be influenced by borefouling (e.g. aquifer and screen clogging) and/or pump fouling (e.g. pipeline and pump impellor). In this system cleaning for pump fouling consists of high pressure jetting of the pump impellors which is a simple procedure while bore fouling would require suction pipe extraction, expensive chemical and mechanical treatment with significant pump downtime.

Pump operating data collected were valve opening (%), pumping hours, flows (m^3), kilowatthours, suction (kPa), discharge pressure (kPa), tubewell water levels (m), flowrate (l/s) and amps. These data were generally collected at intervals of 1 - 6 months. Sites 13 and 17 had been running for over 6 years while site 29-58 had run for 3 years. Since pump data sets on other schemes may not be as comprehensive, the statistical relationships between variables (Table 1) may be of use in determining scheme performance elsewhere.

On suction lift pumps the suction point is often measured at the pump as an indication of pumping water levels. A good correlation was generally found on single tubewell pumps but the relationship was much poorer on the two tubewell systems. The suction point was found to be a better indicator of the combined waterlevels on these systems.

An instantaneous flow rate determined by the pump operator from the flow over a minute was found to be a good approximation of the flow rate calculated from pumping hours and flow (m^3).

The pumps' discharge pressure was found to be very poorly correlated with the flow rate from the pump. Conditions further down the pipeline that would influence discharge pressure include fouling in the pipeline and whether other pumps were operating.

There was a poor correlation between the pump amperage and the flowrate. The variations in flowrate may be too small to affect the pumps energy consumption due to the relatively low flowrates (8-30 l/s).

An often used indicator of pumps and bore efficiency is the amount of power (kilowatt hours) consumed for a given volume of water pumped. This will be influenced by the pumping water level and discharge head. When graphs of kwhr/m^3 vs. time for each site were examined no significant trends within or between fouled and non fouled bores were detected. Difficulties were found using specific capacity or energy per volume of water type indicators on this scheme. The pumping water level has remained fairly stable being limited by the suction capabilities of the pump while the static water level has varied between 0.5 and -1.0 metres over the last 6 years. Although the flowrate is accurately known the specific capacity can not be calculated unless the drawdown is known. Groundwater recharge from irrigation and rainfall change the static water levels and affect measured drawdowns by an unknown amount precluding the use of specific

Table 1. Correlation coefficents of pump data

Site/ Tubewell No.	Suction vs waterlevel	Inst. flowrate vs calc. flowrate	Discharge pressure vs calc. flowrate	Discharge pressure vs inst. flowrate	Amps vs inst. flowrate
13/					
733	0.484	-	-	-	-
1706	0.939	-	-	-	-
combined	0.860	-	-	-	-
17/					
1466	0.874	-	-	-	-
29/					
1450	0.806	0.816	-0.010	-0.500	0.463
30/					
1444	0.915	0.976	-0.122	-0.050	-0.214
34/					
380	0.944	0.942	-0.540	-0.494	-0.086
38/					
1719	-	-	0.045	-0.327	0.372
41/					
1472	0.911	0.842	-0.075	0.703	0.905 P11
43/					
1885	0.965				
1874	0.955				
combined	-	0.932	-0.119	-0.892	0.118
54/					
1954	0.217				
1957	0.408				
combined	0.374	0.818	-0.747	-0.101	0.069
58/					
1981	0.645				
1997	0.781				
combined	0.802	0.605	-0.240	-0.326	0.118

capacity analysis on the currently available data.

An indicator of pump fouling is the ratio of pump discharge to suction pressure. As the pump impellor becomes clogged the tangential velocity of the water can not be maintained and the pressure differential between the suction and discharge drops. This is re-established when the pump impellor is cleaned. In this type of scheme, system performance monitoring needs to be extended from the currently collected data to include specific capacity of tubewells and aquifer storativity to evaluate bore and aquifer clogging. On twin tubewell sites provision needs to be made for individual testing.

4 Hydrochemistry

Water chemistries were measured in the tubewells using a Riha sampling pump (Table 2). These values indicate local aquifer hydrochemistry under non-pumping conditions.Samples for comparison were also taken using a centrifugal suction (Sites 13, 17, 41, 54) or a submersible pump (Site 38) under normal operating conditions (Table 2).

In 5 of the 7 tubewells dissolved oxygen levels were higher under pumping conditions. At Site 38 using a submersible pump the O_2 increased by 19%. This increase under drawdown conditions was accentuated by the low open area of the slots (7%) causing turbulence and oxygenation by agitaton with air at the air watertable interface. At the other 4 tubewells which used suction pumps the increase ranged from 49 - 708%. These increases may be due to drawdown conditions and to leaking seals in the suction lines. This indicates that oxygenation which increases fouling could be reduced by improving pipe connections.

Water temperatures measured using the sampling pump were 1-3°C higher than from the other pumps due to the much lower flow rate and longer travel time from the sampling point to the flow cell.

At Site 54 the 2 tubewells are located within 30 metres of each other These tubewells have significantly different water chemistries although they stratigraphically tap the same aquifer. This shows the localized hydrochemical variations that occur in this environment.

The iron rich waters (T.No. 1957, 1706) are also the most saline at the sites with 2 tubewells. The tubewells with high iron levels also have the most significant fouling problems.

As an indication of the water to chemically precipitate, water analyses from tubewells 733, 1706, 1466 and 1472 were modelled using MINTEQA2 (Brown and Allison, 1987) an aqueous chemical equilibrium program. The measured redox was used to speciate iron. There was no correlation between measured redox and the dissolved oxygen level. Theoretical values of redox calculated from the Nernst equation using the O_2/H_2O couple and based on the level of dissolved oxygen were between 800-850 mv while measured **redox** were between 45 and 193 mv. This suggests there is no

Table 2. Tubewell water chemistry

Site	Tubewell No.	Temperature (°C)	Conductivity (ms/cm)	pH	Eh (mv)	Fe^{2+} (mg/L)	Dissolved oxygen (mg/L)	Tubewell Fouling (a)	Pump Fouling (a)
Samples from pump under operating conditions									
13	733	18.2	11.9	6.29	107	0.2	0.44	I	–
13	1706	18.7	33.1	6.23	68	3.0	0.97	F	–
13	combined	18.6	17.9	6.25	116	0.8	0.53	–	F
17	1466	17.9	28.7	5.98	90	1.2	0.88	F	–
38	1719	18.4	39.9	6.27	197	0.0	0.93	I	C
41	1472	18.3	24.0	6.57	86	0.8	0.10	F	F
54	1954	19.1	30.8	6.23	159	0.0	1.79	C	–
54	1957	18.6	43.5	5.84	112	3.0	0.55	F	–
54	combined	18.9	41.7	6.01	137	1.2	0.82	–	F
Samples from Riha sampling pump									
13	733	21.0	28.0	5.98	140	0.0	0.25	I	–
13	1706	21.7	35.8	6.10	045	5.0	0.12	F	–
17	1466	20.5	36.6	6.05	134	2.0	0.22	F	–
38	1719	19.6	40.4	6.30	161	0.5	0.78	I	C
41	1472	19.0	33.8	6.48	193	2.0	0.12	F	F
54	1954	21.0	40.8	6.04	132	0.0	0.80	C	–
54	1957	21.5	42.5	5.49	120	3.0	0.37	F	–

a. C – clean; I – iron stained; F – fouled.

equilibrium between dissolved oxygen and measured Eh. Disequilibrium may also occur between dissolved oxygen and the ferrous/ferric redox couple (Nordstrom et al 1978). Dissolved oxygen was excluded from model runs using MINTEQA2.

Ferrihydrite and the more poorly ordered Fe-oxides have been associated with filamentous 'iron bacteria' (Chukhrov et al 1973) while Ferrihydrite was identified in precipitates from a groundwater treatment plant (Vuorinen et al 1986). The saturation indices for these minerals were calculated (Table 3.). They show that theoretically all of the bores would not precipitate amorphous Fe-hydroxide or Ferrihydrite but may precipitate Lepidocrocite or Goethite. Although disequilibrium between the measured Eh and the ferrous/ferric couple and chelation with humics would shift the saturation indices. Within the ranges of water qualities measured the saturation indices were more sensitive to changes in Eh than iron levels. This indicates the need to determine the ferrous/ferric couple for accurate modelling. Of interest is that the non fouling tubewell had a similar saturation index to the fouling tubewells. Although the mass of precipitate formed on the fouling tubewells was about an order of magnitude greater. The CCTV survey has shown zones of heavy and light fouling in the slotted interval of the fouled bores. The bulk water chemistry measured by the water sampling may not be representative of the localized zones causing precipitation in the tubewells.

Table 3. Chemical modelling results

Tubewell No.	Saturation Index			Precipitated Goethite (mol/kg)
	Ferrihydrite	Lepidocrocite	Goethite	
733	-3.28	0.24	0.87	3.11×10^{-6}
1706	-3.06	0.46	1.14	5.08×10^{-5}
733 &1706	-2.69	0.83	1.47	1.40×10^{-5}
1472	-2.28	1.24	1.87	1.43×10^{-5}
1466	-3.85	-0.30	0.28	1.05×10^{-5}

5 Deposit Analysis

Examination of tubewell deposits by phase-contrast microscopy revealed many different types of bacteria, the majority of which were filamentous. (eg. Gallionella Sp.).

Heterotrophic iron bacteria were present in all deposits (Table 4). However, their numbers did not correlate with the observed degree of fouling in the bores (CCTV survey) or the quantity of carbohydrate in the residue. Chemical analysis indicates the deposits are predominantly iron.

Table 4. Analysis of Tubewell deposits

Site	Tubewell No.	Viable Cell counts ($x10^3$cells/g)	Heterotrophic cell count ($x10^3$cells/g)	Carbo-hydrate (µg/g)	Fe (%)	Mn (%)
13	1706	20	35.0	987	28.57	0.01
17	1466	18	6.2	629	23.50	0.04
41	1472	2	4.8	1451	30.96	0.06
54	1957	3	4.0	739	12.10	0.03

Electron microscopy showed the deposits to consist of filamentous bacteria, with precipitated iron hydroxides and other inorganic materials accummulated on the extracelluar polymer matrix (Figure 2).

A review of the literature (McLaughlan & Knight 1989) has found no definite evidence that iron related bacteria actively oxidise Fe^{2+} into Fe^{3+}, under the hydrochemical conditions existing at this scheme. In either low pH waters or where Fe^{2+} is bound with humic substances the iron related bacteria may play an active role. It is likely that their role is passive, with the extracellular polymers either accummulating colloidal iron hydroxide or acting as a substrate for precipitation. It would be expected that their presence would increase the rate of fouling.

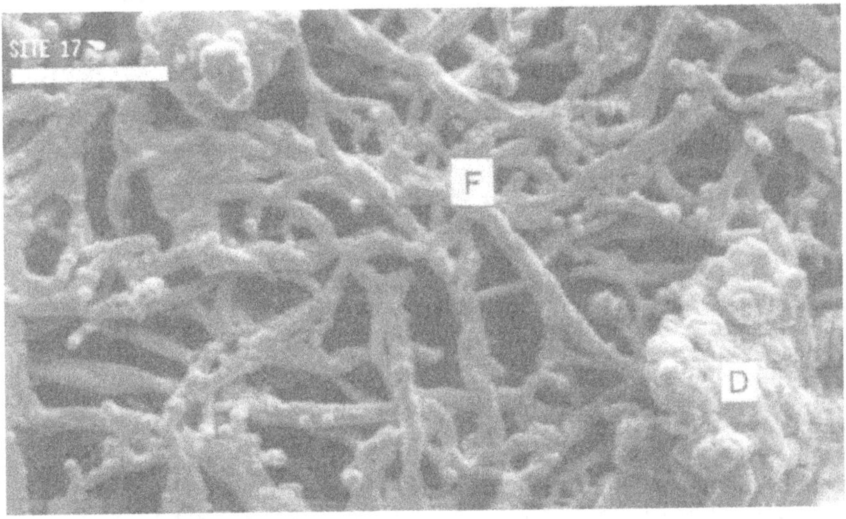

Fig. 2. Scanning electron micrograph of frozen-dehydrated tubewell deposit (bar = 10 µm) D - Iron precipitate F - Filamentous bactaria.

6 CCTV Surveys

A CCTV camera was used to define the location and degree of fouling in 6 tubewells (Table 2). In general the casing of the fouled tubewells was slightly iron stained, from the static water level to the pump inlet. The fouling was heaviest within 2 to 3 metres of the pump inlet, which is the zone of groundwater mixing and aeration. At Site 13, tubewell 733 was slotted from 3 to 8 metres and had a slight discolouration between 4 to 5 metres (Figure 3); whereas tubewell 1706/2 was sloted between 4.5 to 8 and 12 to 15 metres and had heavy fouling from 7 to 8 (Figure 4) and 12 to 15 metres. This was reflected in the hydrochemical data, which showed that the upper part of the aquifer has a lower iron level, while the deeper portion of the aquifer had a higher salinity and iron level. CCTV surveys proved to be useful tool in evaluating hydrochemical zones within the tubewells.

7 Conclusion

The use of specific capacity analysis on the collected pump data was not appropriate to determine tubewell fouling at this drainage scheme. Groundwater recharge from irrigation and rainfall affect the water levels by an unknown amount. A commonly used indicator of pump and tubewell efficiency is the amount of power consumed for a given volume of water pumped. Analysing the pump data indicated no significant relationship between power consumption and flowrate. The ratio of pump discharge to suction pressure was an effective indicator of pump fouling. More specific performance monitoring need to be undertaken to quantify the hydraulic and economic effects of fouling. Differences between measured oxygen levels in the aquifer and the those of the pumping tubewells, show leaking seals may oxygenate the water potentially increasing fouling problems.

CCTV surveys showed that heavy fouling occurred around the pump inlet, where groundwater mixing and aeration occurred. This would have been increased by the low screen open area. At some tubewells, slotted intervals also had localized deposits, indicating that tubewells draw water from discrete hydrochemical zones causing precipitation upon mixing. These deposits are likely to contain iron as an amorphous hydroxide or Ferrihydrite. Chemical modelling of the water indicated that it was undersaturated with regard to these minerals. However, the bulk water chemistry modelled from the tubewells may not reflect the local hydrochemical zones which are causing the fouling. Iron related bacteria were found to be associated with these deposits and will affect the rate of deposit accumulation.

Fig. 3 Photograph of CCTV survey of unfouled tubewell 733

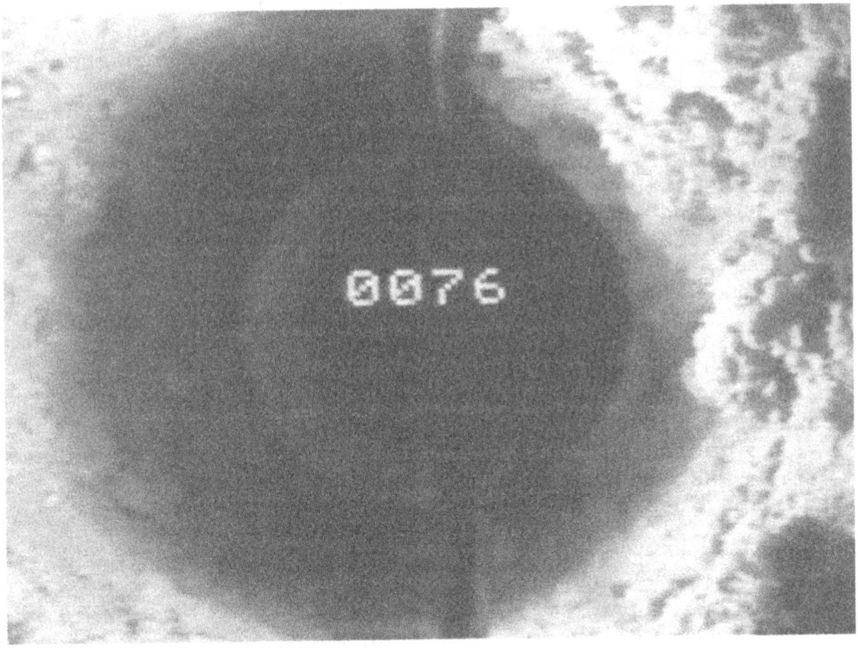

Fig. 4 Photograph of CCTV surveys of fouled tubewell 1706

8 Acknowledgements

The authors would like to thank the Department of Water Resources for their assistance and permission to use Figure 1. Dr. M.J. Knight has critically reviewed the paper. The Federal Government and the State Water Authorities provided funds for this work as part of a grant into corrosion and incrustation in groundwater bores.

9 References

AWWA, APHA, WPCF (1981) Standard methods for the examination of water and wastewater. Fifteenth edition. American public health assn., Washington D.C.

Brown, D.S., and Allison, J.D. (1987) MINTEQA1, an equilibrium metal speciation model :user's manual, EPA/600/3-87/012, United States Environmental Protection Authority, Athens, 92pp.

Chukhrov, F.V., Zvyagin, B.B., Ermilova, L.P. and Gorshkov, A.L. (1973) New data on iron oxides in the weathering zone, in **Proc. 4th Clay Conf. 1972 (Madrid)**, vol. 1, pp.333-341.

Dubois, M. Gilles, K.A. Hamilton, J.K. Rebers, P.A.and Smith, F. (1956) Colourmetric method for determination of sugars and related substances. **Anal. Chem**. 28,350-356.

Kolbel-Boelke, J. Tienken, B and Nehrkorn, A (1988) Microbial communities in the saturated groundwater environment I: Methods of isolation and characterisation of heterotrophic bacteria. **Microb. Ecol**. 16,17-29.

McLaughlan, R.G. and Knight, M.J. (1989) Corrosion and Incrustation in Groundwater Bores : A Critical Review , Research Publication 1/89, Centre for Groundwater Management and Hydrogeology, University of New South Wales,42p

Nordstrom, D.K., Jenne, E.A. and Ball, J.W. (1978) Redox equilibria of iron in acid mine waters, in **Chemical Modelling in Aqueous Systems: Speciation, Sorption, Solubility, and Kinetics**, (ed E.A. Jenne), ACS Symposium Series 93, pp 81-95.

Oxborrow, S. (1987) The Salinity problem of the Wakool Irrigation District and Tullakool Irrigation Area : The wakool- Tullakool sub-surface drainage scheme, B.Eng Thesis, New South Wales Institute of Technology.

Vuorinen, A., Carlson L., and Tuovinen O.H. (1986) Groundwater Biogeochemistry of iron and Manganese in relation to well water quality, in **Inter. Symp. on Biofouled Aquifers: Prevention and Restoration** (ed D.R. Cullimore), American Water Resources Assoc. Tech. Public. Series TPS-87-1, pp 157-168.

7 WELL DETERIORATION IN SINDH, PAKISTAN

C.C. LEAKE
Groundwater Development Consultants Limited,
Cambridge, UK
T. KAMAL
Water and Power Development Authority, Pakistan

Abstract
Since the mid 1960's almost 4000 public tubewells have been constructed in the alluvium of the River Indus in the Province of Sindh, Pakistan. These tubewells are used for irrigation and water table control. The specific capacities of some tubewells declined rapidly although designs were based on sound construction principles. Others, of the same design and installed in similar conditions, maintained their efficiencies. Initial attempts failed to relate simple chemical parameters of the groundwaters to the rate of well deterioration. It now appears that deterioration may be due to hydrochemical processes, possibly assisted by physical and biological factors. The hydrochemical component may have two sub-components: incrustation at the gravel pack/formation interface due to high pressure drop; and derogation of the aquifer near the tubewell due to chemical alteration of abundant mica flakes by reaction with flowing groundwater of varying hydrochemistry. Determination of the significance of these two sub-components is important as remedial measures are possible in the former whereas, if the latter is prevalent, cheaper, less durable well design should be adopted and early replacement planned.
Keywords: Indus Plain, Well Deterioration, Hydrochemistry

1 Introduction

The Province of Sindh, located in the south of Pakistan, consists of three essentially different physiographic units (Figure 1):

- the rocky plains and hills of Kohistan and the Khirtar Range in the west;
- the alluvial Indus Plains in the centre;
- the sand dunes of the Thar Desert in the east.

Much of the alluvial plains is irrigated by canals originating at three barrages (at Gudu, Sukkur and Kotri) on the Indus River; the total commanded cultivable area (CCA) of these structures is some 5.6 million hectares. Irrigated agriculture is the mainstay of Sindh's economy, providing for the livelihood of most of the population.

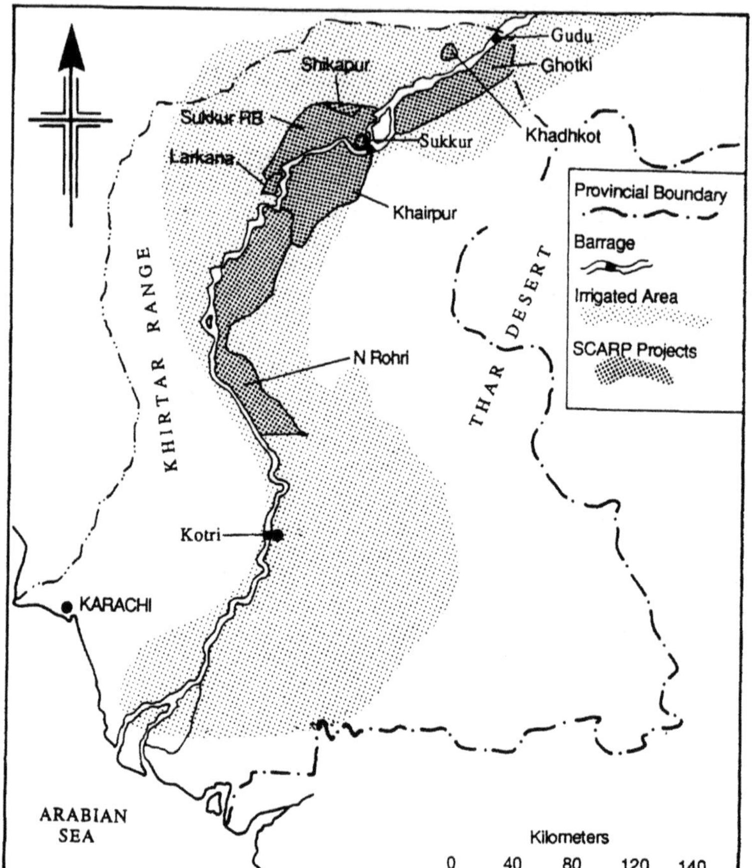

Fig. 1. Location map, Sindh Province, Pakistan

Irrigation of a large, contiguous flat area of these Lower Indus Plains without artificial sub-surface drainage, inevitably resulted in soil waterlogging and salinisation, seriously threatening agricultural production. The optimum remedial measure to this situation was identified as vertical drainage by pumped tubewells, to lower the water table to safe levels below ground surface, that is, to depths below the rooting zone of crops and of low evaporation. The tubewell drainage concept had already been demonstrated as effective in Punjab.

2 Background hydrogeology

2.1 The Lower Indus alluvial aquifer complex

The Indus Plains in Sindh are underlain by predominantly sandy alluvium, except for the extreme south where the subsurface is dominated by deltaic and marine clays and silts. The sands are mainly grey, micaceous, of medium to fine grading and contain thin, lenticular silty clay layers. The thickness of these predominantly sandy deposits is about 70 m in the south of Sindh and considerably more in the north.

Thus the alluvium constitutes a huge essentially unconfined aquifer. The permeability of the sands is higher than their grain size suggests, being mainly in the range of 20 to 50 m/d. The storage coefficient of the aquifer averages about 13 percent but varies from less than 5 to more than 20 percent. Much of the groundwater in this aquifer is brackish or saline, but important fresh groundwater bodies occur along the Indus and at some of the major irrigaton canals.

The alluvial aquifer is suitable for tubewells of high discharges. In the north, test well discharges of up to 150 l/s from cut sections of about 75 m were sustained for several days, showing tendencies of attaining equilibrium after only a few days. In the south, shallower wells with screens of 25 m length and yields of 30 to 60 l/s could be obtained, again with near equilibrium conditions.

2.2 Wells and wellfields

Public tubewell installation started in Punjab under the control of the Water and Power Development Authority (WAPDA). Several hundred deep (about 100 m), high capacity (up to 150 l/s) tubewells were constructed with the dual purpose of water table control (subsurface drainage) and additional supply of irrigation water. The basic well design (Bakiewicz, Milne and Pattle, 1985) was based on reverse circulation drilling at a large diameter (c. 600 mm), mild steel casing and slotted screen surrounded by natural gravel shroud, and turbine pumps driven by electric motors through vertical shafting (Figure 2). Some design deficiencies were quickly identified; some of the wells pumped appreciable amounts of sand and, more seriously, many of the wells deteriorated rapidly in terms of production and specific capacity. Studies carried out (Clarke, 1965) suggested that a relation existed between the chemistry of the pumped water and well deterioration (Figures 3 and 4). The contention was that in aggressive water (defined by pH and Eh (where pH is the redox potential), mild steel corrosion products plugged the slots of well screens, affecting their specific capacities.

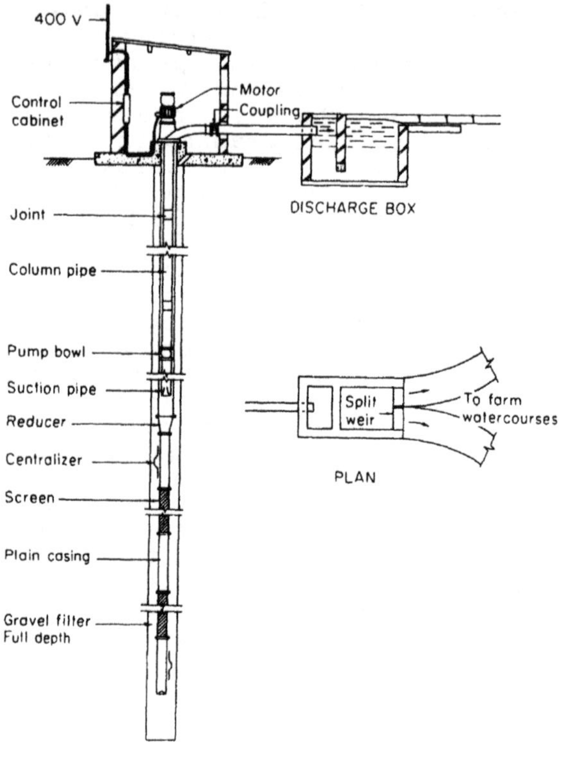

Fig. 2. Typical design of irrigation tubewell

Well design in Sindh took account of the Punjab experience. To overcome the corrosion and associated incrustation problems, inert (glass reinforced plastic or GRP) and resistant (stainless steel) materials were used. To avoid sand pumping, finer graded gravel packs were adopted. These designs proved highly successful and by the end of the 1980s, almost 4000 public tubewells of this kind had been commissioned in Sindh as part of Salinity Control and Reclamation Projects (SCARP), listed in Table 1.

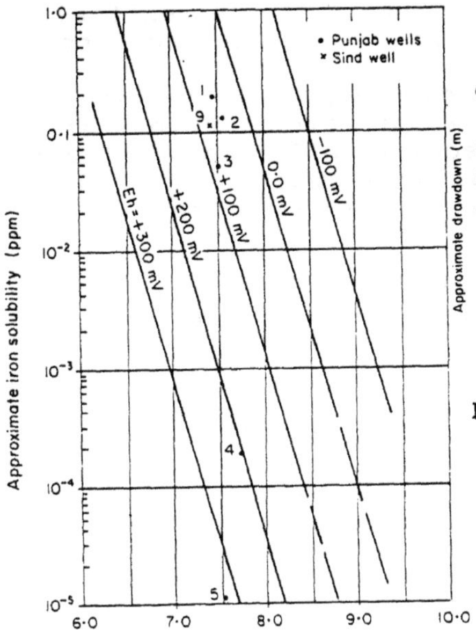

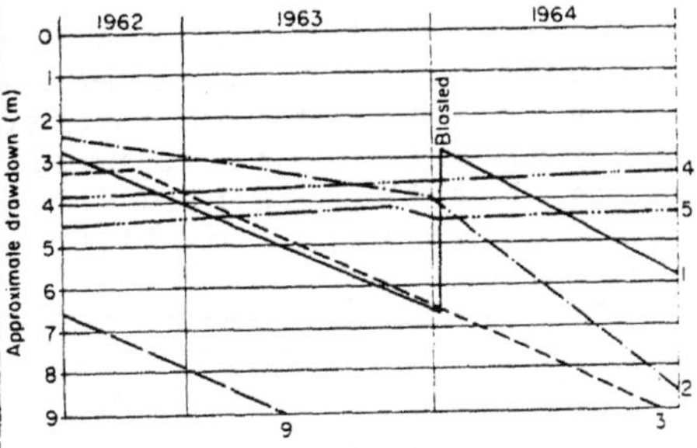

Fig. 4. Well performance (early design)

Fig. 3. Eh-pH relationship

Table 1. SCARP tubewell projects in Sindh

Project	Year of Con-struction	Nr of Tubewells Saline	Fresh	Pumpage (millions m^3/annum) Rabi	Kharif	Total
Khairpur	1963-70	365	175	162.3	198.0	360.3
N Rohri	1969-79		1192	588.2	605.9	1294.1
Shikapur	1973-74		50	13.4	12.7	26.1
Larkana	1974-75		35	5.3	5.1	13.4
Sukkur	1977-78		18	5.5	3.1	8.6
Khadhkot	1977-78		26	7.0	6.4	13.5
SukkurRB	1975-79		400	67.8	76.9	144.8
Ghotki	1976-90		410	80.2	249.3	329.4
S Rohri	1976-89		1200	53.7	37.5	91.1
	Total	365	3506	1083.5	1197.8	2281.3

Source: WAPDA, 1988

All of the wells of the improved design were initially very successful. Design yields were produced with no sand content. In the case of fresh discharges, the pumped water was reused for irrigation; saline effluents were disposed by dilution in the river or in the major canals. The water levels in the wellfield areas declined as predicted providing the required water table lowering.

However, it gradually became apparent that even tubewells of the improved design were deteriorating faster than expected. Effective well life (excluding the pump and motor) is usually assumes as 15 to 20 years, but this is generally viewed as conservative, since older fully productive wells are common throughout the world.

The earliest SCARP Project in Sindh is Khairpur. Comparison of specific capacities of 381 wells in 1967 and 1977 showed a mean deterioration of about 30 percent (Central Monitoring Organisation 1978). This average masked a variation ranging from slight improvement to decline of specifc capacity of 80 percent (Figure 5). This pattern of well deterioration was also apparent in later wellfields.

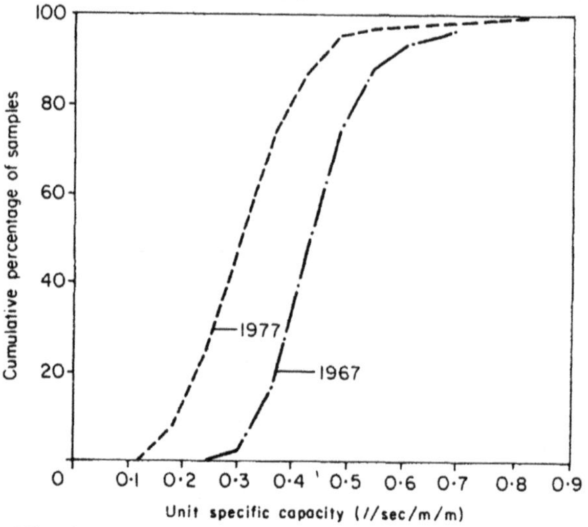

Fig. 5. Khairpur wells: unit specific capacities

3 Studies of Public Tubewell Deterioration

Though the original Khairpur well design are still generally viewed as successful and are being followed in new drainage wellfields, it is apparent that economic advantages are to be gained if specific capacity deterioration can be avoided or slowed down. Consequently, attempts were made to identify the causes and mechanisms responsible.

Review of the Khairpur experience provided some useful clues. This wellfield utilised both slotted GRP and wire-wound stainless steel screens.

Comparison of the performance of the two types showed that the percentage deteriorations had practically identical distributions (Figure 6); in view of the completely different character and open area of the two kinds of screen, this suggested that the mechanism responsible took place away from the screen face. This was confirmed by the circumstantial evidence of the absence of sand pumping in any of the wells; moreover, neither inspection of badly deteriorated wells with an underwater television camera, nor extraction of some GRP casing and screens disclosed any incrustation. Though not absolutely conclusive, this evidence is strongly indicative that whatever was affecting the wells, was happening away from the well screens, either at the formation/gravel interface or in the formation further away from the screen.

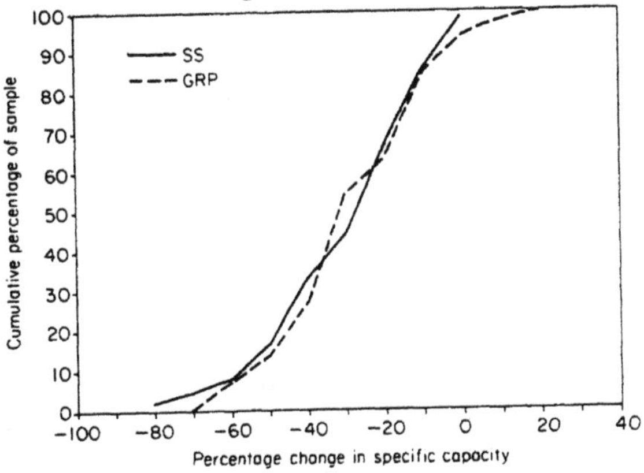

Fig. 6. Khairpur wells: change in specific capacity with respect to screen type

Of course it is possible that the deterioration of the Khairpur wells was the result of mechanical clogging away from the screen face, most likely at the aquifer/gravel pack interface. If this were the case there should be some relation between the decline of specific capacity and grain size properties of the formation at a particular well. Examination of some representative profiles failed to show such a relation (Figure 7).

If the main causes of well deterioration are chemical, then a correlation might be expected between groundwater chemistry and specific capacity decline. This is difficult to establish because of the multiplicity of hydrochemical factors which might be operative. As a first step the influence of overall salinity was investigated; there was a suggestion that saline wells performed better than fresh ones, but the evidence is inconclusive (Table 2).

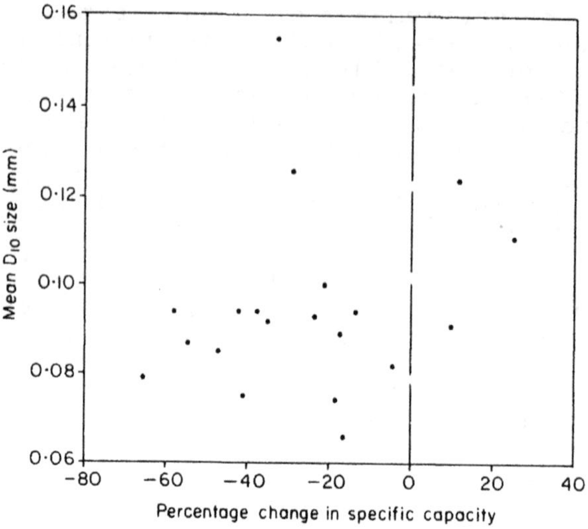

Fig. 7. Khairpur wells: relation between change in specific capacity and effective particle size

Table 2. Khairpur wells: relation between decline in specific capacity and pumped groundwater salinity

Salinity class (EC × 10^6 at 25°C)	*Sample population (No. of wells)*	*Percentage change in specific capacity 1967–1977*				
		Mean	*Standard deviation*	*Lower quartile*	*Median*	*Upper quartile*
EC < 700 and steady	100	−31.2	17.5	−18.0	−33.0	−42.5
EC 700–1500 and steady	37	−22.6	17.6	−11.0	−21.0	−35.0
EC 1500–3000 and steady	27	−26.7	20.7	−7.0	−19.0	−43.0
EC > 3000 and steady	64	−25.1	24.9	−7.5	−23.0	−38.0
EC < 700 and rising	72	−29.4	20.1	−21.0	−33.0	−40.5
EC 700–1500 and rising	40	−31.7	18.6	−20.0	−33.0	−44.0
EC 1500–3000 and rising	15	−26.6	34.4	−8.0	−22.5	−41.5
EC 700–1500 and falling	8	−29.8	22.8	−12.0	−25.0	−47.5
EC 1500–3000 and falling	9	−22.8	16.4	−16.0	−35.0	−52.5
EC > 3000 and falling	9	−27.8	27.9	−5.0	−23.0	−44.0
Overall	381	−29.0	—	—	−29.9	—

Stronger evidence for a hydrochemical cause of well deterioration comes from a more recent tubewell project - South Rohri. Statistical treatment of the performance of 180 wells, in operation for about 3 years, showed a markedly greater specific capacity decline in wells in which the discharge started fresh (electrical conductivity (EC) <1000μS/cm) but increased in salinity by at least 70 percent; other salinity classes gave practically identical results (Figure 8).

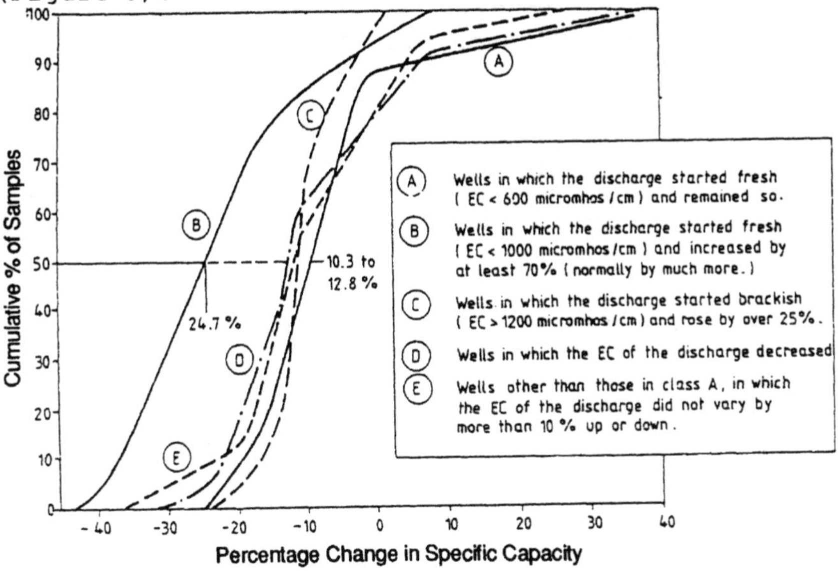

Fig. 8. South Rohri wells: change in specific capacity, 1978-1981

Further attempts to identify the deterioration mechanism more precisely involved monitoring some physical and chemical parameters at selected South Rohri tubewells. A plot relating pH, EH and specific capacity decline is shown in Figure 9. Clearly there is no definite correlation, but allowing for errors in the difficult measurements of Eh and complicating hydrochemical factors, it may be significant that the most severely deteriorated wells pumped relatively high Eh water and the least deteriorated wells pumped low Eh water. This could mean that some metallic constituents (iron, manganese) of the native groundwater may have precipitated and clogged the formation or the formation/gravel interface at some wells. However, it is fair to point out that it is unlikely that the native groundwater is sufficiently rich in such metals to produce the magnitude of well deterioration observed.

Another set of data examined in detail was that from Khadhkot Pilot Project, (Figure 1). Several chemical analyses of water samples were available over a ten year period 1979 to 1988/89. The changes in EC of these wells are summarised in Table 3, together with overall decline in specific capacity. Typical chemical compositions of water from each well is shown in the stiff diagrams in Figure 10.

Table 3. EC and deterioration of Khadhkot wells

Tube-well No	Electrolytic Conductivity of the Pumped Water (1979-1988/89) Initial (μS/cm)	Maximum (percent of initial)	Minimum (percent of initial)	Decline in Specific Capacity of the Well (1979-1988/89) (percent)
KK1	1320	108	68	14
KK2	1640	105	61	7
KK3	2250	110	80	11
KK5	720	247	100	30
KK6	480	134	83	21
KK7	460	141	83	47
KK8	560	177	100	28
KK10	1690	111	96	40
KK11	2440	100	86	70
KK12	1640	119	99	20
KK14	2800	107	79	68
KK15	2990	110	90	20
KK16	1900	142	100	19
KK17	2600	100	85	74
KK18	2350	100	61	21
KK19	700	139	89	39
KK20	1000	100	49	21
KK21	500	184	100	27
KK22	480	190	100	33
KK23	690	123	86	50
KK24	860	116	72	40
KK25	940	150	74	31
KK26	700	136	86	32
Average	-	-	-	33

Source: SCARPs Monitoring 1989

Correlations of various chemical parameters with well deterioration were attempted but no clear pattern emerged. Grouping the results in terms of the magnitude of EC gave the following results:

EC @ 25 C (μS/cm)	Decline in Specific Capacity (percent)
<1000	34.4
1000 to 2000	200.2
>2000	44.0

It is doubtful that these are significant. There is a suggestion that the severely deteriorated tubewells pump water particularly rich in sodium, (Figure 10), but there are sufficient anomalies to make even this contention questionable.

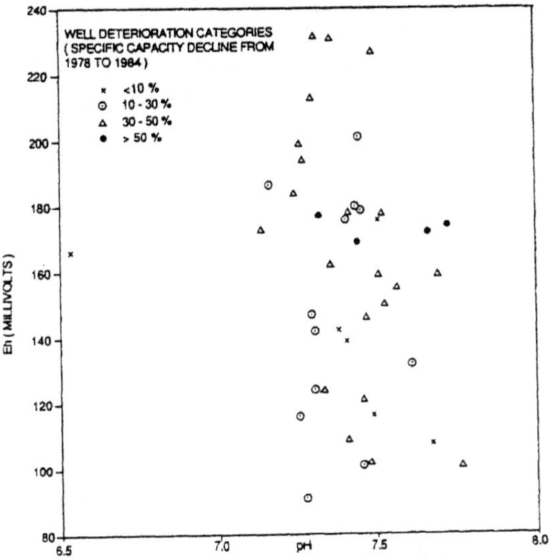

Fig. 9. pH, Eh and decline in specific capacity

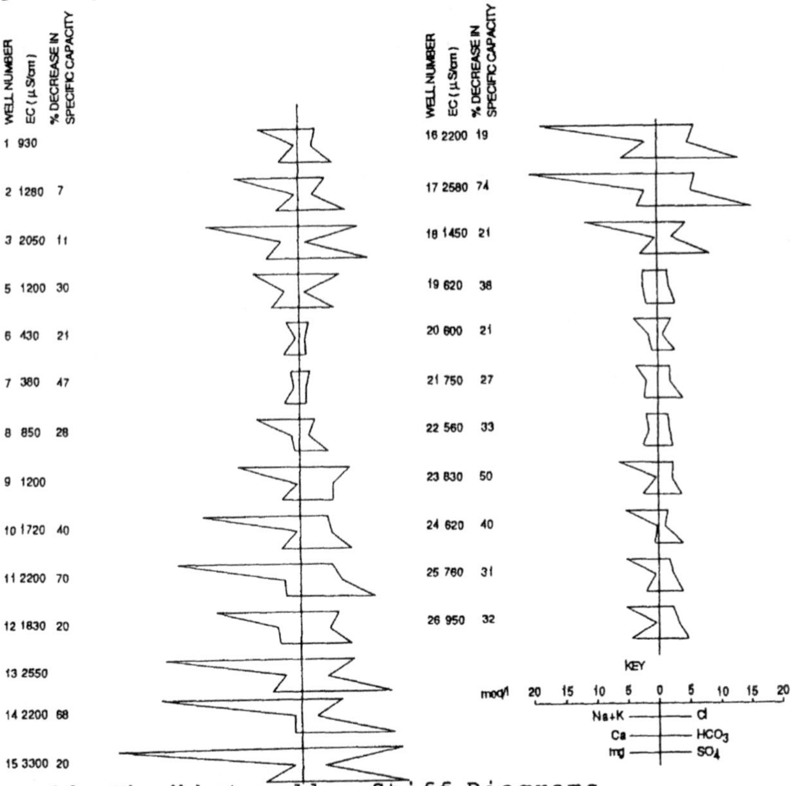

Fig. 10. Khadhkot wells: Stiff Diagrams

The Khadhkot wells demonstrate that the hydrochemical processes responsible for the deterioration of tubewells with time are highly complex and are not understood at present.

4 Discussion

There is considerable circumstantial evidence that hydrochemical processes are partly if not wholly responsible for tubewell deterioration with time in the Indus Plains of Sindh. At least two processes are possible:

- clogging of the aquifer/gravel pack interface by incrustants precipitating out of solution;
- decrease in aquifer permeability by reaction of certain minerals in the aquifer with groundwater of variable chemistry flowing towards the well.

The precipitation mechanism is most likely to take place in the immediate vicinity of the pumped well. The apparent absence of any incrustants of well screens would seem to mitigate against this mechanism, but it is possible that the process is effective at the formation/gravel filter interface where mixing of particles reduces porosities so increasing flow velocities and causing the highest pressure drops.

The process of chemical reaction between the aquifer and the groundwater is most likely to affect mica flakes, which are extremely abundant in the Lower Indus alluvium, constituting up to 50 percent of the formation. The alteration may occur when the potassium of the mica exchanges with one or more of the cations in the groundwater. This process has been documented in the oil industry, where an increase in salinity has been shown to cause potassium removal at the edges of mica particles which then expand and fray. Broken particles migrate down-gradient and clog pores, causing a reduction in permeability. This mechanism will be effective to some degree whenever micaceous sands are invaded or flushed by more highly mineralised water.

On balance, it seems that the second mechanism is more likely to be the dominant one affecting the tubewells in Sindh. Nevertheless, it is most important to establish the causes of well deterioration definitively, as the appropriate measures to deal with the two situations may be quite different.

If incrustation at the formation gravel interface is the main problem standard methods of reworking the wells should be effective. These might include redevelopment by jetting and, in extreme cases, primer cord blasting and acidising followed by cleaning through backwashing and overpumping. It is important that well rehabilitation be undertaken before the deterioration is very advanced; reworking should be carried out when the specific capacity of the well declines by say, 20 percent.

If chemical alteration of the mica flakes is the process leading to well deterioration then well rehabilitation is probably not possible. The appropriate approach may then be to design cheaper, less durable wells and replace them, when their use becomes uneconomic. It may be significant that the private sector has used this design philosophy without modification over the last forty years.

References

Bakiewicz, W., Milne, D.M. and Pattle, A.D. (1985) Development of public tubewell designs in Pakistan. **Q.J.eng.Geol.**, 18, 63-77

Clarke, F.E. (1965) Interim report on evaluation of corrosion and incrustation in tubewells of the Indus Plains. **Unpubl. Rep.** (Pakistan)

PART THREE

DIAGNOSIS AND MONITORING

8 BIOFOULING MONITORING METHODS FOR PREVENTIVE MAINTENANCE OF WATER WELLS

S.A. SMITH
Ada, Ohio, USA
O.H. TUOVINEN
Department of Microbiology, The Ohio State University,
Columbus, Ohio USA
Columbus, Ohio USA

Abstract
Monitoring for biofouling problems in water wells is important in attempts to alleviate or forestall the need for intensive well rehabilitation. In this paper, sampling strategies and analytical methodology for the detection of biofouling are reviewed, with emphasis on problems originating from the microbiological oxidation of Fe(II) and Mn(II) compounds.
Keywords: Biofouling, Iron Precipitation, Manganese Precipitation, Slide Collectors, Well Maintenance.

1 Introduction

Preventive monitoring and maintenance have been weak aspects of routine operation and management of water wells. This neglect of maintenance results in problems of decreased well efficiency and water quality and it stands in contrast to the routine practices in other industrial plant environments, including water treatment plants.

Biofouling in water supply wells and other types of wells is a common, recognized problem although it is not routinely monitored in water wells. Consequently, biofouling problems are usually discovered only when well and pump performance and water quality markedly deteriorate.

The present paper is concerned with biofouling problems which originate from the microbiological oxidation and subsequent precipitation of iron and manganese compounds in water wells and from associated biofilm formation. Several methods have been proposed for the analysis of biofouling problems in water wells but none appear to have gained wide utility for routine maintenance monitoring purposes.

2 Sampling and analysis of biofouling in water wells

Bacteria responsible for biofouling phenomena typically occur in complex biofilm communities. These biofilms range in size and depth depending on a number of physical, chemical, and biological factors which are site-specific. Analysis of these biological communities is essential for understanding the underlying biological nature of the problem and for designing effective well maintenance and control procedures.

Sampling for biofouling can take the form of: (i) pumping and collection of water flow; or (ii) collection of biofilm development on coupons and wellhead filtration devices.

Pumped water samples, when turbid, are often useful for biofouling monitoring purposes. Turbidity may be caused by biofouling or precipitation phenomena including red-water conditions. In water wells, such turbidity is often considered to constitute an advanced stage of biofouling. Therefore, turbid samples should be examined also with microbiological methods of analysis. In turbid water samples, besides free-swimming bacteria which are not representative of biofilm communities, the other bacteria captured are those shed by or sloughed off from the biofilm upon sudden changes in the hydraulic velocity occurring during pumping. Although accurate estimates are not available, these bacteria are generally assumed to constitute only a minor fraction of the entire microbial population associated with the biofilm of concern.

Surfaces immersed in well water readily develop a biofilm which varies in its characteristics depending on the prevailing environmental conditions. Biofilms also develop on in-line cartridge filters and other configurations near the wellhead which can therefore be used for microbiological analysis. Immersed surfaces will also collect biofilm-dwelling organisms from clear water which may show relatively low numbers of organisms in pumped water samples.

2.1 Microscopic examination

Light-microscopic examination has traditionally been the method of choice for identification of **Gallionella, Leptothrix** spp., and other organisms involved in iron encrustation which have distinct morphological features. Historically, light microscopy has been the sole method of examination of Fe- and Mn-bacteria in engineering practice. Although in many instances biofouling phenomena in water wells may not include these filamentous or stalked bacteria, their presence is positive evidence for Fe- and Mn-related biofouling. Microscopic examination of encrustations may show fragments of stalks and sheaths as well as nonfilamentous bacteria. It should be noted that the absence of bacteria is inconclusive evidence.

The thickness and rate of growth of biofilm can be measured with relatively simple microscopic techniques (Bakke and Olson, 1986); several imaging systems are also available at various levels of sophistication. Both types of techniques are sensitive to interference by detritus and other particulate material, including iron and manganese encrustations. These approaches assume relatively uniform distribution of biofilm on the test surface which rarely is the case.

Fluorochrome staining methods, e.g., with acridine orange, can be used to increase the level of resolution between organisms and detritus. In epifluorescence microscopy a wide range of materials can be used as test slides, whereas light-microscopy (bright field, phase-contrast, Nomarsky-interference contrast) necessitates the use of glass slides for optical transparency. Staining techniques combining a vital stain (i.e., fluorochrome) and a respiratory stain have been developed for a number of purposes which also have application for studying bacteria in groundwater samples (King and Parker, 1988). Bacteria may be concentrated by centrifugation or filtration before microscopic enumeration, but a loss of viability occurs to a varying degree, thereby preventing a meaningful count of viable bacteria. Microscopic enumeration is, at best, still qualitative for sedimentary materials because it is not possible to count individual bacterial cells in encrustations and in microcolonies.

Scanning electron microscopy (SEM) has a more powerful level of resolution in providing detail of biofouled surfaces. Preparation is destructive to the sample. With attached microprobe analysis (EDXA), the technique provides an opportunity to analyze elemental composition and distribution with morphological features of sample materials. Robbins device (McCoy et al., 1981) is a system in which SEM stubs are exposed to pressurized flow, followed by their removal for SEM examination. These devices have found application in a variety of industrial settings where monitoring for biofouling problems is desired. These devices can be mounted as a sidearm or outflow device in the same manner as slide box collectors and "moncells." To date, there is no information on their utility or testing in wellheads.

2.2 Cultural methods

Samples for biofouling analysis contain an unknown spectrum and number of microorganisms. In many cases it is not clear which of those organisms should be targeted for cultural methods of recovery. Thus the significance of quantitative enumeration by cultural methods remains elusive. The recovery of organisms is defined by the choice of the nutrient medium and therefore both the qualitative and quantitative aspects of cultural recovery must be addressed in future work.

It should also be noted that threshold values of bacterial counts for engineering practice, such as when a well should be rehabilitated, are arbitrary at best. However, culture methods are useful in providing information on the nature and extent of microbial growth that would then necessitate antimicrobial treatment such as chemical disinfection.

Several media for recovering Fe- and Mn-precipitating bacteria have been described in the literature. In general, these media are based on formulations which lack selective agents. Positive enrichment and recovery by cultural methods is recorded on the basis of Fe (or Mn) precipitation but several other bacteria can also grow in these media. Bacterial enumeration by standard microbiological methods is subject to problems of data interpretation because of the lack of selectivity of these media. If media selective for the growth of specific Fe- and Mn-precipitating target bacteria were developed, they may have limited practical use in view of the large spectrum of bacteria potentially involved in Fe- and Mn-precipitation. Field test kits employing nutrient media are now commercially available, providing presence-absence results for Fe- and Mn-precipitating bacteria. Quantitative interpretation of results may be difficult because of unknown effects of environmental stress on bacterial recovery. For some culture media, microaerophilic conditions are recommended. It has not been established, whether microaerophily is required because of oxygen toxicity, redox conditions, or some other factors important during incubation.

2.3 Collector methods

Several models of in-well and wellhead collectors have been used in previously described studies. These devices are readily installed in wells and wellheads and have been demonstrated to support the development of biofilm communities for subsequent microbiological investigations. With some collection methods it is possible to recover intact biofilm for further analysis.

Some of the primary considerations for the selection and utility of a suitable in-well or wellhead collector system are the following:

(i) access for installation;
(ii) influence of the collector on the technical performance of the well or water system;
(iii) collection of a representative sample of biofilm community.

Well performance is normally not disrupted during these sampling events. However, in-well collectors should be designed and installed carefully to avoid problems near the pump intake and installations of wellhead collectors should preclude back-siphonage.

In-well collectors necessitate access to the well bore through the well cap or piezometers. Some wells, particularly those equipped with surface-mounted lineshaft motors or other obstructive installation, do not provide physical access into the wellbore. In this case, installation through a piezometer or an insert tube is necessary. Piezometers are strongly recommended in new well constructions and required in some European countries (Wojcik and Wojcik, 1987; van Beek, 1987). Proper positioning of in-well collectors is important in order to avoid various submerged obstacles; for example, the collector has to avoid hanging up on submersible pump wire.

Various inserter devices have been designed for tight wells with submersible pumps, but their ability to provide proper conditions for simulating submerged surface for biofilm formation has yet to be established. In these devices the test slides or coupons are mounted within the inserter structure. The exposure of test surfaces to normal well-water conditions may be physically hindered if not properly mounted in the mounted slots.

Representative sampling remains an elusive issue. Glass and polycarbonate slides appear to provide for surface attachment, as also do stainless steel and other metal coupons. A "moncell" collector (Howsam and Tyrrel, 1989) is designed to collect deposits inside a sand-filled chamber, simulating a biofouling mass or other plugging in the filter pack of a well. The device, also collects other particulate materials in addition to the various constituents of biofouling.

Table 1 provides a listing of some collection apparatus applied to well biofouling problems.

Table 1. Well biofouling collection apparatus

Description	Reference
In-well suspended slides, no protective casing	Hässelbarth and Lüdemann (1972), Wojcik and Wojcik (1987)
In-well collector, slides in protective inserted device, several design versions	Smith (1984, 1988) Figure 1
Well-head slide collector, racks of cover slips in flow-through box	Hallbeck and Pedersen (1987)
Well-head slide collector, pressurized flow-through chamber with slides attached on a spool	Figure 2
Well-head sand-filled pressure cell ("moncell")	Howsam and Tyrrel (1989)
Robbins Device, removable coupon stubs exposed to pressurized flow	McCoy et al. (1981)

In-well suspended slides place the collection surface directly in the well bore and have the advantage of simplicity. The slides are exposed to the well environment without physical interference. These devices have proved suitable for large-diameter pumped wells (Hässelbarth and Lüdemann, 1972), mine observation wells (Wojcik and Wojcik, 1987), and pressure-relief wells (Hackett and Lehr, 1986).

Most pumped water wells prevent practical challenges to suspending slides in the well bore. In-well collector apparatus provide a protective inserter for the slides to permit their installation. These apparatus can be easily removed also in tight well clearances obstructed by pump cable and other equipment.

The apparatus have gone through several iterations (Smith, 1984, 1988). Two current inserter arrangements are in use. One consists of two acrylic plastic strips, spaced apart by shorter lengths of strips. The strips are cemented together and the ends ground to a taper to further ease insertion and removal. The slots are protected on two sides, eliminating problems with slides rubbing on the side of the well bore. Slides are clipped in the slots. The assembly is suspended on high-strength monofilament line.

A recent iteration uses a single, thicker acrylic or polycarbonate plastic bar which is slotted and tapered on the ends (Figure 1). The slides (standard microscopic slides, approx. 75 x 25 x 1 mm) are banded or clipped in the slots and the assembly suspended on

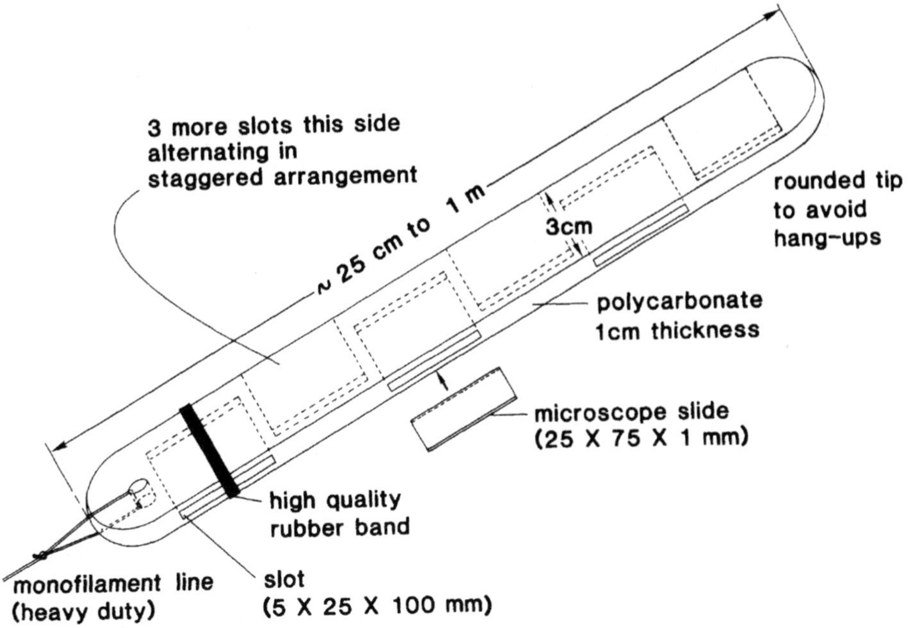

Fig. 1. Schematic of in-well slide biofilm collector (not to scale).

monofilament line. Both acrylic and polycarbonate can be disinfected with a hypochlorite solution. Polycarbonate is also autoclavable. Either device may be assembled in any desired length. Other materials, such as stainless steel, may also be used.

Wellhead slide collectors provide a means of exposing immersed test surfaces at wells where entry is not possible or prudent for a variety of reasons. If properly positioned, wellhead devices can be directly observed without physical disturbance or removal of the apparatus.

Two apparatus have been described for the wellhead application. Hallbeck and Pedersen (1987) described an autoclavable polycarbonate slide box in which cover slips mounted in racks were used as test surfaces. The box was designed as a flow-through system. Figure 2 illustrates a wellhead collector designed to meet a perceived experimental need of controlling flow with minimal oxidation in the present work. This device allows slow flow under pressure past glass slides. The collector keeps the chamber full when the flow is valved or switched off. The internal slide spool can (Figure 2) allows flow around the slides and prevents excessive slide movement. The spool can be easily inserted and retrieved from the pressure chamber. Devices in use or development are constructed from (i) acrylic with nylon and plastic fittings which may be chemically disinfected; and (ii) polycarbonate with metal fittings which can be either disinfected chemically or sterilized by autoclaving.

The "moncell" device (Howsam and Tyrrel, 1989) is a pressure cell with a valved inlet and a screened outlet, filled with sand as a packing material. By varying the particle size of the packing medium, the device has the ability to mimic the filter pack of the well. The acrylic shell of the moncell allows visual observation of discoloration in the sand, which can be documented by time-lapse photography. Changes in the permeability of the packing material can be quantified using falling-head measurements.

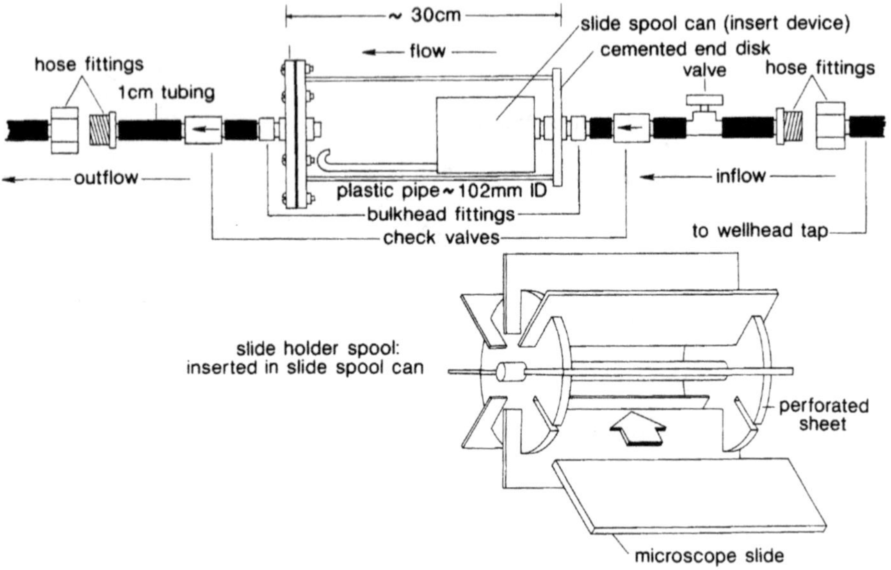

Fig. 2. Schematic of wellhead slide collector apparatus (not to scale).

Robbins Devices have not been applied to water wellhead problems to our knowledge, although they have been used in a variety of industrial settings. These devices could be mounted in a sidearm or outflow configuration in the same manner as slide box collectors and moncells.

3 Integrating biofouling monitoring into routine operations and maintenance

A definitive recommendation for devices and approaches is not possible at the present stage of the research. Each of the devices and approaches has both merits and drawbacks. Pumped sampling and analysis of sloughed biofilm using appropriate methods can provide a useful indication of biofouling. Slide-collected methods allow quantification of biofilm development. Moncell-collected materials provide a model of plugging processes in a filter pack and a means of qualitative assessment of well plugging.

The approach considered should fit the level of sophistication appropriate and the information needed. Such monitoring and analysis must be integrated into a comprehensive maintenance program for wells supportable by available technical and financial resources. How the water supply industry uses such methodologies in well maintenance will depend on further research and education on their value in the operating environment.

Acknowledgment and disclaimer: Research support from the American Water Works Research Foundation (AWWARF) is gratefully appreciated. AWWARF assumes no responsibility for the content of this work or for the opinions or statements of fact expressed.

4 References

Bakke, R. and Olson, P.Q. (1986) Biofilm thickness measurements by light microscopy. **J. Microbiol. Meth.**, 5, 93-98.

Hackett, G. and Lehr, J.H. (1986) **Iron Bacteria Occurrence, Problems and Control Methods in Water Wells.** National Water Well Association, Worthington, OH.

Hallbeck, E.-V. and Pedersen, K. (1987) The biology of **Gallionella**, in **International Symposium on Biofouled Aquifers: Prevention and Restoration** (ed. D.R. Cullimore), American Water Resources Association, Bethesda, MD, pp. 87-95.

Hässelbarth, U. and Lüdemann, D. (1972) Biological incrustation of wells due to mass development of iron and manganese bacteria. **Water Treat. Exam.**, 21, 20-29.

Howsam, P., and Tyrrel, S. (1989) Diagnosis and monitoring of biofouling in enclosed flow systems -- experience in groundwater systems. **Biofouling**, 1, 343-351.

King, L.K. and Parker, B.C. (1988) A simple, rapid method for enumerating total viable and metabolically active bacteria in groundwater. **Appl. Environ. Microbiol.**, 54, 1630-1631.

McCoy, W.F., Bryers, J.D., Robbins, J., Costerton, J.W. (1981) Observations of fouling biofilm formation. **Can. J. Microbiol.**, 27, 910-917.

Smith, S. (1984) An investigation of tools and field techniques for the detection of iron-precipitating bacteria in ground water and wells. MS Thesis, The Ohio State University, Columbus, OH.

Smith, S.A. (1988) Low-tech well microbiology research at SASCS and Ohio Northern University. **Biofilm Bull.**, 1(5), 7-8.

van Beek, C.G.E.M. (1987) Clogging of discharge wells in the Netherlands II: causes and prevention, in **International Symposium on Biofouled Aquifers: Prevention and Restoration** (ed. D.R. Cullimore), American Water Resources Association, Bethesda, MD, pp. 43-56.

Wojcik, W. and Wojcik, M. (1987) Monitoring biofouling, in **International Symposium on Biofouled Aquifers: Prevention and Restoration** (ed. D.R. Cullimore), American Water Resources Association, Bethesda, MD, pp. 109-119.

9 THE USE OF BOREHOLE CCTV SURVEYS

T. COSGROVE
Wessex Water Plc, Poole, UK

Abstract
The use of closed circuit television (CCTV) surveys in boreholes and wells has become common over the past twenty years. They provide a quick and relatively inexpensive means to obtain a visual record of the downhole construction and condition.

Technological advances have resulted in improved picture quality, more rugged construction and miniaturisation. Colour systems are now widely available and have demonstrated their superiority over black and white systems.

Video recording capability is essential if CCTV surveys are undertaken as part of long term monitoring, maintenance and rehabilitation programmes.

CCTV surveys are particularly useful in the investigation of biofouling phenomena and assessing the efficacy of any treatment or rehabilitation process which may have been undertaken to reduce encrusted deposits.
Keywords: Miniaturisation, Colour, Video Recording, Rehabilitation.

1 Introduction

Remote inspection by CCTV has long been established in industry, with applications ranging from nuclear fuel rod inspection to underwater pipe line surveys. The technology is proven and there is a plethora of systems, adapted for specific uses, commercially available.

Within the Water Industry the technique gained wide acceptance in sewer inspection initially but its benefits for water well investigation were recognised and it has now become a standard part of in-house downhole logging units.

In common with downhole geophysical logging systems, CCTV systems comprise four main elements: downhole camera and lights, cable and winch, surface power supply, data processing modules and data display/recording units. Recent technological advances have produced a miniaturisation of these components (with the exception of the cable) and modern systems can be highly portable. The development of reliable and rugged colour camera systems has greatly improved the interpretative potential of CCTV surveys and whilst they are by nature non-quantitative, they are an invaluable complement to more standard downhole logging investigations.

2 Equipment description

Numerous systems are commercially available and are designed according to specific survey environments. In the UK, borehole CCTV systems have developed in parallel to sewer CCTV systems and the principal requirement of the equipment is to produce good quality visual images within a low light intensity underwater environment.

The CCTV system is generally operated separately to standard downhole geophysical systems, primarily because the CCTV cable requires more conductors. The cable as a consequence is thicker (approximately 10mm) than standard logging cable and in deep hole applications may require extra reinforcement, commonly achieved by Kevlar coating.

Camera sizes vary but are generally in the order of 300-400mm in length, 50-75mm in diameter. They may have either built in lighting arrangements or bolt-on lighting attachments. The advantage of the former is a reduction in overall size and improved access capability. As a general rule, the larger the borehole the greater the light requirement. However, excessive light can reduce image quality by flaring and back reflection.

Cameras are fitted with a variety of lenses to provide either normal or wide angle forward views or side (radial) views. The latter is commonly achieved by a rotating mirror assembly. Some cameras are capable of scanning by remote control in both 180 degrees in the forward plane, and 360 degrees in the radial plane. Remote focussing is an advantage in situations where the camera cannot be centrally placed.

Colour camera systems are a recent introduction in borehole investigations and have added new interpretative dimensions such as defining lithological boundaries and detecting ferric staining.

The surface system comprises power and camera remote control unit, monitor screen and preferably a TV writer to annotate the images of log depth. Video cassette is essential to record the survey. Power supply is 110/24 v A.C.

3 Survey methodology

Downhole turbidity is the principal limitation to CCTV survey quality and every effort should be made to minimise downhole disturbance. Should this be unavoidable, as in instances where inhole pumping plant has to be removed to facilitate access, then ideally 24 hours should elapse between the removal and commencement of survey.

The actual survey can produce disturbance, particularly in small boreholes and it is common to run a forward view survey first. From experience this view provides the greatest amount of information. The radial view survey is then used to examine side wall features more closely.

Picture orientation is invaluable and can be achieved by simply lowering a marker line, such as a weighted and calibrated tape, in the borehole prior to the survey. Features such as fissures and encrusted zones can then be located relative to this and should visibility allow, the tape calibratation can be used to measure their size.

Where the borehole has been cleared of plant and the diameter is large enough, tripod mounted winched surveys are recommended at speeds between 1-3 metres per minute. Camera centralisers can be used but may themselves cause disturbance by brushing the borehole sides. When there is insitu plant or access is limited, manual lowering is necessary and the operator must also be able to see the TV monitor. Logging speed will be determined by access in these instances. The survey log should ideally be recorded on video tape which has the obvious advantage of providing reproducible archive data.

4 Survey applications and limitations

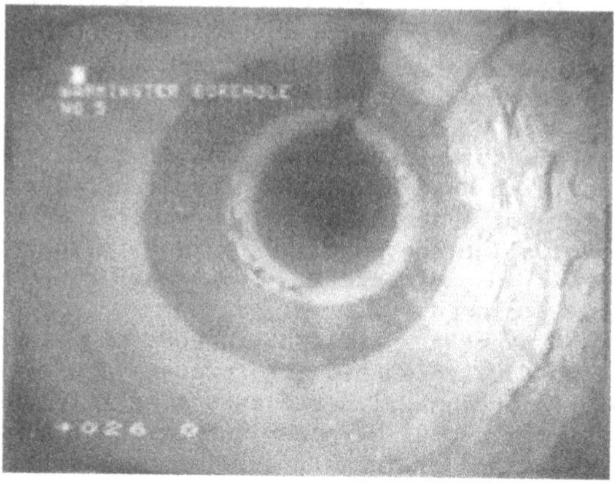

Fig.1
Lithological boundaries in Lower Chalk strata at Warminster, Wiltshire. This particular zone, probably a marl horizon, was not indicated by gamma logging of the borehole.

Until recently, downhole CCTV surveys were considered a novelty and at the best regarded as simply a visual confirmation of the information obtained from more standard logs, such as caliper and occasionally flow meter when particle movement was observed.

The quality and therefore the interpretative value of modern systems is higher and although their intepretation is still dependent on the skill of the operator, their value has increased to the extent that arguably in rehabilitation programmes they are the best means to evaluate the efficacy of any treatment undertaken.

As a simple visual record of physical downhole features, they provide confirmation of casing location and type, the presence of fissures, of collapses and diameter variations. These features may have been revealed by caliper logging but it is improbable that a caliper log would reveal whether the notation "slotted casing" on an old borehole record meant for instance, bridge slotted or flame cut slots and whether they were encrusted or clear. Furthermore a caliper log may not detect features such as verticle fissures, inclined bedding planes and changes in lithology (Fig.1). The latter is particularly evident in colour surveys.

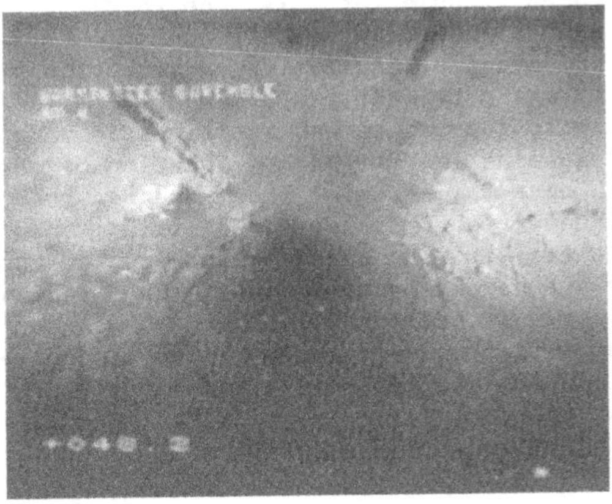

Fig.2

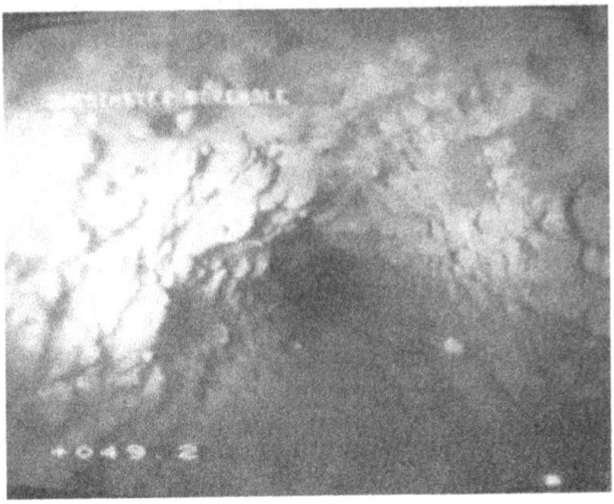

Fig.3

Fig.2 and Fig.3 show variation in encrustation on flame slotted casing over 1 metre depth in the same borehole. This was undefined by caliper logging.

Accepting that turbidity is a major limitation to CCTV surveys, the presence of turbid horizons or observation of particle movement does also provide evidence of flow phenomena within a borehole which may assist the interpretation of standard flow logs.

The examination of downhole plant can only be undertaken by CCTV survey and, providing access is available and good quality images can be obtained, is a non-disruptive and inexpensive means of examination.

Arguably, however, the greatest contribution of CCTV surveys is the insitu examination of downhole conditions with regard to the nature and extent of encrustations (Figs.2 and 3). Other downhole logging methods are unsuitable for this, though caliper logs may occasionally reveal heavily encrusted zones.

Biofouling and chemical precipitation in boreholes can be a major problem with regard to well performance. Before any rehabilitation work to remove the encrustations is undertaken, it is important to determine the extent and location of these deposits, inasmuch that the remedial programme can then be designed to ensure that whatever the process used (jetting, chemical or biocidal treatment etc) it is applied to the problem zone. Post treatment surveys are necessary to evaluate the effectiveness of the treatment in physical terms i.e., what percentage of material is removed or what percentage of slots/perforations were cleared. Used in conjunction with pre and post treatment yield and performance tests, greater confidence can then be applied to future remedial programmes.

5 Comments

The use and methodology applied in downhole CCTV surveys has evolved in an individualistic and haphazard manner. Evaluation of their worth is difficult as individual operators will hold varying opinions dependent on their own experiences and equipment capability.

The full benefit of modern systems, particularly colour ones, will not be achieved unless a more consistent approach is applied.

By sharing experiences and standardising survey procedures, users will be in a better position to direct manufacturers efforts to meet their requirements and not remain in the present position of accepting and adapting what is on offer.

6 Acknowledgements

The views in this paper are those of the Author, and do not necessarily coincide with those of Wessex Water Plc.

7 References

British Standards Institution, (1988). Geophysical logging of boreholes for hydrogeological purposes. BS 7022: 1988.

10 WELL PROBLEM IDENTIFICATION AND ITS IMPORTANCE IN WELL REHABILITATION

N. MANSUY and C. NUZMAN
Layne Western Co. Inc., Mission Woods, Kansas, USA
D.R. CULLIMORE
Regina Water Research Institute, University of Regina, Regina, Canada

Abstract

Problems associated with water wells can be placed into three categories including physical, chemical and biological. Biological problems are usually more common and often create difficult conditions to remediate. Biological Activity Reaction Test Systems (BARTS) have been developed to monitor the presence and activity of three important groups of bacteria found in groundwater systems. The three groups which have been generally identified include Iron Related Bacteria, Slime Forming Bacteria and Sulfate Reducing Bacteria. These three groups contribute to problems relating to quantity and quality. Field results have given valuable information with regard to environmental growth conditions and relative population size of each important population. Timed interval sampling has shown different incumbent populations at different zones within an aquifer using the BARTS. Bacterial concentrations and zones of growth are important considerations when designing effective remediation. Different sequences of chemical treatment and methods of application are used when combatting different problems. Many wells have been effectively rehabilitated using different chemical and physical techniques following the proper identification of the problem.
Keywords: Biofouling, Iron Related Bacteria, slime Forming Bacteria, Sulfate Reducing Bacteria, Rehabilitation, BARTS.

1 Introduction

The process of plugging of aquifers is a complex phenomenon which can be caused by a variety of factors functioning alone or in combination with each other. The causes of production losses and water quality deterioration during the aging process of wells are often difficult to define. This is partially as a result of the lack of knowledge of the physical, chemical and biological processes taking place and limitations in testing techniques.

Well rehabilitation is often made very difficult and ineffective in the event that the problem is not successfully identified. Historically acids have been used extensively in well rehabilitation and are often inhibited by extensive blankets of extra cellular polysaccharides (ECPS).

2 Forms of plugging

Plugging can be of a physical, chemical or biological nature or as a result of the conjugation of several phenomena. Plugging will usually result from the combination of several factors.

2.1 Physical plugging

Plugging of a physical nature may originate from either the partial collapse of the screen, infiltration of sand or silt into the interior of the screen, or through particulates lodging in the slots of the screen and the associated interstitial spaces of the sand and gravel.

2.2 Chemical plugging

Plugging of a chemical nature may result from the deposition of chemical precipitates which are often initiated by the presence of oxygen, pH shifts to the alkaline range or increases in the redox potential. Many hard brittle chemical deposits are actually precipitation byproducts of bacterial origin. Many of these chemical deposits are not recognized as bacterial origin. Singer and Stumm (1970) compared the rates of oxidation of ferrous iron under sterilized and natural conditions in water and found that the microbial mediation accelerated the reaction by a factor larger than 100,000 times.

2.3 Biological plugging

Biological plugging is the most common form of plugging

of water wells. From experience about 80% of all wells that are experiencing plugging, have a high level of biological activity. The presence of the biological component is not the only form of plugging resulting in the reduction of specific capacity. The presence of the bacteria will usually aggravate the problem. The production of large amounts of extra cellular polysaccharides (ECPS) will act as a glue or cement which will hold the chemical precipitates and silt, sand or clay together. It is the combination of this total mass which causes reduced aquifer holding capacity or causes entrance resistance.

2.4 Bacterial groups

There are three major groups of bacteria which have been generally identified as contributing to problems related to water quantity and quality. These three groups have been identified as Iron Related Bacteria (IRB's), Slime Forming Bacteria (SFB's) and Sulfate Reducing Bacteria (SRB's).

IRB's are normally considered nuisance microorganisms often causing problems with massive deposits of iron and manganese oxides and hydroxides. Mansuy (1988) lists the complete classification of the iron related bacteria. The majority of iron related bacteria are also slime forming bacteria which are very capable of precipitating and accumulating many chemical precipitates.

SFB's are the most common group of microorganisms found in aquifers. The majority of slime forming bacteria are also responsible for large deposits of chemical precipitates leading to the overall volume of biomass.

SRB's are a group of obligately anaerobic heterotrophic organisms which utilize organic matter via the reduction of sulfate to sulfide (eg. **Desulfovibrio** and **Desulfotomaculum**). The products of sulfate reduction are hydrogen sulfide and sulfuric acid which lead to corrosion. The production of hydrogen sulfide also creates problems with water treatment.

3 Biological Activity Reaction Test Systems (BARTS)

Biological Activity Reaction Test Systems (BARTS) have been developed to monitor the presence and activity of the three important groups of bacteria in groundwater systems. BARTS were originally designed for the diagnosis of biofouled wells. They have more recently been used for more varied applications, wherever there is a need to assess the bacterial population type or size. BARTS are simple in design and interpretation and can be used without the need for expensive equipment (eg. incubator).

The BARTS test works by creating a variety of different environmental growth conditions, allowing the growth of different microorganisms with different physiological requirements. this varied growth environment is achieved with a nutrient gradient diffusing from the bottom of the tube upward and an oxygen gradient diffusing from the top of the tube downward. The oxygen gradient is created by a floating intercedent device (Ball) restricting the passage of dissolved oxygen into the liquid filled BART. Refer to the diagram in Figure 6 for a description on how the BARTS test works.

4 Methods

Prior to sampling it is desirable to allow the well to rest for a period of at least 12 hours to concentrate themselves and detach for sampling. Water samples are collected according to standard methods at timed intervals corresponding to 5 minutes, 15 minutes, 30 minutes and 60 minutes. The samples are then transferred to the IRB-BART, SLYM-BART and the SRB-BART according to the instructions supplied with the tests. The tests are transported to an office or laboratory facility and incubated at room temperature. It is important to keep the tubes in an upright position and do not allow the temperature to get above 30 degrees celsius or below freezing during transport.

The day the test is initiated is noted on the tube. The tubes are then observed every day for any sign of a reaction. The first day that a reaction occurs is referred to as the days of delay. The reaction is then matched with the reaction patterns on the DESI poster. Refer to Figure 5 for a color interpretation of the BARTS reactions.

5 Results and discussion

Timed interval sampling has proven to be a valuable tool in the diagnosis of biological zones of growth. The BARTS tests have proven to be sensitive enough to detect variable indigenous populations of bacteria at different timed intervals after initiation of pumping.

5.1 Quantitative interpretation of BARTS.

The BARTS tests for detection of specific bacterial groups was originally designed to be a presence/absence test. It then later became semi-quantitative with "days of delay" corresponding to relative size of the incumbent bacterial populations. From one to three days was considered to be a major bacterial activity, from four to six days was considered a moderate bacterial activity and seven to ten days was considered a light bacterial activity.

From recent comparative studies using spreadplates and appropriate agar media the tests appear to be semi-quantitative from reactions using "days of delay'. The following results have been obtained by the manufacturer of the BARTS.

< 6 hours	- Very large (>10^6) and aggressive bacterial populations.
< 24 hours	- Very significant (>10^5) and active population.
48 to 96 hours	- Active population >10^4 cfu/ml.
Less than 6 days	- > 500 cfu/ml.
Greater than 6 days	- < 500 cfu/ml
Negative at 45 days	- Bacterial group not detected.

5.2 BARTS results form pumping station #24 Well #5, Kalamazoo, Michigan.

In Kalamazoo, Michigan, 3 wells were tested in the same well field using the above described procedure of timed interval sampling. Table 1 and Figure 1 show the days of delay and the reaction type from Well #5

Table 1. BARTS results from pumping station #24 Well #5

Timed interval	IRB-BART		SLYM-BART		SRB-BART	
	Days	Reaction	Days	Reaction	Days	Reaction
5 min	7	2	3	2	7	1
15 min	neg	-	4	2	neg	-
30 min	10	5	3	2	7	1
60 min	9	5	2	2	7	1

A rapid reaction in the SLYM-BART at all different timed intervals shows that a severe problem of slime forming bacteria exists and probably extends considerable distances into the aquifer. From the comparative spreadplate results there are >10,000 cfu.ml which represents a significant and active population. The days

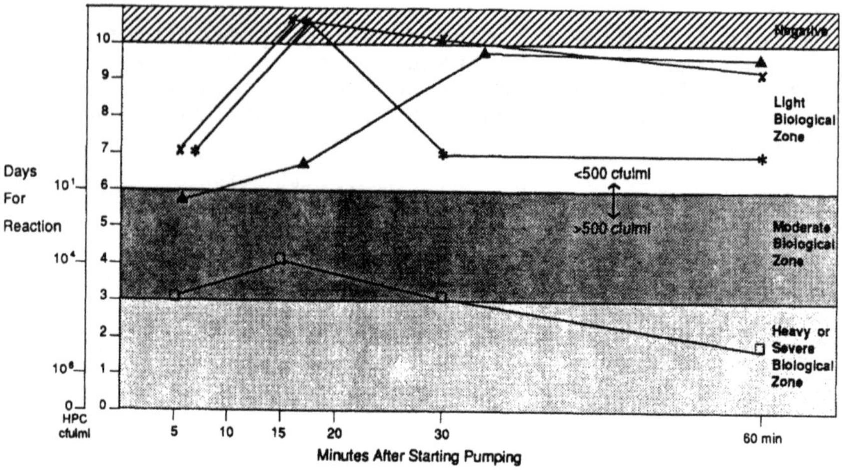

Figure 1: **Timed Interval BARTS Results for Pumping Station 24, Well #5**

✗ – Well #5 Kalamazoo IRB's
☐ – Well #5 Kalamazoo Slime Forming Bacteria
✱ – Well #5 Kalamazoo Sulfate Reducing Bacteria
▲ – Heterotrophic Plate Count (colony forming units [m])

of delay in the IRB-BART and the SRB-BART indicate population size to be < 500 cfu/ml. Refer to Figure 1 for corresponding days of delay and the heterotrophic plate count results.

The reaction type in the IRB-BART is most likely due to Gram-Negative aerobic rods and cocci in the 5 and 15 minute sample changing to possibly facultatively anaerobic gram-negative rods or anaerobic gram-negative straight curved and helical rods in the 30 and 60 minute samples. This would indicate that there is greater availability of oxygen closer to the well screen with reduced oxygen conditions further out into the aquifer. The difference in reaction may also be related to facultatively anaerobic or anaerobic microorganisms detaching from depths of the biofilm as pumping progresses. Reaction #2 in the SLYM-BART is most likely due to a consortial population allowing the growth at different zones within the BART tube.

5.3 BARTS results from pumping station #24 Well #16 at Kalamozoo, Michigan

Well #16 is in close proximity to well #5 and exhibits a few differences in BARTS results. Refer to Table 2 and Figure 2 for results from well #16 at Kalamazoo, Michigan.

Table 2. BARTS results from pumping station #24 Well #16

Timed interval	IRB-BART		SLYM-BART		SRB-BART	
	Days	Reaction	Days	Reaction	Days	Reaction
5 min	6	2	5	2	6	1
15 min	5	2	3	2	6	1
30 min	10	2	3	2	6	1
60 min	neg	-	2	2	6	1

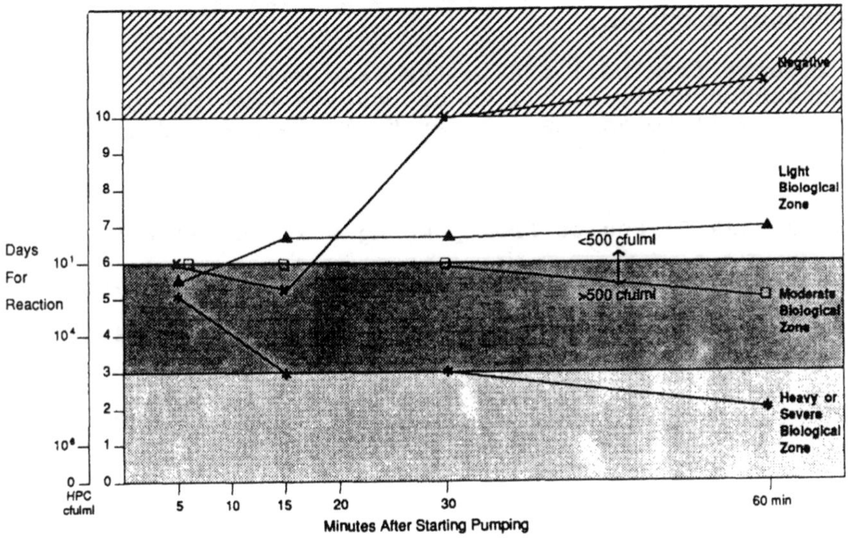

Figure 2: **Timed Interval BARTS Results at Pumping Station 24, Well #16 Kalamazoo, Michigan**

✘ – Iron Related Bacteria
✱ – Slime Forming Bacteria
□ – Sulfate Reducing Bacteria
▲ – Heterotrophic Plate Count (colony forming units [m])

IRB-BART demonstrated higher bacterial numbers in the first two timed interval samples with reduced bacterial numbers in the next two subsequent samples. This gives an indication that the major iron bacterial activity is occurring closer to the well. There was no indication of bacterial species succession as pumping progressed which indicates that environmental growth conditions are most likely more constant as the aquifer is penetrated. Reaction #2 again corresponds with Gram-negative aerobic rods and cocci. **Pseudomonas** being the most commonly found microorganism in this group.

The SLYM-BART showed a more rapid reaction as pumping progressed which demonstrates that there are more slime forming bacteria further out into the aquifer. Another possible explanation could be a higher rate of detachment as pumping of the well continued.

A reaction in the SRB-BART supports the fact that sulfate reducing bacteria are present. The most commonly isolated genera belong to **Desulfovibrio**. From the days of delay it appears that there are approximately 500 cfu/ml in all the different timed interval samples. There appears to be a constant detachment of SRB's into the water supply or they are constant into the aquifer.

5.4 BARTS results from pumping station #24 Well #18 at Kalamazoo, Michigan.

The results from Well #18 in the same well field demonstrates the variability of results that can occur from one well to another. Refer to Table 3 and Figure 3 for results from this well.

Table 3. BARTS results from pumping station #24 Well #18

Timed interval	IRB-BART		SLYM-BART		SRB-BART	
	Days	Reaction	Days	Reaction	Days	Reaction
5 min	5	2	5	3	neg	
15 min	neg	-	5	3	neg	
30 min	neg	-	6	1	neg	
60 min	7	4	6	1	neg	

The iron bacterial results were found to be > 500 cfu.ml at the 5 minute sample, negative at the 15 and 30 minute samples and < 500 cfu/ml at the 60 minute sample. There appears to be two very distinct and separate zones of biological activity isolated from each other by a zone of low bacterial activity. The 5 minute sample is most likely dominated by gram-negative aerobic rods and cocci. The reaction from the 60 minute sample could possibly be

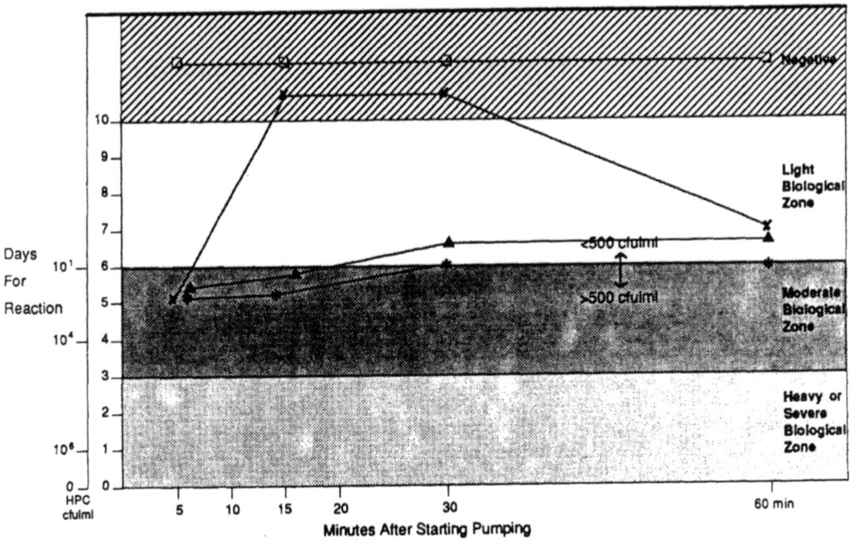

Figure 3: **Timed Interval BARTS Results at Pumping Station 24, Well #18 Kalamazoo, Michigan**

✘ – Iron Related Bacteria
✱ – Slime Forming Bacteria
▢ – Sulfate Reducing Bacteria
▲ – Heterotrophic Plate Count (colony forming units [ml])

dominated by gram-negative aerobic rods and cocci, facultatively anaerobic gram-negatives, gram-positive cocci, endospore forming gram-positive rods and cocci, regular non-sporing gram-positive rods or the sheathed and stalked iron bacteria. This indicates that the zone of higher oxygen concentrations is closer to the well and the reduced oxygen conditions around the well supports a different consortial growth.

The differences in environmental growth conditions around this particular well is also supported by the results in the SLYM-BART. the numbers of slime forming bacteria is constant throughout the aquifer as indicated by the similar days of delay at the different timed intervals.

Another interesting finding in this well compared to the other two wells at the same pumping station is the lack of sulfate reducing bacteria.

5.5 Results from Gainesville, Florida

The microbial changes that can take place can be seen by different reaction type at each timed interval. Refer to Table 4 and Figure 4 for timed interval results from Gainesville, Florida.

Table 4. Timed interval BARTS results at Gainesville, Florida

Timed Interval	BARTS Reaction #	BART Test	Possible Major population
5 minutes	4	IRB	Filamentous IRB,s
	-	SLYM	no reaction
	2	SRB	Desulfovibrio
15 minutes	7	IRB	Enterobacter
	-	SLYM	no reaction
	1	SRB	Desulfovibrio
30 minutes	6	IRB	Citrobacter
	-	SLYM	no reaction
	1	SRB	Desulfovibrio

The differences in incumbent populations as a well is being pumped will correspond to either different concentric growth zones into the aquifer or differences in detachment intervals. The results from Gainesville showed possibly filamentous iron related bacteria close into the well followed by Enterobacter and Citrobacter in concentric growth rings around the well. The day of delay on the results given in Table 1 was day 6 which for simplistic reasons has been given a designation of moderate levels of bacterial activity. Figure 4 shows a graph of the results from Gainesville. The same days of delay at the different timed intervals indicates that there is most likely a constant detachment of bacteria and the bacteria extend considerable distances into the aquifer. Figure 4 shows the same days of delay for the IRB-BART and the SRB-BART but no reaction in the SLYM-BART. The detection of sulfate reducing bacteria was expected in this case because hydrogen sulfide is a major problem in this particular well. The SLYM-BART was negative, although the reaction may have been missed if the tube was not swirled and the slimy growth occurred in the bottom of the tube.

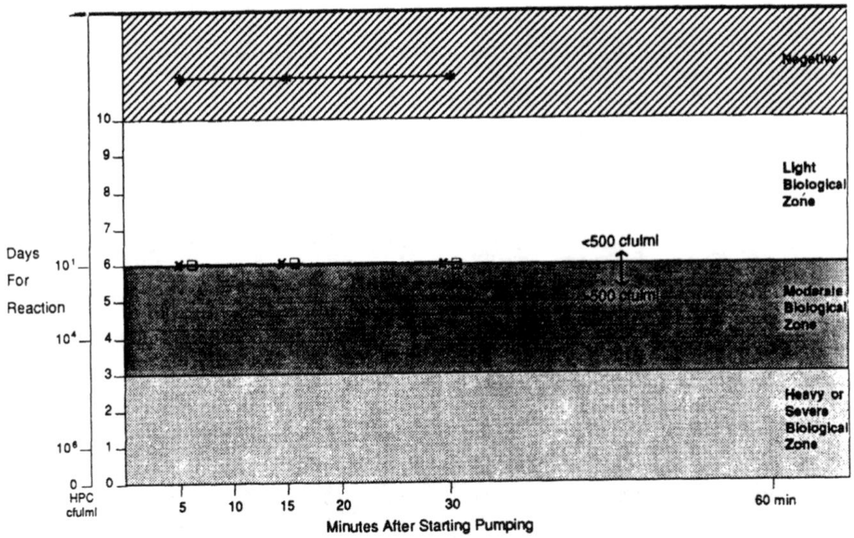

Figure 4: **Timed Interval BARTS Results from Well #3 at Gainesville, Florida**

x - Iron Related Bacteria
* - Slime Forming Bacteria
□ - Sulfate Reducing Bacteria

5.6 Chemical treatment variations

The detection and assessment of bacterial populations and activity is very important when designing a chemical treatment scenario. The major differences in treatment options as a result of the presence or absence of bacterial problems relate to;

1. Chemical volume.
2. Chemical concentration.
3. Contact time.
4. Chemical sequences.
5. Chemical combinations.
6. Methods of application.
7. Choice of chemicals.
8. Physical treatment option.

Due to the number of options and the variability of treatment, each recommendation is different on a case by case basis. These results represent a few case studies and do not represent all of the timed interval BARTS results. Following the identification of the specific groups of bacteria, the treatments have been altered and have been shown to be much more effective over previous treatments.

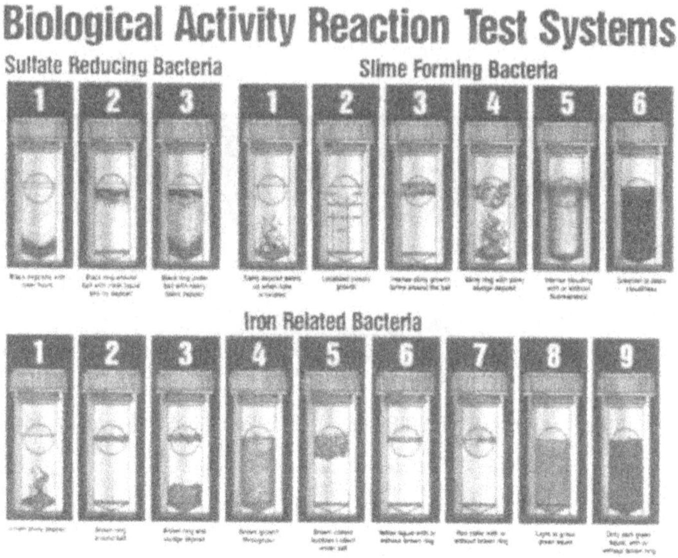

Figure 5. Interpretation photograph for BARTS.

6 Conclusions

The following conclusions can be drawn from the results obtained during these field trials.

1. BARTS offer a valuable tool for the diagnosis of bacteriological problems in water environments.

2. BARTS are semi - quantitative and qualitative in nature and are a valuable source of information.

3. Timed interval sampling is a useful sampling method to determine zones of bacterial growth.

4. It is important to watch for days of delay and reaction type when interpreting the BARTS.

5. Proper well problem identification is essential to effective rehabilitation strategies.

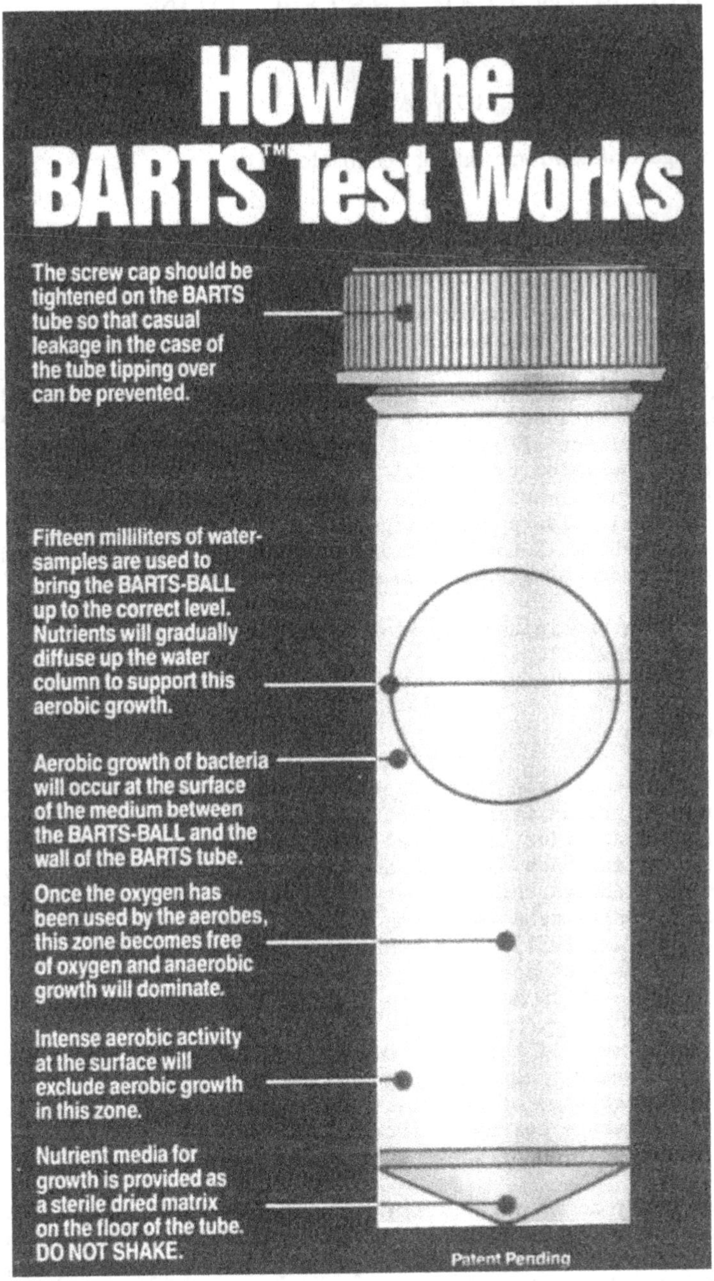

Figure 6. Illustration on how the BARTS test works.

11 MONITORING AND PREVENTION OF IRON BIOFOULING IN GROUNDWATER ABSTRACTION SYSTEMS

S.F. TYRREL and P. HOWSAM
Department of Agricultural Water Management, Silsoe College, Cranfield Institute of Technsology UK

Abstract
Iron biofouling is a complex process influenced by the interaction of the engineering and management aspects of the borehole (design and operation) with the environmental characteristics of the aquifer (hydrogeology, hydrochemistry and microbiology). A simple model is proposed which outlines the key factors governing the iron biofouling process. Experience of biofouled boreholes in the UK suggests that oxygen and flow velocity control iron biofouling in the subsurface environment. Basic variables are suggested which, if monitored regularly, will allow prompt diagnosis of the onset and extent of a biofouling problem. Recommendations are made for the design and operation of boreholes to reduce the problems associated with iron biofouling.
Keywords: Iron Precipitation, Biofilm Formation, Biofouling, Monitoring, Prevention.

1 Introduction

The development of iron-based deposits is a familiar and unwelcome occurrence in groundwater abstraction systems. The deposits consist of mainly iron hydroxides, microbial matter (cells and extracellular slime) and water. Such deposits are commonly of a slimy/sludgey consistency although this may vary especially if the deposit is quite old (iron deposits tend to become harder, brittle and more compact as they age)(Howsam, 1987). Iron biofouling deposits accumulate on surfaces and as such are capable of forming coatings in boreholes and well components which may lead to a variety of deleterious effects:

(a) impairment of hydraulic efficiency due to clogging filter packs, screens, pumps and pipework.
(b) deterioration of materials due to the creation of environments which encourage corrosion processes.
(c) deterioration of water quality.

It is necessary to identify suitable variables which may be monitored in order to assess the state of iron biofouling in a borehole or well. Monitoring should be carried out regularly so that problems can be dealt with as they arise. It is generally accepted that remedial measures are more likely to succeed if carried out before iron biofouling deposits begin to age harden. Effective monitoring and prevention of iron biofouling depend upon an understanding of the basic principles of the processes involved. This paper explains how the engineering and management aspects of a groundwater abstraction system can interact with the environmental characteristics of the aquifer. Armed with an understanding of these interactions it is possible to identify the key variables to be monitored on a regular basis and the design/operational features which can slow down or even prevent the deleterious processes of iron biofouling.

2 Iron biofouling processes and their control

The two key processes which are responsible for the development of an iron biofilm on a submerged borehole component surface are firstly, biofilm formation and secondly, iron precipitation.

2.1 Biofilm formation

Biofilm formation is a multi-stage process whereby a biological population develops on a submerged surface fed by nutrients carried along by fluid movement. A biofilm is typically composed of cells (predominantly bacteria and protozoa in the case of a borehole biofilm), slime produced by the cells and particulate matter trapped by the layer of slime and cells. Living an attached lifestyle is advantageous to the sessile organisms as conditions for nutrient uptake, protection from anti-microbial agents and inter-species cooperation are all enhanced. It is frequently the case that the attached population is substantially more active and greater in size than the free-floating (planktonic) population as a result of these ecological advantages (Costerton and Laschen, 1984).

2.2 Iron precipitation

Insoluble iron compounds make up approximately 90% of the dry weight of iron biofouling deposits (Hallbeck and Pedersen, 1985) and so an understanding of their precipitation processes is of value when attempting to prevent subsurface clogging. Unfortunately, the chemical and biological mechanisms by which iron may be transformed from the soluble to the insoluble state are not easy to distinguish. Iron is dissolved from iron-containing formations under anaerobic/acidic conditions. The iron will remain in the soluble ferrous form until there is a rise in the pH or Eh (redox potential) of the water. Because of increased oxygenation, an increase in Eh may be encountered as the groundwater approaches a pumping well, initiating chemical iron precipitation at the aerobic/ anaerobic interface. Some bacteria (notably Gallionella) can derive energy from the oxidation of ferrous to ferric iron and tend to thrive at the aerobic/anaerobic interface. Distinguishing between the chemical and biological process of iron precipitation taking place in this region is difficult. Iron may also be precipitated by the utilisation of organic iron complexes by heterotrophic bacteria (McCrae et al, 1973).

2.3 Factors controlling iron biofouling processes

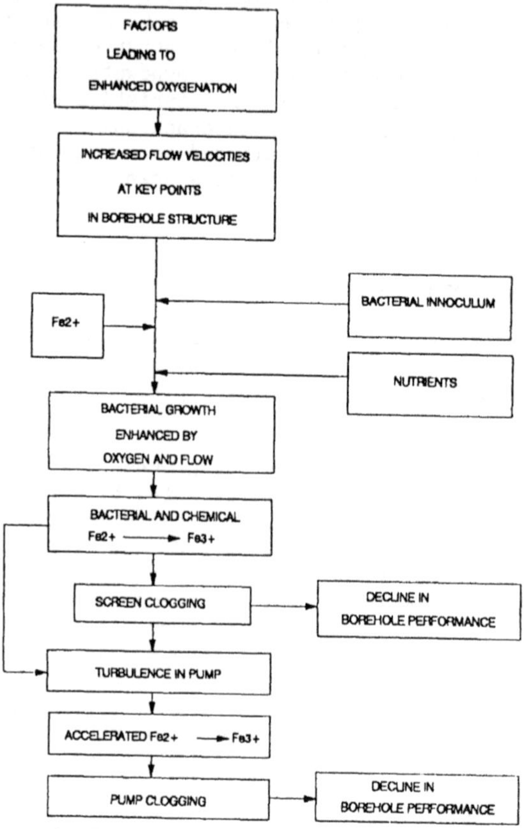

Figure 1. Proposed iron biofouling process model.

Iron biofouling in the groundwater environment is a complex process dependent upon the interaction of a number of key factors. A simple process model (Figure 1) has been formulated which identifies those factors considered to govern the initiation and the rate of iron biofouling.

Oxygen is considered to be the key factor in iron biofouling as it is responsible for the growth of aerobic iron precipitating bacteria and the oxidation of soluble ferrous to insoluble ferric iron. As already mentioned, flow is a prerequisite of biofilm development and it has been demonstrated that in certain circumstances, higher flow velocities will increase the rate of biofilm development (Pedersen, 1982, Caldwell, 1986). In addition it might be expected that turbulence caused by higher flow velocities would enhance nutrient uptake (Cullimore, 1986) and also promote nucleation. Iron biofouling requires a groundwater with a 'significant' dissolved iron content, a microbial inocculum and concentration of nutrients sufficient to support an active biofilm. The premise of the model is that the nutrient supply and microbial inocculum is omnipresent. Where the groundwater contains a dissolved iron load, it is proposed that oxygenation and flow are the key controlling factors without which iron biofouling will not proceed irrespective of the other elements of the model.

3 Iron biofouling in practice

During the course of a study of iron biofouling in forty four UK groundwater abstraction systems a large body of data was collected concerning design, operation, hydrogeology, hydrochemistry and microbiology (Tyrrel, 1990). These data were used to identify where and why biofouling was occurring and which parameters might be the most appropriate to monitor.

In practice, evidence of iron biofouling patterns go some way to confirming the general principles of the model and the ways in which the design and operation of a groundwater abstraction system can interact with the aquifer environment to influence iron biofouling processes.

3.1 Screen biofouling patterns

The most commonly encountered screen biofouling pattern is characterised by the development of deposits from the top of the screen down to a point (variable) near the bottom of the screen. Where this type of pattern is occurring the bottom section of the screen (approximately 25%) remains free from biofouling deposits (in some cases the clean part of the screen may appear as good as new) in contrast to the upper sections which may be unrecognisable due to the massive development of iron biofilms. This may be explained if the water entering the lower part of the borehole is anaerobic. The anaerobic water will hold any iron in the soluble form as it passes through the screen. As the water passes up the screen towards the pump it encounters oxygenated water entering the borehole from the near surface. At this point an aerobic/anaerobic interface occurs at which point aerobic bacterial development and chemical iron precipitation may be inititated.

There are several possible explanations for this upper layer of oxygenated groundwater. Firstly, it is to be expected that the upper layers of groundwater would contain more oxygen beause they are in closer proximity to the atmosphere. Secondly, it is suggested that design and operational factors may interact with hydrogeology to enhance the oxygenation of the upper aquifer layers. For example, if the pumping water level is allowed to fall below the top of the screen, a seepage face can develop which will promote cascading. Intermittent pumping is also thought to promote oxygenation because of water table fluctuations. Higher flow velocities may also contribute to this pattern of deposition on the screen. The highest flow velocities are often encountered at the top of the screen corresponding to the layer of oxygenated water and the main area of iron/biofilm deposition. It is suggested that the combination of an aerobic/anaerobic interface and high flow velocities leads to enhanced biofouling conditions. Indeed CCTV evidence suggests that in some boreholes biofouling deposits are significantly worse in the high flow regions of the screen close to the pump.

3.2 Biofouling of the pump and pipework

From the study of UK boreholes it became clear that whenever biofouling began on the screen the problem became transmitted into the pump and up the rising main. Flow velocities are at their highest in the pump and therefore any flow related enhancement of iron biofouling would be expected to peak at the inlet, within the pump bowls or up the riser. Experience has shown that this is in fact the case. Pumps choked with iron biofouling deposits cause disruption of supply and require costly refurbishment. In practice the higher flow velocity theory has been upheld in cases where low revving pumps have been replaced with higher revving models. This has led to significant reductions in the length of time a pump can be run before cleaning is required.

The idea that higher flow velocities can enhance biofouling is not always easy to believe as it might be expected that scour would be predominant over deposition under these conditions. The tenacity of iron biofilms was demonstrated well in a case where a dewatering system was performing inefficiently due to the development of biofouling deposits. Biofouling was occurring throughout the system but the zone of maximum fouling corresponded to the parts of the system subjected to velocities of up to 4m/s (a situation which might be expected to promote scouring).

3.3 Other environmental factors influencing iron biofouling

It was suggested in section 2.3 that dissolved iron, nutrients and a microbial inocculum are required for iron biofouling to be initiated. A number of interesting points emerged concerning these elements of the process model during the course of the UK borehole study. Monitoring of the groundwater using continuously running monitoring cells (moncells)(Howsam and Tyrrel, 1989) showed that bacteria were present in every borehole. More importantly, iron precipitating bacteria were identified at every borehole site studied, even those without known biofouling problems, suggesting that the necessary innoculum is always available.

The study highlighted the fact that iron biofouling occurs across many of the major UK aquifers (Greensand, Bunter sandstone, Chalk and alluvium). Significant biofilm based microbial populations appear to be able to obtain sufficient nutrients to develop in the relatively unpolluted water supply aquifers in the UK. The level of iron in the groundwater required for iron-based biofilms to develop was shown to vary between 0.03 and 6.50 mg/l. Thus it would seem that even very small amounts of iron in groundwater can lead to iron biofouling problems given the appropriate environmental conditions.

4 Monitoring

Experience of the UK water industry suggests that monitoring of borehole efficiency in general and borehole biofouling in particular is at times inadequate. During the course of the UK borehole survey it became clear that the level and sophistication of monitoring varied quite considerably. In some areas, boreholes are being systematically investigated using pumping tests and CCTV surveys. However it came as some surprise that a large number of boreholes did not even have regular water level or yield data available. Historical operational information is often at best anecdotal and the cases where it would be possible for a well operator to access a comprehensive dossier of information concerning the state, past and present, of their underground assets would seem to be few and far between.

The aim of this paper is not to suggest new techniques of borehole monitoring but rather to reinforce the need for basic monitoring practice. Even simple variables checked on a regular basis can provide sufficient information to diagnose borehole biofouling and perhaps more importantly to assess how much of a problem it is causing.

4.1 Recommended monitoring practice

Yields and water levels are fundamental variables providing information on borehole/aquifer performance which can be interpreted easily. The increasing application of remote monitoring in the UK should make these measurements more convenient in the future. The monitoring of the chemical and microbiological variables should be carried out regularly and the results checked for trends which may indicate the onset of iron biofouling. CCTV surveys are invaluable as they provide solid evidence of the extent of biofouling in addition to information concerning the general state of repair of the well or borehole. The monitoring periods suggested are for guidance and may vary with circumstances.

a) Pumping and rest water levels (monthly).
b) Yield (monthly).
c) Standard bacterial plate count (monthly).
d) TOC, nitrogenous compounds, phosphate, ferrous and total iron (monthly).
e) CCTV survey (every two years).

5 Design and operation

Leading on from the study of borehole biofouling processes, features have been identified which should be borne in mind when designing and operating a groundwater abstraction system. Again, the suggestions may not be revolutionary but are worth re-emphasizing.

5.1 Recommended design/operational practice

a) Operate boreholes continuously at the appropriate lower abstraction rate. Additional storage capacity may be required to cope with surges in demand.
b) Where possible, ensure that pumping water levels do not fall below the top of the screen. Pumping continuously at lower abstraction rates would aid this.
c) Increase the path length of aerobic groundwater into borehole by setting the screen deeper.
d) Use pumps which result in as low flow velocities as possible in all parts of the system.
e) Avoid the mixing of oxygenated waters with iron-containing waters in the borehole.

6 Acknowledgements

The authors would like to ackowledge the cooperation of the following organisations:

Allied Colloids Limited
Anglian Water
Droycon Bioconcepts Inc.
East Anglian Water Company
Luton College of Higher Education
National Rivers Authority
North West Water
Rofe Kennard and Lapworth
Slough Estates Limited
South West Water
Tendring Hundred Water Company
Thames Water
Water Research Center
W.J. Engineering

This research was funded by a grant from the Overseas Development Administration (Engineering Division).

7 References

Caldwell, D.E. (1986) Microbial colonisation of surfaces,in **Int. Symp. on Biofouled Aquifers: Prevention and Restoration. AWWA**, pp. 7-9.

Costerton, J.W. and Laschen, E.S. (1984) Influence of biofilms on efficacy of biocides on corrosion causing bacteria. **Materials Performance.**, 23, 34-37.

Cullimore, D.R. (1986) Physicochemical factors in influencing the biofouling of groundwater, in **Int. Symp. on Biofouled Aquifers: Prevention and Restoration.** AWWA, pp. 23-36.

Howsam, P. (1987) Biofouling in Wells and Aquifers. **J. IWEM.**, 1, 209-215.

Howsam, P. and Tyrrel, S.F. (1989) Diagnosis and monitoring of biofouling in enclosed flow systems - experience in groundwater systems. **J. Biofouling.**, 1, 343-341.

McCrae, I.C., Edwards, J.F., and Davis, N. (1973) Utilisation of iron gallate and other iron complexes by bacteria from water supplies. **Applied Microbiology**, 25, 991-995.

Pedersen, K. (1982) Factors regulating microbial biofilm development in a system with slowly flowing seawater. **Appl. and Env. Microbiology.**, 44, 5, 1196-1204.

Pedersen, K. and Hallbeck, E. (1985) Rapid biofilm development in deep groundwater by **Gallionella ferruginea. Vatten.** 41, 263-265.

Tyrrel, S.F. (1990) Unpublished MPhil thesis. Silsoe College, Cranfield Institute of Technology.

12 DIAGNOSTIC ANALYSIS OF WELL PERFORMANCE DATA

Ch. ATA–UR–REHMAN
Water and Power Development Authority, Lahore, Pakistan

Abstract
In the Indus Plain over 10,000 large capacity tubewells have been installed by WAPDA to combat waterlogging and to generate additional water for leaching salinized soils. A tubewell monitoring programme was also initiated to identify deteriorating tubewells requiring remedial measures. Major causes of discharge reduction include choking of the well screen/gravel pack, decreased efficiency of the pump and lowering of the regional water table. A graphical procedure is described which was developed to analyse well performance data and determine the causes of discharge reduction and their extent.
Keywords: Indus Plain, Tubewells, Pumps, Well Performance

1 Introduction

As a result of canal irrigation without contemporary drainage, waterlogging and salinity have made their gradual appearance in irrigated areas of the Indus Plain, throwing considerable acreage of land out of cultivation. Many unsuccessful attempts to reclaim the affected areas were made through different measures and ultimately these led to large scale development of groundwater through a public tubewell programme, both to control subsoil water levels and to generate additional water for leaching soils. Successful use of groundwater for irrigation triggered its large scale development in the private sector also. Currently, over 10,000 large capacity public tubewells and 200,000 small capacity private tubewells pump nearly 40 - 45 million acre feet (Maf) (Public 10 Maf, Private 30 Maf) of groundwater annually.

To protect the large investment in the public tubewell programme, and also to improve tubewell design, periodic monitoring of well performance was introduced. Discharge and specific capacity were adopted as key indicators and the benchmark for each well was established through an acceptance test performed on the completed well. This consists of a one hour single step specific capacity test. Prior to this, during the development stage, contractors also performed a multi-step specific capacity test at 75%, 100%, 125% and 150% of the design discharge to ensure completed development indicated by stabilization of the specific capacity.

Post project monitoring consisted of closing the well and observing the recovery of water levels followed by a one hour single step specific capacity test. Data thus observed were compared with the benchmark information.

2 Theory and Case History

2.1 The problem

In an irrigation-cum-drainage project the success of a scheme is based on maintaining the pumpage as close to design as possible. However, over a period of time the discharge of a well is affected by:

reduction in specific capacity,
lowering of groundwater level,
decrease in pump efficiency.

An understanding of how much effect each of these causes has on the reduction in discharge is required. To demonstrate the procedure take the case of tubewell LN 151 in Lalian Scheme (Pakistan), the relevant data from which are given in Table 1.

Table 1. Data from tubewell LN 151, Lalian Scheme, Pakistan

Item	Acceptance test	Subsequent test
Date	8/63	11/70
Q(gpm)	2390	1651
Specific capacity (gpm/ft)	112	69
Depth to Water (ft)	16.18	21.83

The specific capacity of the well has reduced from 112 to 69 gpm/ft, indicating a substantial decrease in efficiency. Also the water level has declined by 5.67 feet. What percentage of the reduction in well discharge is attributable to lowering of the water level, reduction in efficiency of well screen and gravel pack, and poor maintenance of pumping equipment?

2.2 General approach

The total head on the pump obtained from the pump characteristic curve at the acceptance test is:

$$H_o = sQ_o + D_o + C'Q_o^2$$

Where:

H_o = total pump head at acceptance (ft)
s = specific drawdown (ft^3/s)
Q_o = discharge at acceptance (ft^3/s)
D_o = depth to watertable at acceptance (ft)
C' = coefficient for pump losses

If the drawdown (sQ_o) is subtracted from the total head (H_o) the remaining head ("net head (H'_o)") is:

$$H'_o = H_o - sQ_o = D_o + C'Q_o^2$$

Assuming that during the subsequent test the well and pump are in the same original condition, then any lowering of the regional water table would increase the net head "H'_o" by an equal amount, thereby increasing the total head and so reducing discharge for a constant power input. In reality the well and pump also deteriorate with time and cause a reduction in discharge. Now to find out this new discharge under lowered water table conditions, the new total head is known but the drawdown component is not. This component depends on specific capacity and the new discharge of the well. Thus drawdown and discharge, two inter-dependent variables, require evaluation.

2.3 Pump characteristic curve

The first step in the analysis is the preparation of a field characteristic curve of the pump using data from acceptance tests on a number of wells with similar pumps. Each set of data consists of a pumping water level below a known measuring point and the corresponding discharge, together with other head losses. The sum of the various losses is plotted against discharge on the same graph as the manufacturer's characteristic curve, to give a smooth curve parallel to that of the manufacturer's and called the "field characteristic curve" (see Figure 1 for tubewell LN 151).

2.4 Well characteristic graph

The second step is to prepare a well characteristic graph at the same scale as that of the pump. This correlates discharge and drawdown, and consists of a series of straight lines representing the constant specific capacity for each well. The graph includes a "net head" scale (see Figure 2).

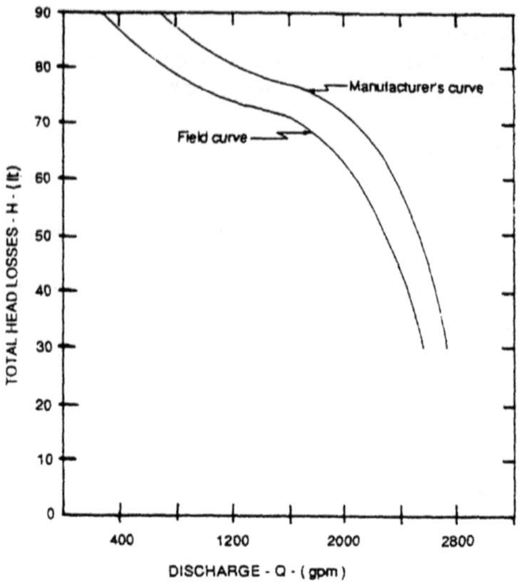

Fig.1. Characteristic curves for pump at tubewell LN 151

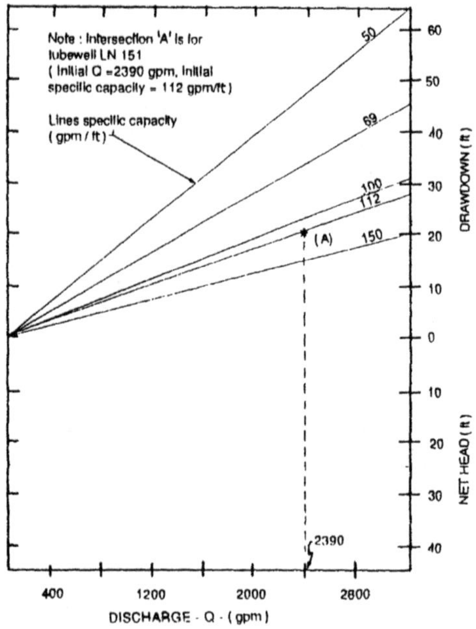

Fig.2. Well characteristic graph

2.5 Analytical Procedure

1. On the well characteristic graph mark the initial discharge on the initial specific capacity line (point "A" on Figure 2 for tubewell LN 151, corresponding to initial values of 2390 gpm and 112 gpm/ft respectively.)
2. Overlay this graph on the pump characteristic graph keeping the Y axes coincident so that point "A" lies over the field characteristic curve of the pump (Figure 3). Now mark line "a", which corresponds to the zero axis of the pump graph. This line is offset from the zero axis of the well graph by an amount equal to the net head (23 ft for tubewell LN 151).
3. To account for changes with time add the increased depth to the water table by drawing another line "a'" below the line "a" (5.67 ft for tubewell LN 151).
4. Now slide the well graph vertically up, keeping the Y axes coincident until the line "a'" coincides with the zero head axis of the pump graph (see Figure 4).
5. Read the values of discharge corresponding to the points of intersection of the initial and subsequent specific capacity lines with the field characteristic curve, marked "B" and "C" on Figure 4. These are the discharges the pump would have drawn from the lowered water table under the initial and the subsequent well conditions, assuming the pump to be in order. For tubewell LN 151, values of "B" and "C" are 2290 and 2080 gpm respectively. The actual well discharges at the times of the initial and subsequent tests are "A" and "D", (2390 and 1651 gpm respectively).

Now:

A-B is the reduction in discharge due to lowering of the water table.
B-C is the reduction in discharge due to the well problem.
C-D is the reduction in discharge due to the pump problem.

Thus in the case of tubewell LN 151, the deficiency in discharge has components as follows:

a) lowering of water table	= 2390 - 2290 = 100 gpm
b) well problem	= 2290 - 2080 = 210 gpm
c) pump problem	= 2080 - 1651 = 429 gpm

To refine the analysis, a correction can be made to account for the decreased head loss due to the

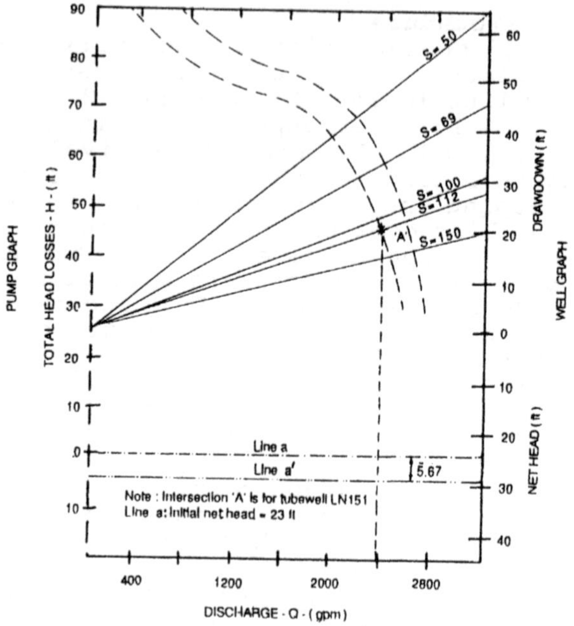

Fig.3. Graphical procedure to determine initial net head and additional decline in the water table

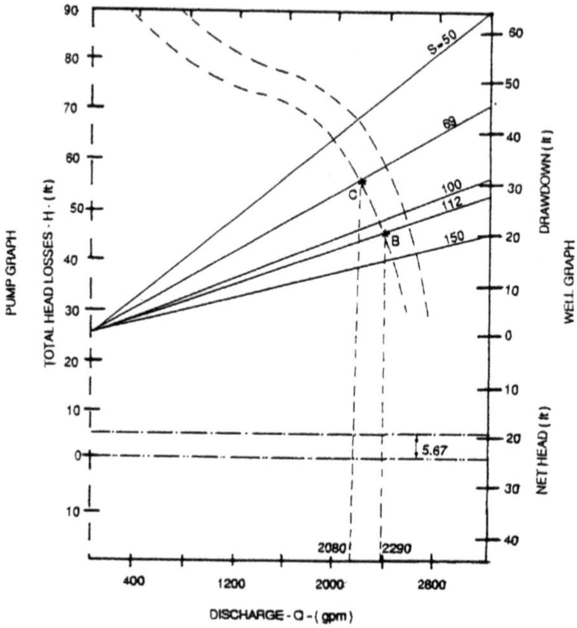

Fig.4. Graphical determination of reduced discharges

reduction in discharge compared to the initial value (2390 gpm). In the first approximation, the pump and the well in their original condition but with lowered water table would have pumped 2290 gpm instead of 2390 gpm. This reduction in discharge of 100 gpm causes say 0.7 feet less head loss. Therefore, in the second approximation the lowering of the water table by 5.67 feet causes an actual increase in net head of 4.97 feet (5.67 - 0.70). Now using 4.97 feet as the net head and repeating the process the value of discharge under the lowered water table condition marked as "B" would be 2305 gpm.

Had there been no pump problem, the decrease in discharge according to the first approximation would have been 310 gpm (2390-2080). This decrease requires a reduction in head loss of say 1.4 feet. To determine the discharge with lowered water table and deteriorated well condition, but no pump problem, the original assumed increase in the net head of 5.67 feet will now be 4.27 feet (5.67 - 1.4), and the discharge designated by point "C" would be 2110 gpm, so that the components reducing the discharge are:

a) lowering of water table	= 2390 - 2305 = 85 gpm
b) well problem	= 2305 - 2110 = 195 gpm
c) pump problem	= 2110 - 1651 = 459 gpm

3 Discussion

The analysis in the worked example indicates that the major reduction in discharge is due to the pump problem (62%), followed by the well efficiency (26%). The remedial measures required are cleaning of the pump and its accessories and the rehabilitation of the well screen and gravel pack by appropriate methods. The discharge reduction due to lowering of the water table is only 85 gpm (12%) and thus does not warrant the addition of any pump stage at this time.

4 Conclusion

The method of analysis described above can be applied to all operational tubewells for which benchmark data are available, thereby allowing the components of any discharge reduction to be identified. This information is the basis on which suitable rehabilitation measures can then be planned.

13 PREDICTING AND INTERPRETING WELL BEHAVIOUR AND WELL EFFICIENCY

R. HERBERT and J.A. BARKER
Hydrogeology Unit, British Geological Survey,
Wallingford, UK

Abstract
The effects on well performance of well geometry and materials, as well as aquifer properties, can be estimated using simple theoretical formulae along with laboratory-determined hydraulic parameters for well screens. The validity of this theory is demonstrated using step-test results from experimental wells constructed in Bangladesh. Comparison of field and theoretical results for such tests can now be used in a rational and diagnostic manner to assess the extent to which gravel pack and screen blocking and deterioration are reducing well efficiency. It is our hope that these findings will help to remove some of the mystery surrounding the topic of well losses.

1 Introduction

The costs of delivering water to the surface from a borehole are in two parts: (a) the capital costs of borehole construction and (b) the running costs such as pumping and maintenance. Optimising well design relies on being able to predict the combined capital and running costs. The pumping costs are largely dependent on the performance of the well, that is to say the well discharge-drawdown ratio, Q/s_w, (also known as the 'specific capacity'). A well with a high Q/s_w will deliver water more cheaply to the surface than one with a low Q/s_w. Campbell and Lehr (1973) describe how to calculate running costs due to pumping. If a well is badly designed or if encrustation or some other deterioration occurs, Q/s_w will be unnecessarily low. It is important, therefore, to be able to predict well performance, Q/s_w, for different well designs and also to monitor deterioration in Q/s_w in order that a decision can be made as to when it is economical to rehabilitate a well or when to replace it with a new well.

Section 2 of this paper shows how to calculate a theoretical value for Q/s_w for many simple situations and how complex situations can affect well performance. Section 3 describes the step-drawdown test used to determine Q/s_w, and how to interpret it in the light of the new theory. Many of the techniques described are taken from Herbert et al. (1989).

2 Predicting Specific Capacity, Q/s_w

The introduction demonstrates the desirability of being able to predict the relationship between well discharge, Q, and drawdown, s_w. There are many factors which affect well performance and a completely generalised solution does not exist. This section begins with a fairly common, easily solved situation and then goes on to identify other less simple cases. Q/s_w is measured in the field by means of a step-drawdown test (see Section 3). This section concentrates on predicting the relationship for a perfectly constructed well in pristine condition.

2.1 Steady state flow to a well having uniform inflow

Consider the steady-state flow system shown in Figure 1. The quantity of interest is the total head loss between the outer aquifer boundary and the head in the casing just below the pump. For a given discharge rate, head losses between this point and the discharge end of the surface pipe do not depend on the well design and, although not unimportant, are not relevant here.

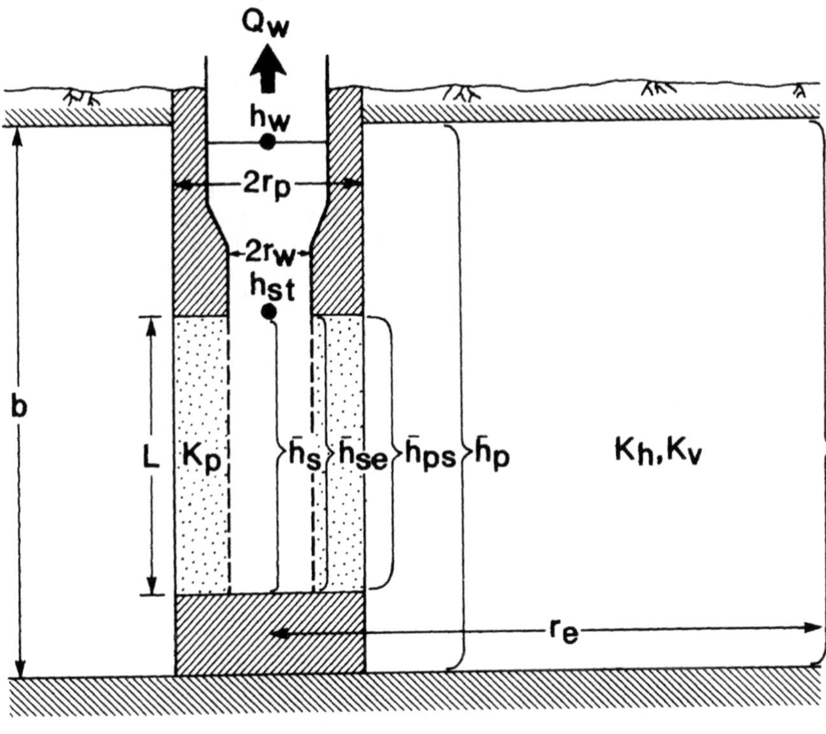

Figure 1. Simple model of a well showing intervals over which average heads apply.

The approach adopted here is to break the total head-loss problem down into a number of simpler head-loss problems (Figure 1). The total head loss can be split up as follows

$$
\begin{aligned}
\text{Total head loss } (\bar{h}_e - h_w) = \; & \text{Aquifer loss } (\bar{h}_e - \bar{h}_p) \\
& + \text{'Penetration' loss } (\bar{h}_p - \bar{h}_{ps}) \\
& + \text{Gravel-pack loss } (\bar{h}_{ps} - \bar{h}_{se}) \\
& + \text{Slot losses } (\bar{h}_{se} - \bar{h}_s) \\
& + \text{Well-screen loss } (\bar{h}_s - h_{st}) \\
& + \text{Losses above screen } (h_{st} - h_w) \qquad (1)
\end{aligned}
$$

where:

- $\bar{h}_e$ is the average vertical head over the exterior boundary,
- $\bar{h}_p$ is the average vertical head over the whole aquifer depth at the radius of the gravel pack,
- $\bar{h}_{ps}$ is the average vertical head over the screened interval of the well at the radius of the gravel pack,
- $\bar{h}_{se}$ is the average vertical head over the screened interval of the well on the outside of the screen,
- $\bar{h}_s$ is the average vertical head over the screened interval of the well on the inside of the screen,
- h_{st} is the head at the top of the screen inside the well, and
- h_w is the head indicated by the water level in the well.

Each of the components of head loss in equation (1) will be dealt with separately in the following sections.

2.1.1 Aquifer and Gravel-Pack Loss

The components of loss in the aquifer and in the gravel pack will both be estimated using a generalisation of the Thiem formula for radial flow derived by Barker and Herbert (1989) in which heads are replaced by their vertical averages. This formula is based on the assumption of Darcian flow between horizontal confining boundaries in a homogeneous (but possibly anisotropic) aquifer.

The head loss through the aquifer is

$$\bar{h}_e - \bar{h}_p = \frac{Qw}{2\pi b K_h} \ln \frac{r_e}{r_p} \qquad (2)$$

where b is the full aquifer thickness and K_h is the horizontal hydraulic conductivity of the aquifer and radii are defined in Figure 1. Similarly, the head loss across the gravel pack is

$$\bar{h}_{ps} - \bar{h}_{se} = \frac{Qw}{2\pi L K_p} \ln \frac{r_p}{r_s} \qquad (3)$$

where L is the length of the screen and K_p is the horizontal hydraulic conductivity of the gravel pack.

Note the implicit assumption that no water flows vertically through the gravel pack opposite the casing or in the region beneath the gravel pack and well (Figure 1).

The head, $\bar{h}_{se}$, in (3) is the average over the outside of the screen. In reality, the flow in the gravel pack must converge to the slots, and this introduces an extra component of head loss (which is analogous to the penetration loss discussed in section 2.1.2). Unfortunately, this loss will depend on: the slot design, the percentage open area, the slot spacing, and the gravel-pack. However, simple calculations indicate that this loss is small in relation to other losses provided total open area is greater than 5%, so it has been ignored in this study.

2.1.2 Penetration Loss ($\bar{h}_p - \bar{h}_{ps}$)

A previously developed formula (Huisman, 1972; Anon, 1964) was simplified (Barker, 1989) to give:

$$\bar{h}_p - \bar{h}_{ps} = \frac{Qw}{2\pi K_h L} \frac{(1-p)}{p} \ln \left[\frac{p(1-p)}{(2-\eta^2)} \frac{b}{r_p} \left(\frac{K_h}{K_v} \right)^{\frac{1}{2}} \right] \qquad (4)$$

where p is the fractional penetration:

$$p = \frac{L}{b} \qquad (5)$$

and η is a parameter which represents the eccentricity of the screen with respect to the vertical centre of the aquifer:

$$\eta = \frac{2z_c}{b(1-p)} \qquad (6)$$

and z_c is the vertical distance from the centre of the screen to the centre of the aquifer.

2.1.3 Slot Losses ($\bar{h}_{se} - \bar{h}_s$)

There is some head loss as water flows through a slot. This loss is normally assumed to be given by the **orifice law**, which relates the head loss to the square of the inflow velocity. For uniform flow to a screen of length L, this law takes the form

$$\bar{h}_{se} - \bar{h}_s = \left[\frac{Qw}{4\pi g L r_w C_v C_c A_s} \right]^2 \qquad (7)$$

where: C_v is the velocity coefficient (a value of about 0.97 is appropriate for a slot); C_c is the contraction coefficient (typically 0.63); and A_s is the fractional open area through which flow occurs (normally taken as either half the fractional slot area or the production of the gravel-pack porosity and the fractional slot area).

2.1.4 Well-Screen Loss ($\bar{h}_s - h_{st}$)
Barker and Herbert (1989b) show that the hydraulic characteristics of a wide variety of well screens can be described in terms of the following equation for the rate of head loss with depth, z, inside the screen.

$$\frac{dh_s}{dz} = \alpha Q^2 - \beta Q \frac{dQ}{dz} \quad (8)$$

The two parameters α and β are empirical constants which vary from one screen to another (although β can be closely estimated from the screen radius alone). Table 1 lists values for α and β determined from hydraulic tests on a wide range of screens (described in Hydraulics Research, 1986). In order to evaluate the screen-loss term from this equation, it is necessary to make an assumption about the variation of Q with vertical distance along the screen. The simplest assumption is that the inflow to the well is uniform so Q is linear with depth. With this assumption integration of (8) gives the average head:

$$\bar{h}_s - h_{st} = Q_w^2 \left[\frac{\alpha L}{4} + \frac{\beta}{3} \right] \quad (9)$$

which is the required screen-loss formula.

Equation (9) has appeared previously in the work of Chen (1975, 1985), although its derivation and form are quite different here.

Sixteen wells of different design were constructed in Bangladesh and their performance was closely monitored over a wide range of discharges. In only one case was significantly non-uniform inflow observed to the screen (Davies et al., 1988).

2.1.5 Losses Above the Top of the Screen ($h_{st}-h_w$)
Head losses above the top of the screen must also be taken into consideration in well design. There will be frictional head loss inside the casings, and extra losses at the joints. Head losses across the joints would require special study, and are probably not very significant. The total head loss in all sections of (unperforated) casing can be calculated from a formula of the form:

$$h_{casing} = \frac{8Q_w^2}{\pi^2 g} \sum_s \frac{f_s l_s}{D_s^5} \quad (10)$$

where the summation is taken over all screen sections, and section s has friction factor f_s, length l_s and diameter D_s.

There will often be at least one reducer (at the bottom of the pump casing). There will be a head increase across the reducer (from Bernoulli's equation), but also frictional and turbulent head loss. An estimate of the net loss in pressure head can be made using:

Table 1. Principal results of the screen tests.

Test No.	Make*	Material+	Nom. Diam. (in)	I.D. (m)	α (s^2/m^5)	β (s^2/m^5)
1	Jo	SS	6	0.158	26.00	540.0
2	BC	GRP	6	0.153	18.50	652.0
3	Jo	SS	6	0.158	31.00	523.0
4	No	SS	6	0.150	31.20	637.0
5	No	SS	6	0.151	23.20	662.0
6	Du	P	6	0.147	18.80	732.0
7	Du	P	6	0.148	16.00	746.0
8	Du	P	6	0.148	17.70	730.0
9	Hy	P	6	0.147	23.70	759.0
10	Pr	P	6	0.148	15.80	712.0
11	Pr	P	6	0.151	11.80	686.0
12	De	P	6	0.149	17.00	813.0
13	De	P	6	0.147	20.30	738.0
14	De	GRP	6	0.168	20.00	492.0
15	Jo	SS	4	0.112	138.00	2240.0
16	No	SS	4	0.101	76.40	3610.0
18	Du	P	4	0.102	72.90	3490.0
19	Du	P	4	0.102	123.00	3530.0
20	Hy	P	4	0.101	156.00	3750.0
21	Jo	SS	8	0.201	6.14	173.0
22	BC	GRP	8	0.203	3.49	201.0
23	Du	P	8	0.197	3.47	220.0
24	Hy	P	8	0.200	3.25	229.0
25	Jo	SS	10	0.259	2.46	78.3
26	No	SS	12	0.298	0.81	41.6
27	Du	P	12	0.301	0.71	42.8
28	Hy	P	12	0.300	0.71	43.7
29	BC	GRP	6	0.152	13.10	646.0

+ SS - Stainless steel
P - Plastic
GRP - Glass-reinforced plastic

* Jo - Johnson
Bc - Bristol Composite
No - Nold
Du - Durapipe
Hy - Hydrotech
Pr - Preussag
De - Demco

$$h_{reducer} = \frac{8Q_w^2}{\pi^2 g}\left[\left(\frac{1}{D_2^4} - \frac{1}{D_1^4}\right) - k'\left(\frac{1}{D_2^2} - \frac{1}{D_1^2}\right)^2\right] \quad (11)$$

where D_1 and D_2 are the internal diameters at the lower (upstream) and upper (downstream) ends of the reducer, respectively (see, for example, Daugherty and Franzini (1977)). For a well-designed reducer - one with a cone angle less than about 15° - the value of k' should be about 0.2.

There must be a further significant head loss as water enters the pump intake. This will be roughly proportional to the square of the well discharge rate and must also depend on the diameter of the casing in which the pump is housed. However, for a given discharge rate the loss will not depend on the design of the well below the pump casing.

As a first estimate of this additional loss in head, it has been assumed that it will be roughly equal and opposite in sign to that induced by the expansion term of equation (11). This would be exact if the pump intake were of diameter D_1 and were connected directly to a reducer of the same but inversed geometry of the expansion joint.

Use of this approximation reduces the right-hand side of equation (11) to

$$\frac{8Q_w^2 k'}{\pi^2 g}\left(\frac{1}{D_1^2} - \frac{1}{D_2^2}\right)^2 \quad (12)$$

The drawdown in the well, s_w, can then be usefully thought of as recording the piezometric level of the water at the pump intake.

2.2 Some field results

If equations (2), (3), (4), (7), (9), (10) and (12) are substituted in equation (1), a general expression relating s_w (= initial head - h_w) and Q_w is obtained which is appropriate for a well pumping at a steady-state and for which uniform inflow occurs along the length of the screen. This expression can be simplified to the form:

$$s_w = BQ_w + CQ_w^2 \quad (13)$$

where B and C are constants which may be calculated from equations (2) to (12) listed above. This expression is exactly that used by Rorabaugh (1953) in analysis of step-drawdown tests. Using the simple theory presented earlier, the theoretical value of C is given by equation (14) for a single reducer and where slot losses have been ignored.

$$C = \left(\frac{\alpha L}{4} + \frac{\beta}{3}\right) + \frac{8}{\pi^2 g}\left[\sum_s \frac{f_s l_s}{D_s^5} + k'\left(\frac{1}{D_1^2} - \frac{1}{D_2^2}\right)^2\right] \quad (14)$$

Seventeen step-drawdown tests were carried out in Bangladesh on wells with varying screen materials, design and geometry (Davies et al., 1988). The results of these tests are summarised in Table 2

Table 2. Comparison of the step test results and values (of C) predicted by the theory.

Well	Material*	Diameter (nominal) (in)	Length (feet)	B (field) (s/m^2)	C (field) (s^2/m^5)	C (theory) (s^2/m^5)	Error in theory C (%)
1c	SS	10	90	62	240	73	70
2	SS	6	80	30	920	958	-4
3	SS	6	60	110	960	857	11
4	SS	8	80	32	264	187	29
5	SS	8	120	30	238	185	22
6	SS	6	100	30	360	554	-54
7	Gtex.	6	80	63	1000	487	51
8	Gtex.	6	120	25	586	655	-12
9	Gtex.	8	90	33	168	127	25
10/4	Gtex.	4	120	43	1600	2521	-58
10/6	SS	6	80	23	740	700	5
11	GRP	6	120	28	328	550	-68
12	GRP	6	80	75	540	584	-8
13	GRP	8	80	32	184	105	43
14	GRP	10	80	32	56	52	8
15	GRP	8	120	24	126	125	1
16	GRP	6	60	56	380	510	-34

* SS - Stainless steel, wire wound
Gtex. - Mesh-wrapped plastic
GRP - Glass-reinforced plastic

where comparisons are made between the values derived graphically from the field data and from the theory presented earlier. The average contribution to C from the three terms in equation (14) were: 67% (screen loss), 30% (casing loss), and 3% (reducer loss).

Figure 2 is a plot of C values derived from theory and field tests. It can be seen that an unexpectedly good agreement is obtained, suggesting the theory presented in Section 2.1 is reasonably accurate.

Aquifer properties contribute to the value of B. However, these properties are variable in space and the average values 'seen' by the well vary in time. Consequently there was no similar rigorous attempt to compare theoretically predicted values for B with those observed in the field.

2.2.1 A Particular Field Result

Figure 3 shows the constructional details and lithology of a test well drilled near to Dhaka, Bangladesh. Figure 4 plots the results of two step tests carried out on the well. The field values, deduced graphically, for B and C on Figure 4 have units of metres.seconds/litre and metres.seconds^2/litre^2 respectively. (These values of B and C should be multiplied by 10^3 and 10^6 to conform with the values given in Table 2).

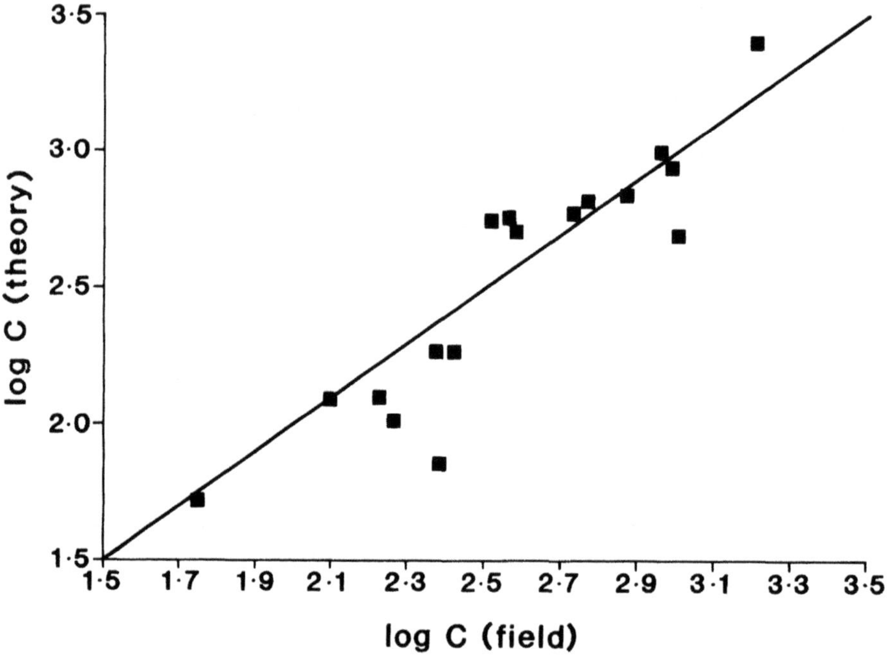

Figure 2. Comparison of theoretical and experimental (step-test) values for the coefficient of nonlinear head loss, C.

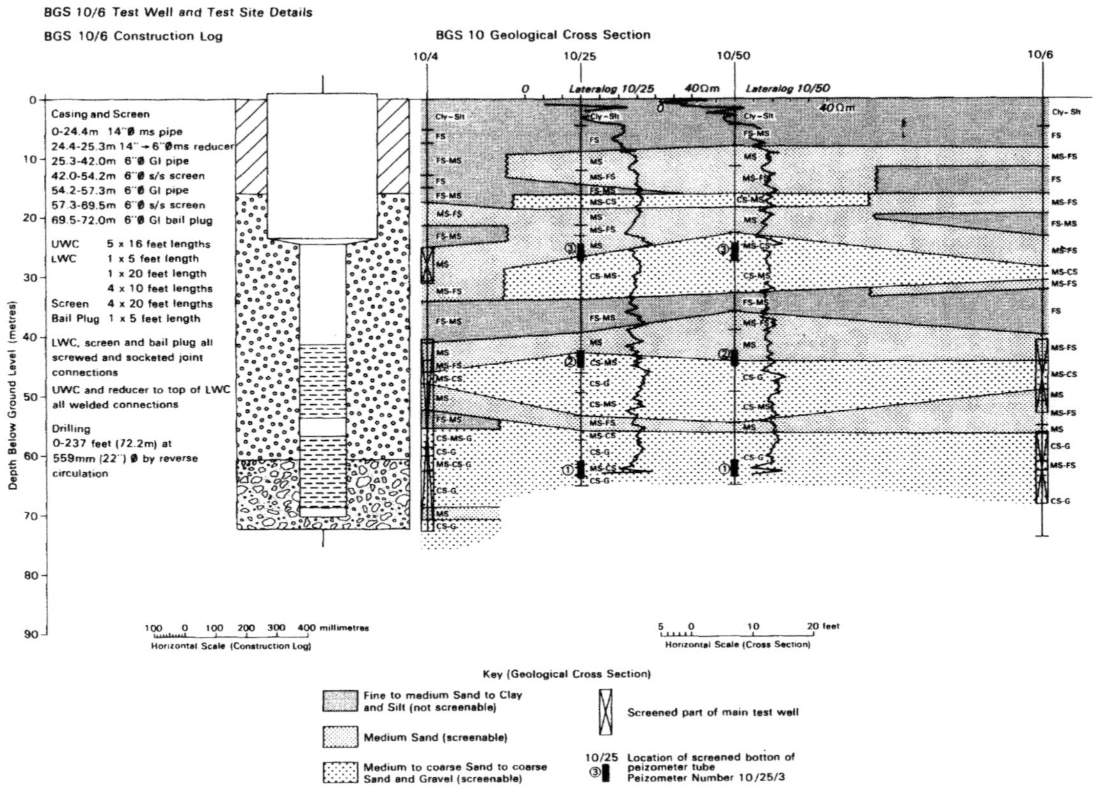

Figure 3. Details of one of experimental-well sites in Bangladesh.

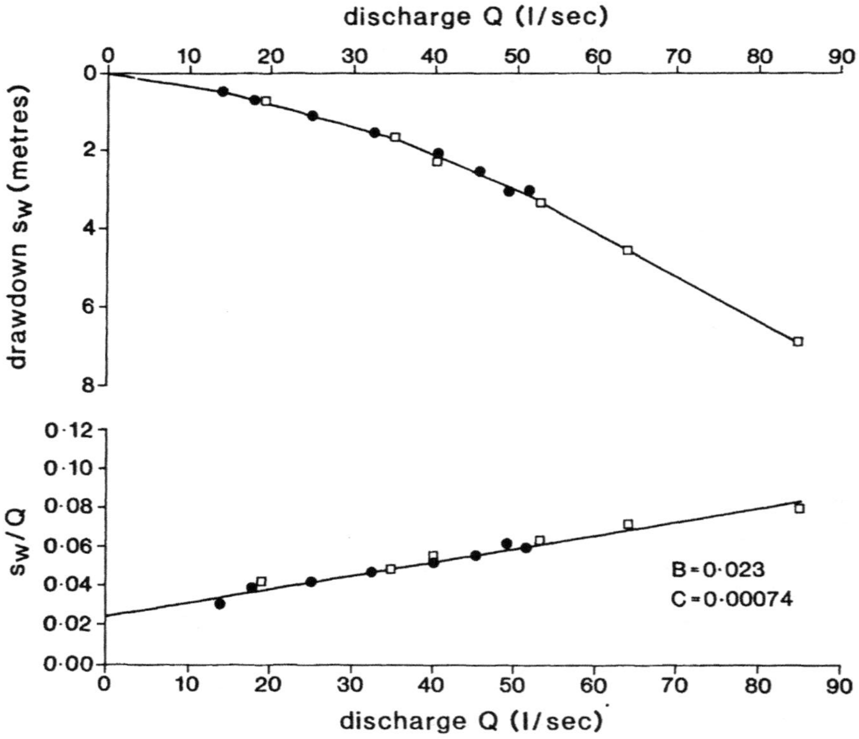

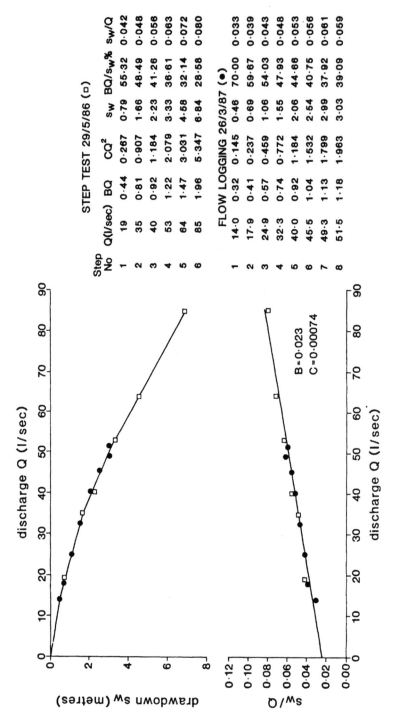

Step No	Q(l/sec)	BQ	CQ^2	s_w	BQ/s_w%	s_w/Q
STEP TEST 29/5/86 (□)						
1	19	0·44	0·267	0·79	55·32	0·042
2	35	0·81	0·907	1·66	48·49	0·048
3	40	0·92	1·184	2·23	41·26	0·056
4	53	1·22	2·079	3·33	36·61	0·063
5	64	1·47	3·031	4·58	32·14	0·072
6	85	1·96	5·347	6·84	28·58	0·080
FLOW LOGGING 26/3/87 (●)						
1	14·0	0·32	0·145	0·46	70·00	0·033
2	17·9	0·41	0·237	0·69	59·67	0·039
3	24·9	0·57	0·459	1·06	54·03	0·043
4	32·3	0·74	0·772	1·55	47·93	0·048
5	40·0	0·92	1·184	2·06	44·66	0·053
6	45·5	1·04	1·532	2·54	40·75	0·056
7	49·3	1·13	1·799	2·99	37·92	0·061
8	51·5	1·18	1·963	3·03	39·09	0·059

Figure 4. Results of pumping tests on the well depicted in Figure 3.

The theoretical value of C is calculated using equation (14) which sums equations (9), (10) and (12). In equation (9), α and β are to be found in Table 1 as 31 s^2/m^6 and 523 s^2/m^5 respectively. In equation (10), a value of f_s appropriate to 6" nominal diameter galvanised iron pipe, was taken from published tables. The theoretically calculated value of C was 700 s^2/m^5 which is surprisingly close to the equivalent field value of 740 s^2/m^5.

Figure 5 presents the results of the step drawdown test carried out on 29/5/86 and also the predicted step drawdown curve using the theoretical values of C and an assumed value of B of .023. The agreement is particularly good demonstrating the accuracy that can be obtained in predictions of well performance using the theory developed in this section.

2.3 More complex situations

2.3.1 Transient Flow

The analysis given above in 2.1 assumes there is steady-state flow to a well. In practice, all wells are pumped intermittently so their

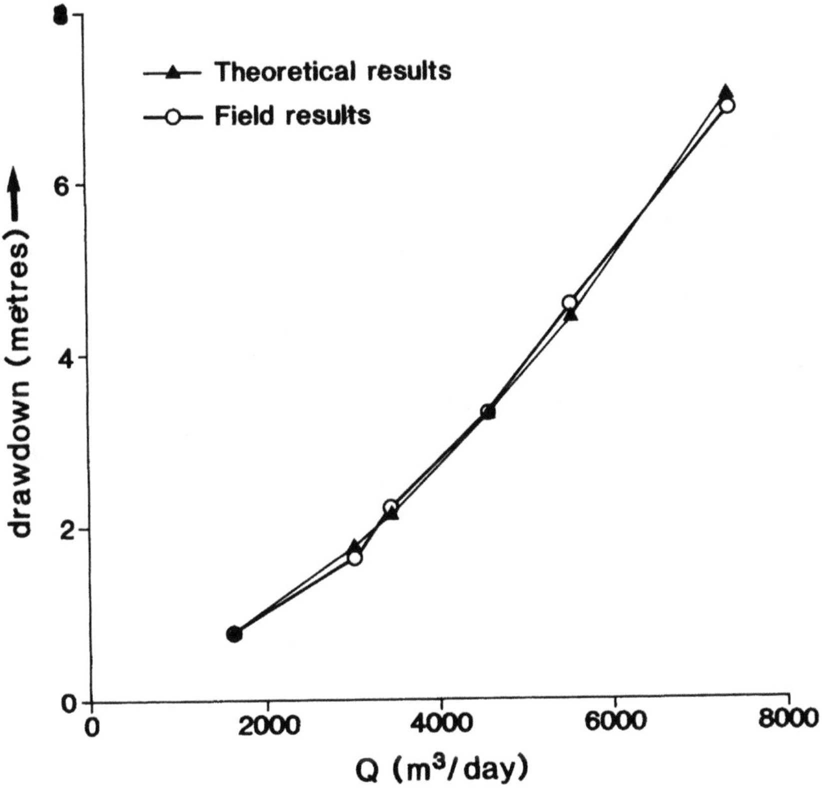

Figure 5. Comparison of observed and predicted results from a step-test on the well depicted in Figure 3.

behaviour is transient. The general transient problem of flow to wells with well losses is far more complicated than the steady-state problem.

Consider transient flow in the immediate neighbourhood of a well. When pumping begins the heads at points close to the well decline rapidly relative to heads at distant points. However, after some period of steady pumping, all of the heads within this neighbourhood begin to decline at similar rates, so the differences between them are near constant. The flow is then said to be in a quasi-steady-state (q.s.s.), meaning that at any instant the head distribution appears to represent steady flow.

Let $s_0(t)$ be the time-dependent drawdown at the outer boundary of the neighbourhood under consideration, and let Δs be the q.s.s. head loss between this boundary and some reference point such as the top of the screen. The average drawdown at the reference point during period t_1 to t_2 will therefore be

$$\langle s_p \rangle = \Delta s + \frac{1}{t_2 - t_1} \int_{t_1}^{t_2} s_0(t)dt \tag{15}$$

While the quantity Δs will depend on the well design, the integral term will not, provided the boundary of the neighbour-hood is sufficiently far from the well that the head variation there depends only on the total flow rate. So, in comparative modelling of well designs it is only necessary to consider the Δs (q.s.s.) term. In practical terms this means that well drawdowns measured in the field during a step-drawdown test should be those drawdowns which occur after the same period of pumping and from the same original hydraulic state (usually undisturbed).

The problem that remains is that of establishing the conditions under which the q.s.s. assumption is justified. Firstly, consideration is given to how large the neighbourhood of the well should be that the behaviour on its outer boundary would not be affected by the well design. Two extreme cases of well design are those where the screen fully penetrates and where the screen is very short, so the flow has to converge rapidly near the well. The effects of this convergent flow are significant out to a distance equal to a few times the aquifer thickness. Therefore a reasonable choise of q.s.s. neighbourhood size is twice the aquifer thickness.

The next problem to consider is how long it takes for q.s.s. conditions to be established within that neighbourhood. One approach to this problem is to determine how long it takes for the radius of influence of the well to extend beyond the neighbourhood. This gives a characteristic time of about

$$t_c = \frac{Sb^2}{T} \tag{16}$$

where S is the total storage coefficient, T is the transmissivity and b is the aquifer thickness. If this characteristic time is significantly shorter than the typical period of pumping then the q.s.s. approach is justified. It should be noted that in very thick aquifers with low transmissivities, t_c can be longer than days and it will be impracticable to run the test such that q.s.s. obtains.

Very often it is not necessary to carry out the above calculation because the period it takes to establish q.s.s. conditions can be estimated from long-term tests. When the drawdowns in all of the piezometers begin to vary in unison (with fixed differences between them) q.s.s. conditions have been established within the radius of the piezometers.

2.3.2 Complex Design Situations

The design of a well need not be restricted to a single screen type or diameter. A well with screen sections of different diameters is often described as **telescopic**. No simple analysis of this problem has been found. Consequently, telescopic designs have to be analysed using more complex numerical models of well flow.

If screen entry velocities are too high, turbulent flow will occur in the aquifer, there will be non-linear head losses in the aquifer and the flow into the well screen will become markedly non-uniform up its length. The exact point at which this occurs requires further research, but there is an indication that if the standard criterion for limiting screen slot entry velocities to less than 0.1 ft/sec applies, then these difficulties will not arise and the methods given in this paper can be used.

Similarly, the presence of hydraulic boundaries and layered complex aquifers can seriously affect well performance and care must be taken in its interpretation.

There is one final, frequently occurring, cause of confusion in the interpretation of well performance. Occasionally, step-tests are carried out before full development of the well has occurred. Thus during the test the aquifer properties near the well can change and make interpretation almost impossible.

3 Measuring and Interpreting Well Behaviour

3.1 The step-drawdown test

In a normal step-drawdown test the well is pumped at successively higher pumping rates and the drawdown for each rate is recorded. The entire test usually lasts no longer than one day and interpretation is simplified if the pumping times are the same for each rate. If time permits, the water level is allowed to recover fully between each step. A plot is made of s_w/Q against Q. Figs. 4 and 5 show the results of one such test.

As shown in Section 2 if the aquifer system and well design are relatively simple (section 2.3.2), if the pumping time is long enough (equation 16), the velocities of entry do not exceed the recommended screen slot entry limit of 0.1 ft/sec and turbulence in the aquifer is avoided then a plot of s_w/Q vs. Q will give a straight line. The intercept will be B and BQ is the total head loss of flow through the aquifer and gravel pack. C is the slope of the plot and CQ^2 describes all the head losses within the well to the pump intake (equation 14).

It is often erroneously quoted (Driscoll, 1986) that CQ^2 results from turbulence of flow through the aquifer, it can now be seen that this is not so and that in most cases the CQ^2 component of drawdown is a result of predictable hydraulic head losses inside the well.

3.2 Interpretation and use of step-drawdown tests

Step-drawdown tests can be run at regular intervals through a well's life. If B increases with time, it can be deduced that gravel pack or slot losses are increasing with time. Also, if C increases with time then it is likely that either internal well losses have increased, probably due to a roughening of the well interior due to rusting, or flow into the well has become turbulent due to an effective reduction of screen length due to gravel pack blocking or encrustation.

Finally, it should be mentioned that if the plot of s_w/Q is not a straight line then one of the more complex situations described in section 2.3 may pertain and great care must be taken in its interpretation.

3.3 Well efficiency - rehabilitate or not?

Well efficiency is defined here as the actual specific capacity, $(Q/s_w)_a$, divided by the theoretical specific capacity $(Q/s_w)_t$, usually expressed as a percentage. $(Q/s_w)_t$ can be calculated by the methods described in Section 2 and $(Q/s_w)_a$ can be deduced from observation of well behaviour using step-drawdown tests. $(Q/s_w)_a$ will get smaller with passage of time, as will well efficiency, due to filter blockage, encrustation or corrosion. As efficiency falls, pumping costs will rise. It is a relatively simple matter to compare the present values of these increased running costs with the cost of rehabilitation and so deduce when rehabilitation of a well becomes economical. When making the decision to rehabilitate or not another factor must also be considered. Ahrens (1975) suggests that if a tubewell has not lost more than 25% of its original specific capacity there is a 90% chance that it can be fully restored by proper maintenance and rehabilitation. However, if it has lost more than 50% there is only about 10% chance that it can be successfully rehabilitated.

For completeness, it should be pointed out that some authors define well efficiency differently to the above. This arises from a lack of appreciation that even the most perfectly designed and constructed well must have internal well losses, i.e. $C > 0$.

Acknowledgement
This paper is published by permission of the Director of the British Geological Survey (NERC).

References

Ahrens, T.P. (1975) Investigation of tubewell deterioration. **Harza Engineering Co. Report (April).**

Anon. (1964) Steady flow of ground water towards wells. **Committee for Hydrological Research T.N.O., Proceedings and Information No. 10,** The Hague.

Barker, J.A. (1989) A study of steady-state flow to partially-penetrating wells. **BGS Report WD/89/23.**

Barker, J.A. and Herbert, R. (1989a) Pilot study into optimum well design, Vol. 4, part 2, A simple theory for approximating well losses. **BGS Report WD/89/12**, pp 22-24.

Barker, J.A. and Herbert, R. (1989b) Pilot study into optimum well design, Vol. 3, Hydraulic tests on well screens. **BGS Report WD/89/10**, p. 8.

Campbell, M.D. and Lehr, J.H. (1973) **Water Well Technology**. McGraw-Hill, 600 pp.

Chen, Yu-Sun (1975) Well hydraulics. **China Building Industry Press** (in Chinese).

Chen, Yu-Sun (1985) Hydraulic head field induced by pumping a well in a confined aquifer. **Mem. 18th Congress Int. Assoc. Hydrogeol., Cambridge**, UK.

Daugherty, R.L. and Franzini, J.B. (1977) **Fluid Mechanics with Engineering Applications**. McGraw-Hill, New York, 564 pp.

Davies, J. Herbert, R. Nuruzzaman, N.D. Shedlock, S.L. Marks, R.J. and Barker, J.A. (1988) Pilot study into optimum well design, Vol. 1, Fieldwork - Results, **BGS Report WD/88/21**.

Driscoll, F.G. (1986) **Groundwater and Wells**. Johnson Division, St Paul, Minnesota, 556 pp.

Herbert, R. Barker, J.A. and Davies, J. (1989) Pilot study into optimum well design, Vol. 6, Summary of the programme and results, **BGS Report WD/89/14**.

Huisman, L. (1972) **Groundwater Recovery**. Winchester Press, New York, 336 pp.

Hydraulics Research (1986) Hydraulic roughness of well screens. **Report No. EX1388, Hydraulics Research Ltd.**

Rorabaugh, M.I. (1953) Graphical and theoretical analysis of step-drawdown test of artesian well. **Proc. Amer. Soc. Civil Engrs., 79, Separate No. 362**, 23 pp.

14 IDENTIFYING CAUSES FOR LARGE PUMPED DRAWDOWNS IN AN ALLUVIAL WADI AQUIFER

M.W. GROUT
National Rivers Authority, Peterborough, UK
K.R. RUSHTON
University of Birmingham, Birmingham, UK

ABSTRACT
Large pumped drawdowns are often attributed to well losses. However, in situations where the vertical components of flow are significant, large drawdowns can result from the natural flow mechanism operating in the vicinity of the pumped borehole. Three idealised examples are considered which show that conditions in the vicinity of the borehole affect both the pumped drawdown and the aquifer flow mechanism and produce a different apparent 'well loss'. Analysis of a pumping test from a wadi aquifer with a numerical model which allows for vertical flows highlights the effect of the seepage face in causing large pumped drawdowns. The examples demonstrate the importance of observation piezometer data collected at different levels within the aquifer system for both identifying the vertical flow mechanism and helping to understand the flow processes in the vicinity of the pumped borehole.

1 INTRODUCTION

The term 'well loss' is generally used to describe the component of the drawdown in a pumped borehole which is additional to the drawdown required to draw the water from source to the vicinity of the borehole. For a 'perfect' borehole there would be no well loss. When estimating the drawdown in a borehole due to the flow of water through the aquifer, equations such as Theis or leaky aquifer theory are generally used which assume a uniform velocity of approach to the borehole over the whole depth of the aquifer. The well losses are then assumed to depend solely on the energy required to move the water through the gravel pack and well screen and into the pump. This ignores another crucially important cause of head loss due to the groundwater flow patterns in the vicinity of the borehole.

For a partially penetrating borehole, the drawdowns in the borehole are larger than if the borehole is fully penetrating. The additional

---

* The author is now with the National Rivers Authority but the work on which the paper is based was carried out at Birmingham University.

drawdown is, in effect a well loss. Whenever a seepage face occurs, the borehole is less efficient than when the approach velocity is uniform over the full depth of the aquifer; this additional drawdown is also a well loss. Whenever vertical components occur in the vicinity of the borehole due, for instance, to layering of the aquifer or the presence of blank casing, additional drawdowns above the horizontal flow assumption result in components of well loss.

This paper considers certain examples of additional drawdowns in the vicinity of pumped boreholes. The discussion is mainly concerned with unconfined aquifers and considers both an idealised situation and an actual pumping test in a wadi aquifer.

2 COMPONENTS OF DRAWDOWN IN A PUMPED BOREHOLE

The drawdown in a pumped borehole is the result of hydraulic head losses associated with the flow of water through the aquifer and into the borehole. The observed drawdown is usually visualised as being made up of two components, a loss due to flow through the aquifer, and an additional loss which depends on the conditions in the vicinity of the borehole, referred to as the well loss. This can be described by the following equation:

$$s_t = s_a + s_w \tag{1}$$

where: s_t is the observed drawdown in the pumped borehole,
s_a is the loss due to the aquifer and
s_w is the well loss.

The aquifer loss, or 'formation loss', depends primarily on the aquifer permeability and the discharge, it is commonly estimated using conventional pumping test models. These models assume that flow to the borehole is horizontal, or near horizontal, and that the discharge is uniformly distributed along the well-face (Theis, 1935; Hantush and Jacob, 1955; Neuman, 1972).

In many instances, the drawdown at a pumped borehole is significantly larger than that which is predicted by the conventional models above. These large pumped drawdowns are usually attributed to well losses. Well losses can be attributed to a number of causes, for example:

- resistance to turbulent flow in the immediate vicinity of borehole and through the well screen (Clark, 1977),

- deterioration in the zone around the borehole due to artificial recharge (Rushton and Srivastava, 1988),

- invasion of the aquifer by drilling mud,

- encrustation of the well-screen causing a decrease in the open area.

In certain cases the additional drawdown due to well losses can be a significant proportion of the observed drawdown.

However, there are additional reasons why the pumped water level is far lower than the drawdown due to horizontal flows through the aquifer. A component of the large pumped drawdowns can also result from the natural flow mechanism operating around a borehole. For example, in unconfined aquifers a seepage face forms at the well-face unless solid casing extends below the well water level. This causes a non-uniform distribution of flow with most of the discharge drawn from deeper horizons within the aquifer. Alternatively, casing out overlying aquifer units can itself cause the discharge to be drawn from deeper horizons. In these circumstances, the vertical flows are significant and substantial vertical head gradients are induced to draw water down from the water table. The vertical flows result in increased drawdowns at the pumped borehole.

This paper considers a number of examples which demonstrate the mechanisms described above. However, prior to this, the flow mechanism around a pumping borehole in an unconfined aquifer is described in more detail.

3 FLOW MECHANISM IN AN UNCONFINED AQUIFER

This section describes the flow conditions in an unconfined aquifer in the period following the start of pumping. An idealised cross-section showing the general form of the boundary conditions which frequently occur is given, Fig.1. Reference is also made to Fig.3 which although does not strictly relate to the example described below, does show the time-drawdown curves which are typically observed in the pumped borehole and at different depths within the aquifer. Descriptions of the flow conditions similar to those presented here are also given in Rushton and Rathod (1988) and US Department of the Interior (1977).

Initially, as pumping starts, most of the discharge is derived from well storage. This causes the pumped water level to drop relatively quickly, Fig.3(a). A seepage face develops at the well-face between the pumped water level and the point at which the free surface intersects the borehole, Fig.1.

The seepage face is inefficient at drawing water into the borehole and a relatively large proportion of the discharge flows into the borehole below the pumped water level. This causes the water levels at depth to fall as the confined storage is depleted, and sets up vertical gradients from the lower aquifer horizons to the free surface, Fig.3(b). This flow pattern is illustrated in Fig.1.

Once the vertical gradients are set up, water starts to flow from the free surface to deeper aquifer horizons. The free surface also starts to fall, although only slowly, as water is released from unconfined

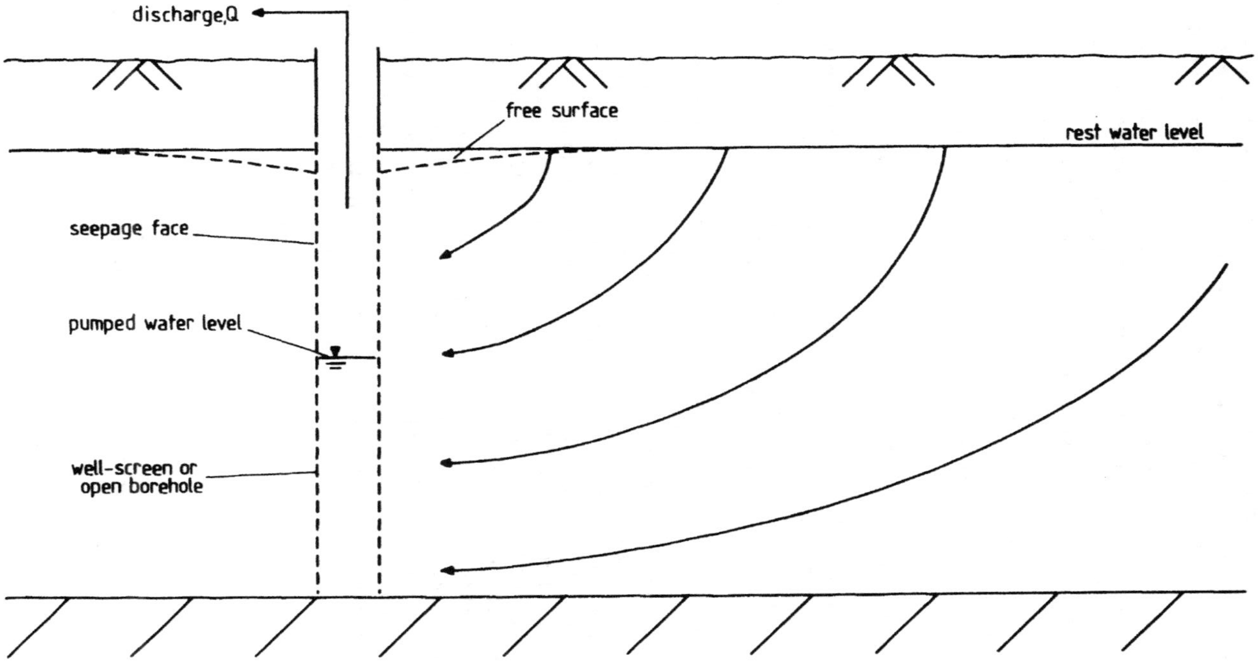

Fig.1. Flow pattern in an unconfined aquifer.

storage, Fig.3(b). The free surface continues to fall at a relatively steady rate following the first day of pumping and this steady fall is also seen in the piezometric levels for deeper aquifer horizons. The response described above was confirmed by a pumping test undertaken in a sandstone aquifer where a nest of piezometers was installed in an observation borehole about 30 m from the pumped borehole (Rushton and Howard, 1982). Relatively large pumped drawdowns can, therefore, often be established in order to draw water down from the free surface.

The pumped borehole drawdown can be even further increased where layers of low vertical permeability are present. Because a relatively large proportion of the discharge flows vertically into the deeper aquifer horizons the water levels at depth are sensitive to the vertical permeability. This is frequently much lower than the horizontal permeability due to the occurrence of layers of fine material. Low vertical permeabilities mean that strong vertical gradients are needed to draw water from the free surface.

4 SIGNIFICANCE OF FLOW COMPONENTS OF PUMPED DRAWDOWNS

This section considers the effect of conditions in the borehole and in the vicinity of boreholes on the pumped well drawdowns. A representative aquifer is considered which is unconfined and has two horizontal zones of lower hydraulic conductivity.

The total depth of the aquifer is taken as 86 m. The flow domain is divided into three horizons, the upper horizon has an initial saturated thickness of 40 m while the middle and lower horizons each have thicknesses of 20 m (Fig.2(a)). The three horizons are assumed to be homogeneous, each having a hydraulic conductivity of 1.5 m/d. Between the upper and middle horizons and between the middle and lower horizons, layers of low permeability are represented. These layers each have a thickness of 3 m and a vertical hydraulic conductivity of 0.015 m/d, or 1/100 of the aquifer hydraulic conductivity; flows are predominantly vertical through these layers. Values assumed for the aquifer storage coefficients are given in Fig. 2(a). The base of the aquifer is impermeable, while the upper boundary is taken to be a free surface. The outer boundary is set at a radial distance of 10 km; the effect of pumping does not reach this boundary. A discharge of 1000 m^3/d continues for 3 days, recovery is monitored for a further 3 days.

Three alternative conditions are represented in or close to the borehole:

(A) the borehole is open to the middle and lower zones only,
(B) the borehole is open to the middle zone only,
(C) the borehole is open to the middle and lower zones and the horizontal hydraulic conductivity of the middle zone is higher within one metre of the borehole.

These conditions are illustrated in Fig. 2(b).

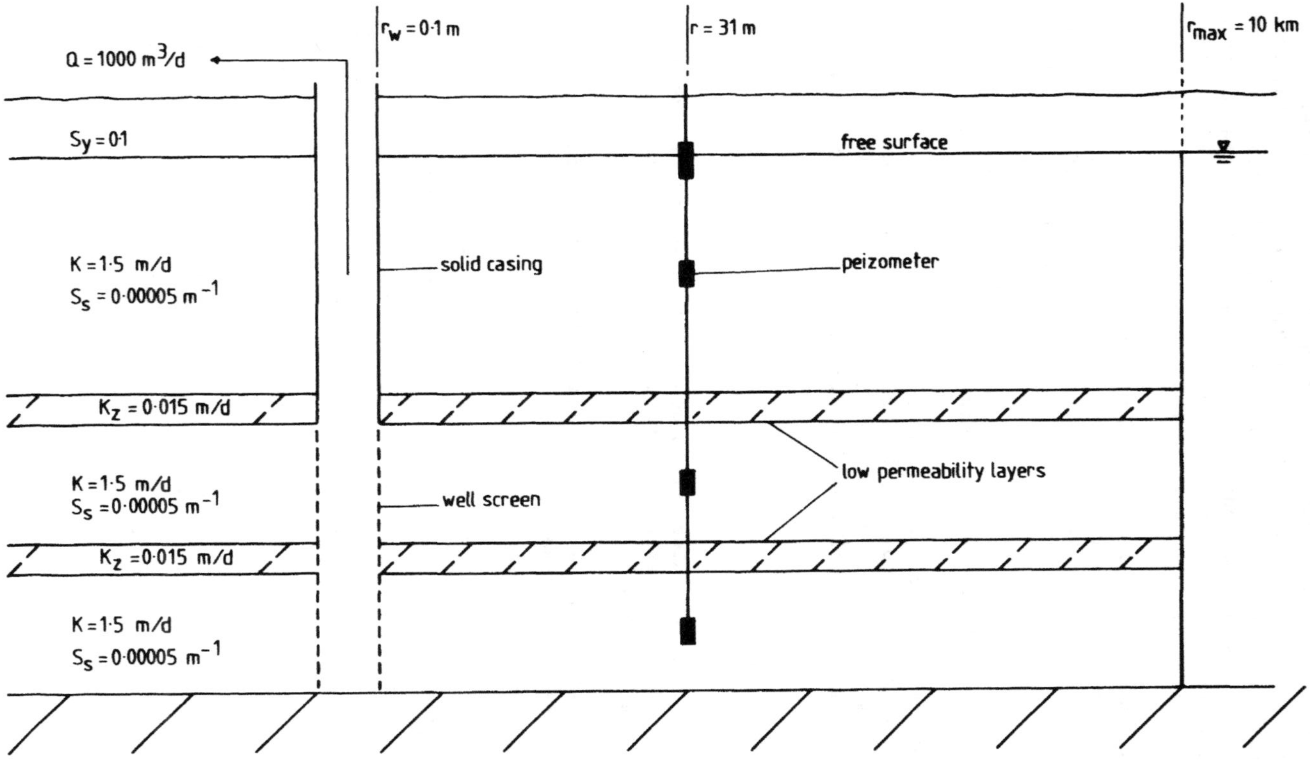

Fig.2(a). General layout and aquifer parameters for idealised examples.

Numerical solutions to these aquifer problems are obtained using a radial flow model which represents flow to a borehole in three aquifer horizons. The model was developed by Grout (1988) and is based on earlier radial flow models described by Rushton and Redshaw (1979). The model provides finite-difference solutions using a direct solution technique for the resultant simultaneous equations. The radial mesh has six mesh intervals for every tenfold increase in radius. Vertical flows are included in the model, the drawdown is calculated at four different levels. Three of the levels correspond to the three flow horizons while the fourth level is the free surface.

In Case A, water is drawn from both the middle and lower zones of the aquifer. Time-drawdown curves in the pumped well are indicated by the full lines of Fig. 3(a). Initially the drawdown is rapid but it tends to level off and after 3 days the pumped drawdown is 17.14 m. Initial recovery is rapid but the residual drawdown is 0.25 m after 1 day. Certain of the pumped drawdowns are recorded in Table 1; these will subsequently be compared with the drawdowns for Cases B and C.

The pumped drawdowns do not provide a clear insight into the flow mechanisms within the aquifer. However, from the numerical model results it is possible to determine the contribution from the middle and lower zones to the well. At 1.0 day the contributions are as follows:

middle zone 548 m^3/d
lower zone 452 m^3/d

This distribution indicates that a considerable amount of energy is used in drawing water vertically through the lower low permeability zone into the pumped well.

Table 1. Pumped drawdowns in metres at different times during the pumping and recovery phase.

Time (d)	Case A	Case B	Case C
Pumping Phase			
0.001	9.84	15.68	7.06
0.01	14.24	27.58	10.73
0.1	16.24	29.85	12.42
1.0	16.91	30.17	12.95
3.0	17.14	30.54	13.21
Recovery Phase			
0.001	7.29	14.86	6.14
0.01	2.90	2.98	2.48
0.1	0.92	0.76	0.81
1.0	0.33	0.55	0.37
3.0	0.25	0.39	0.28

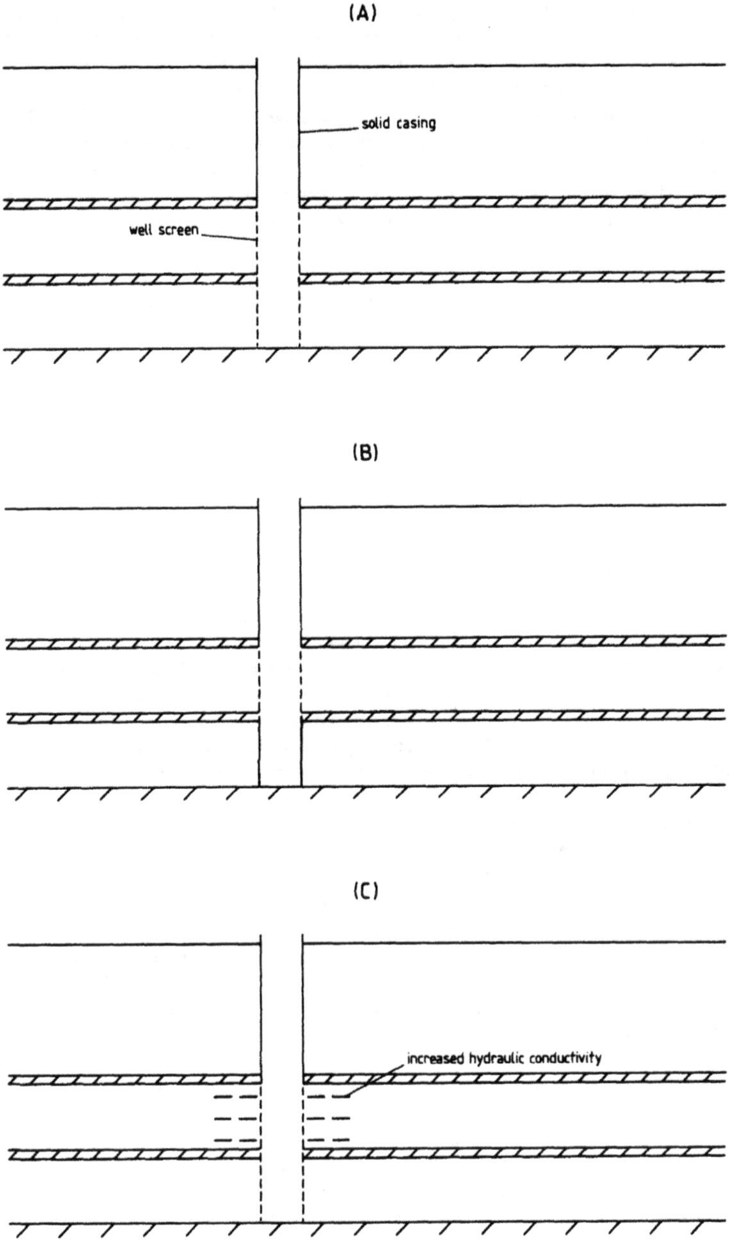

Fig.2(b). Conditions close to the pumped borehole for examples A, B and C.

Further information can be gained from the plots of Fig. 3(b), which refer to the drawdowns for different aquifer horizons at a radial distance of 31 m as indicated in Fig 2(a). Just the curves for the drawdown phase are given in this plot. In a field situation, this type of data would be collected using point piezometers positioned at different levels over the saturated thickness (see Rushton and Howard, 1982; Walthall and Ingram, 1984). The difference in heads between the upper and middle and the middle and lower zones are proportional to the vertical components of flow across thes two low permeability zones.

Table 2. Comparison of drawdowns in metres for the different aquifer horizons at r = 31 m after 3 days of pumping.

	Case A	Case B	Case C
Free Surface	0.74	1.07	0.82
Upper Horizon	0.99	1.41	1.10
Middle Horizon	1.63	2.30	1.79
Lower Horizon	3.34	0.81	2.72

Information for Case B, in which water is only withdrawn from the middle zone, is recorded in Fig. 3(a) by the chain-dotted line; values are also included in the second column of Tables 1 and 2. The pumped drawdowns on the drawdown phase are almost two times those for Case A. This is due to the reduction in the length of the open well-screen. Initial recovery is again rapid. After 0.01 days (14.4 mins) the recovery curve is roughly similar to that for Case A.

An examination of the first two columns of Table 2 indicates that the vertical flow components are different for Cases A and B. In Case B the head difference between the upper and middle zones is 40% greater than the difference for Case A. This indicates that the vertical flows from the free surface at r=31 m are much greater. Drawdowns for the free surface and the upper and middle zones are greater than Case A as a result. However, partial penetration of the pumped well in Case B results in less drawdown in the lower zone with water drawn up into the middle zone at this radius.

Results for Case C, in which the hydraulic conductivity of the middle zone is increased between the well and a radial distance of 1.0 m, are included as the broken line in Fig. 3(a); water is withdrawn from the middle and lower zones. Pumped drawdowns are significantly less than those for Case A. An examination of the contributions to the pumped well indicates why there are such differences; at 1 day the contributions are as follows:

middle zone 659 m^3/d
lower zone 341 m^3/d

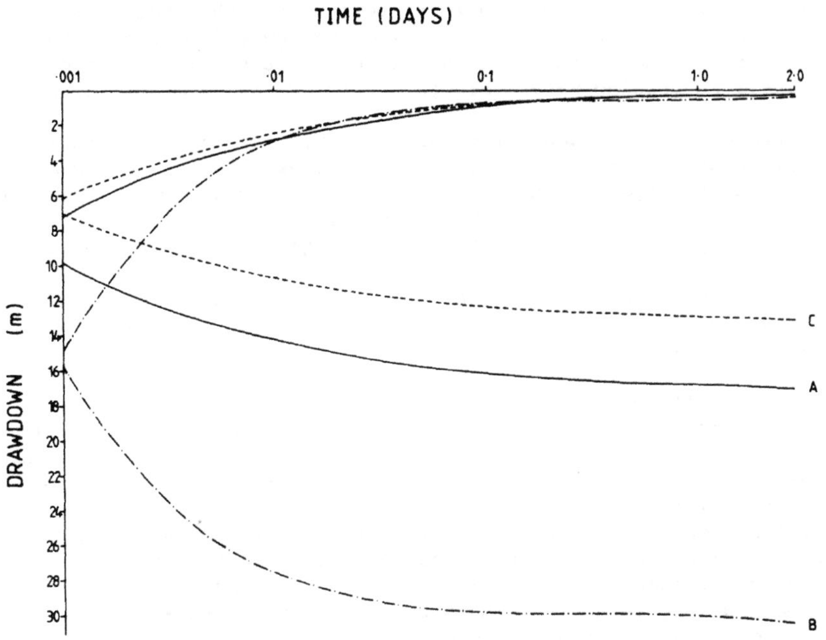

Fig.3(a). Drawdown and recovery curves for the pumped boreholes, Cases A, B and C.

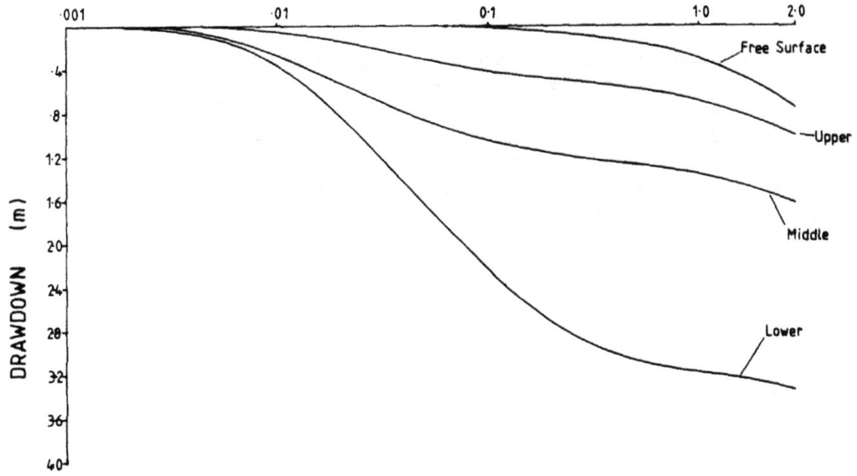

Fig.3(b). Drawdown curves at r=31 m for Case A.

The higher inflow to the middle zone results from the increased hydraulic conductivity in the vicinity of the pumped borehole; this leads to smaller head losses in the vicinty of the well and less energy used in drawing water through the lower low permeability zone.

The increased flow through the middle flow horizon is reflected in the vertical head distribution in Table 2. The drawdowns for the free surface and the upper and middle zones are greater for Case C compared to Case A. The increased flow through the middle flow horizon results in a greater proportion of the total discharge being drawn from the free surface closer to the well. The head difference between the middle and lower zones for Case C is only 54% of the difference for Case A reflecting the smaller vertical flows across the lower low permeability layer.

These three different cases have shown that with the same aquifer, different pumped drawdowns can be obtained dependent on conditions at, or close to the borehole. Since there is no change in the aquifer and the quantity of water withdrawn from the aquifer is identical in each case, these differences in pumped drawdown would normally be classed as 'well losses'. However, the results of Table 2 demonstrate that there are differences in vertical flows at this particular radius and this indicates that there are differences in the 'aquifer loss'. It would be possible to use the Theis theory to calculate a 'theoretical' aquifer loss based on the transmissivity and the specific yield. Alternatively a more sophisticated theory could be used to predict the aquifer drawdown, but there is no theory that can adequately represent the complex flow interactions in this example. Consequently the concept of separating out the aquifer loss and the well loss is not realistic.

Indeed by considering an example of a representative aquifer, it has been possible to show that, even if there are no losses due to the effect of gravel packs, well screens or turbulent losses, the pumped and aquifer drawdowns are dependent on conditions in the vicinity of the well. If an estimate of the well loss or well efficiency is required, a good approximation can be obtained from the change in the pumped drawdown during the first 0.001 day of recovery.

5 FIELD PUMPING TESTS IN AN ALLUVIAL WADI AQUIFER

The idealised examples of the previous section highlighted the dependence of the pumped drawdowns on conditions in the vicinity of the borehole. This section describes a study in a wadi alluvial aquifer in which difficulty was encountered in achieving the expected yields from the boreholes. The existence of a seepage face had a significant effect on the ability of the borehole to withdraw water from the aquifer.

The aquifer system consists of two distinct units. The shallower unconfined unit is an alluvial fan, or wadi deposit, of recent age consisting of sand and silt layers containing sequences of pebbles and cobbles. In the area of the tests the thickness of this unit is approximately 30 m. The deeper unit is a compact sandstone of Cretaceous age which is a predominantly dense and fractured formation and which is generally well-cemented. In the well field the sandstone is more than 100 m thick. Between the wadi deposits and the sandstone there is a conglomerate layer which is approximately 3 m thick and is continuous over the study area. Since the conglomerate is very well-cemented, it has a low hydraulic conductivity; this has a strong influence on the yield of boreholes which penetrate into the sandstone unit.

Three boreholes were drilled into this aquifer system in the test area; additional piezometers (1P and 2P) were also constructed at two of the boreholes. The location and details of the borehole construction can be found in Fig. 4; summary details are recorded in Table 3.

Table 3. Summary of boreholes in test area.

No.	dia. (m)	distance from no 1 (m)	total depth (m)	open to aquifer
1	0.322	-	75	alluvium and sandstone
1P	0.025	0.01	30	alluvium (in gravel pack)
2	0.304	50	37	alluvium
2P	0.025	50	45	top of sandstone
3	0.304	80	70	sandstone

5.1 Pumping Test

In the main pumping test a discharge of 1050 m^3/d was withdrawn from No. 1 borehole for a period of 2 days, recovery was monitored for a further 0.21 d. Results for all five monitoring points are presented as the discrete symbols in Figs 5 (a)-(f). The continuous lines which are obtained from a numerical solution are discussed later.

(i) Pumped well, No. 1, Fig. 5(a) and 5(b); the water levels fall rapidly to about 26 m; readings during the first few minutes are not available due to the difficulty in achieving a steady discharge rate in the early stages of the test. Recovery is rapid with 90% recovery occuring in 0.01 d (14.4 mins).

(ii) Observation piezometer in gravel pack of pumped well, No. 1P, Fig. 5(a) and 5(b); these results show a fall in head to about 11 m

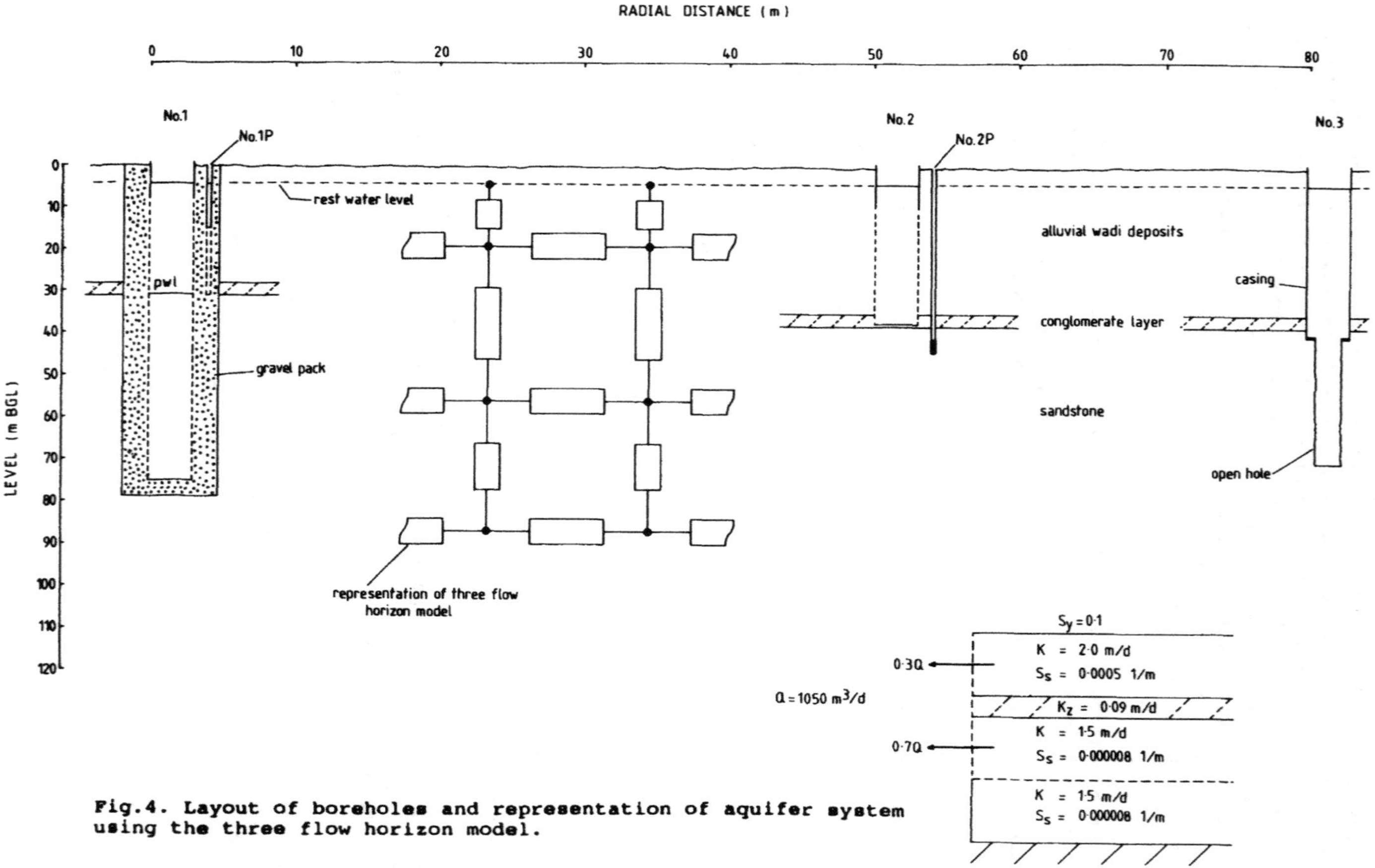

Fig.4. Layout of boreholes and representation of aquifer system using the three flow horizon model.

and then a steady reading throughout the remainder of the test. These readings reflect the conditions in the aquifer adjacent to the borehole and demonstrates that the fall of the water table at the edge of the well is much smaller than the fall in the pumped water level. The difference between the pumped water level and the water table is due to the seepage face; an understanding of the effect of the seepage face is crucial to the potential yield of the borehole. Because this piezometer is in the gravel pack rather than in the aquifer, it will only give an approximate representation of the actual conditions in the aquifer.

(iii) Observation well in alluvium, No. 2, Fig. 5(c) and 5(d); the drawdowns in the alluvium occur only slowly as water is released from storage at the water table.

(iv) Observation piezometer in top of sandstone, No. 2P, Fig. 5(c) and 5(d); the drawdowns in the sandstone show a very different response to the alluvium; not only do they reflect the rapid response of the pumped well, but they are larger than the drawdowns in the overlying alluvium. This confirms that vertical flows do occur.

(v) Observation well in sandstone, No. 3, Fig. 5(e) and 5(f); results are similar to (iv) above but because this borehole is more distant from the pumped well the drawdowns are smaller.

5.2 Interpretation Using a Model

The drawdown and recovery curves of Fig. 5 provide valuable information, but the significance of these results, in terms of flows within the aquifer and losses in the vicinity of the well, can best be gained by developing a numerical model which reproduces the observed drawdowns.

The radial flow numerical model of Grout (1988), was adapted to represent the wadi aquifer. Table 4 shows how the alluvial and sandstone zones are included in the numerical model. The conglomerate is represented as a low permeability layer between the upper and middle horizons. Figure 4 indicates how the numerical model relates to the aquifer horizons.

The major development to the model which was introduced in the previous section is that, instead of no flow occuring from the upper horizon of the aquifer, flow does enter into the pumped borehole from the alluvium. However, because the pumped water level is towards the bottom of the alluvial horizon, most of the water leaves the alluvial horizon of the aquifer along the seepage face. The seepage face is less efficient at drawing water from the aquifer than when it is fully submerged. In fact, to obtain a reasonable simulation of all five of the field drawdown and recovery curves, the flow from the alluvium is set to be only 30% of the total abstraction even though the hydraulic conductivity of the alluvium is higher than the sandstone. Other aquifer parameters as deduced from the model are shown inset on Fig.4.

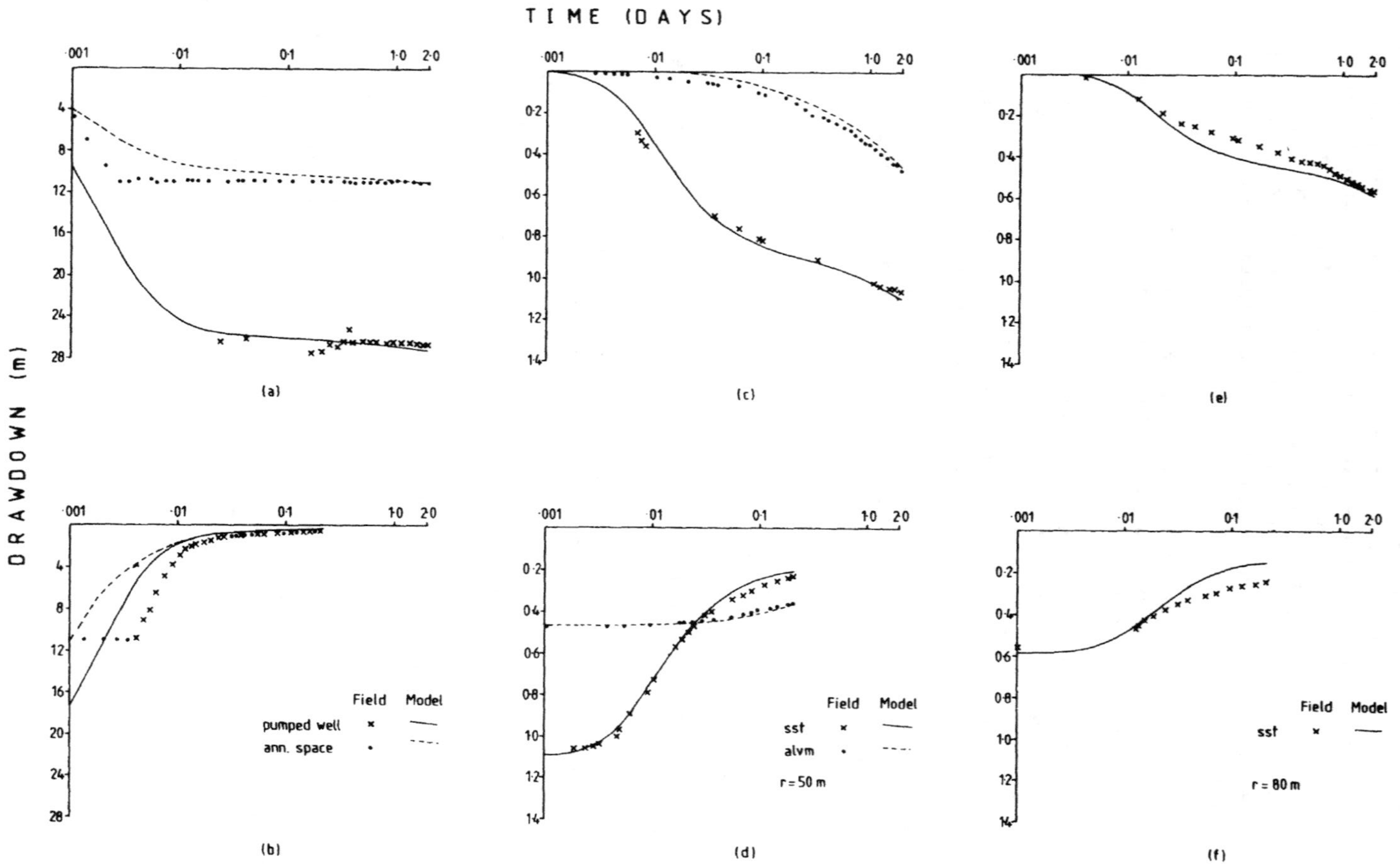

Fig.5. Comparison of field and model drawdowns for field pumping test.

Table 4. Aquifer units represented by the different model flow horizons.

Aquifer Unit	Saturated Thickness (m)	Model Flow Horizon
Alluvium	33 (initial)	Upper
Top of sandstone to base of pumped well	35	Middle
Sandstone below base of pumped well	29	Lower

The saturated thickness of the alluvium varies as drawdown occurs at the free surface.

Although the numerical modelling is not the main aspect of this paper, it is instructive to compare the model predicitions, which are indicated by continuous lines, with the field results. In this discussion it should be recalled that the numerical model results are obtained from a single simulation with the recovery phase following immediately after the pumping phase.

- the agreement between field and modelled results in the observation wells and piezometers, No. 2, No. 2P and No. 3, which are 50 m or 80 m from the pumped borehole are good. Not only are the general trends reproduced, but detailed features such as the shape of the curves at the end of the pumping phase are also modelled.

- the representation of the pumped well drawdowns are encouraging, although the start of the recovery phase is more rapid in the model than observed in the field. This may be due to small errors in the timing of the field records.

- the field results for the piezometer in the gravel pack are of the same order as the modelled results for the aquifer levels at the face of the borehole. The location of the piezometer in the gravel pack is not ideal and also the model simulation of the conditions on the seepage face is only approximate. Consequently differences can be anticipated.

5.3 Discussion

No. 1 borehole was typical of the design used for the development of this aquifer. The initial design assumed individual borehole yields of at least 2000 m^3/d but in the test described above the yield was only 1050 m^3/d with the pumping level just above the conglomerate layer. The situation was even more serious during a test on borehole No. 2 which is constructed totally in the alluvium. With a discharge rate of 500 m^3/d, the pumped drawdowns approached the maximum permissible.

Various suggestions were made to explain the low yields and large drawdowns. These included:

- the aquifer in the vicinity of the borehole was clogged with drilling mud,

- the 15 m head difference between the observation piezometer in the gravel pack and the pumped water level was due to losses across the well screen.

Certain well losses undoubtedly did occur in the vicinity of the pumped borehole during the test. However, the main cause of the large pumped drawdowns was the existence of the seepage face and the vertical components of flow within the aquifer. There is no way in which these 'well losses' can be removed; they are implicit in the groundwater flow mechanisms.

6 CONCLUSIONS

This paper discusses certain reasons for the large drawdowns which can be observed in pumped boreholes. In unconfined aquifers with moderate to low permeability the vertical components of flow are significant. A substantial seepage face can develop at the well-face and most of the discharge flows into the borehole from deeper aquifer horizons below the pumped water level. This flow mechanism produces very much larger pumped drawdowns compared to the theoretical situation where discharge is uniformly distributed along the well-face, particularly when low permeability layers are present which reduce the vertical permeability of the aquifer. The case of an unconfined aquifer has been considered here but large pumped drawdowns due to vertical flows can also occur in leaky aquifer situations.

Three idealised examples have been discussed which indicate that the conditions in the vicinity of the pumped borehole have a significant effect on the aquifer flow mechanism and lead to different apparent 'well losses'. Misinterpretation of pumped borehole data can occur if the analysis does not allow for vertical flow. The assumption of only horizontal flows makes it necessary to invoke large well losses to explain the large pumped borehole drawdowns. This demonstrates the importance of collecting observation borehole data for both identifying the vertical flow mechanism and helping to understand the flow processes in the vicinity of the pumped borehole. The idealised problems show that observation borehole data should be collected at the same radial distance but at different levels over the full depth of the aquifer system.

The field example highlights the effect of the seepage face in causing large pumped drawdowns. The seepage face developed as soon as pumping commenced. The identification of the seepage face is important because the design consultant for the well-field argued that the large pumped drawdowns and low well yields were due solely to well losses. Undoubtedly well losses did occur but the large pumped drawdowns were primarily due to the relatively low permeability of the aquifer system, the vertical flow mechanism and the seepage face. The failure to appreciate these features led to a serious overestimate of the yield of the well-field.

7 REFERENCES

Clark, L. (1977) The analysis and planning of step drawdown tests. **Quart. J. Eng. Geol., London**, 10, 125-143.

Grout, M.W. (1988) **Numerical Models for Interpreting Complex Pumping Test Response.** Unpubl. Ph.D. Thesis, University of Birmingham, England.

Hantush, M.S. and Jacob, C.E. (1955) Non-steady radial flow in an infinite leaky aquifer. **Trans. Am. Geophys. Un.**, 36(1), 95-100.

Neuman, S.P. (1972) Theory of flow in unconfined aquifers considering delayed response of the water table. **Wat. Resourc. Res.**, 8(4), 1031-1045

Rushton, K.R. and Howard, K.W.F. (1982) The unreliability of open observation boreholes in unconfined aquifer pumping tests. **Ground Water**, 20(5), 546-550.

Rushton, K.R. and Rathod, K.S. (1988) Causes of non-linear step pumping test responses. **Quart. J. Eng. Geol.**, London, 21, 147-158.

Rushton K.R. and Redshaw, S.C. (1979) **Seepage and Groundwater Flow.** Wiley, Chichester.

Rushton K.R. and Srivastava, (1988) Interpreting injection well tests in an alluvial aquifer. **J.Hydrol.**, 99, 49-60.

Theis, C.V. (1935) The relation between the lowering of the piezometric surface and the rate and duration of discharge of a well using ground water storage. **Trans. Am. Geophys. Un.**, 16, 519-524.

U.S. Department of the Interior (1977) **Ground Water Manual**. Bureau of Reclamation, Washington D.C.

Walthall, S. and Ingram, J.A. (1984) The investigation of aquifer parameters using multiple piezometers. **Ground Water**, 22(1), 25-30.

PART FOUR

CURES

15 WELL PERFORMANCE DETERIORATION: AN INTRODUCTION TO CURE PROCESSES

P. HOWSAM
Department of Agricultural Water Management,
Silsoe College, CIT, UK

Abstract
Whilst successful cures to well performance deterioration are dependent on a correct diagnosis and a full understanding of the cause process(es) involved, it is important that the cure processes applied, are also equally well understood. A lack of success with maintenance or rehabilitation projects often stems from the application of an inappropriate technique or the inappropriate application of a potentially suitable technique. Much of this approach is summarised in the expression 'suck-it-and-see'; an all too common approach in groundwater engineering in the past. This paper outlines the main types of process involved in well development and rehabilitation.
Keywords: Acid, Chemical, Hydrodynamical, Jetting, Pasterisation, Physical, Repair, Replace, Reline, Structural, Surging

1 Introduction

In general 'cures' can be categorised in a similar way to that for 'causes', ie primary processes:

Physical
Chemical
Hydrodynamical

and secondary procedures:

Operational
Structural/Mechanical

In practice it is likely that a combination of several processes /procedures will be required in order to achieve optimum results. Again as with 'cures' they may be applicable to various parts (as previously defined) of the groundwater abstraction system:

Aquifer
Well
Pump
Distribution

For the sake of clarity the word well shall be used to cover both water supply boreholes and irrigation tubewells. Development shall refer to the action taken to improve well performance immediately after construction. Rehabilitation shall refer to the action taken to improve/recover well performance, after a period in operation, and shall be used to cover the terms refurbishment and stimulation.

2 Cure processes

2.1 Physical processes

(**P1**) Simple physical processes include brushing, whereby a wire brush, is lowered and rotated on the end of a drill string inside the well lining. It is a simple and effective process for removing soft to medium encrustations but becomes much less so where internal casing/screen diameters vary or are not precisely known.

For harder encrustations, processes which induce shock and/or vibration are required (ie to give the same effect as when an iron object is banged with a hammer in order to remove rust scale). Explosives or dry ice are used in this type of process. In the latter case blocks of dry carbon dioxide are crudely dropped down the well. The rapid change from solid to gaseous state in the confines of the well produces an explosive like reaction. As with the use of explosives the condition of the well needs to be known if damage is to be avoided with this relatively uncontrolled process.

Explosives have been used in low yielding fractured rock aquifers in order to extend the fracture system and hence improve the near-well transmissivity. For well screen cleaning only very small amounts of explosive are required, eg 2 or 3 detonators and a single strand of the high velocity detonating fuse ('cordtex'), commonly used in mining and quarrying. THe use of delay detonators will produce a series of shocks in very rapid succession. To avoid damage to the well lining, the explosive must be carefully centred inside the well, by the use of a steel cage. Explosives should never be used in plastic lined wells but can be used in steel and glass reinforced plastic linings, as a quick and effective treatment for stubborn encrustations.

(**P2**) Sophisticated physical processes include pasteurisation, irradiation, and ultrasonics. The latter is a process taken from industrial/laboratory practices of using ultrasonic vibrations to descale vessels. As well rehabilitation process it does not appear to be widely adopted. There are grounds for believing that the process can lead to the breaking of the cement/casing bond and possibly other joints/seals, leading to the ingress of non-aquifer waters into the well.

Pasteurisation and irradiation are common laboratory/industrial sterilisation techniques. Both processes have been adopted for trials in the treatment of biofouled wells. In the case of irradiation it is understood that the doses required are at a similar level to those used for food irradiation. Whilst the process could be effective, public fear of radioactive processes and the high degree of sophistication, would suggest that this process will not be widely utilised. In the case of pasteurisation, the temperature of the water in the well needs to be raised to 60 Celsius for about 30 minutes to be sure that all bacteria have been killed. At this temperature a biofilm also tends to break up and slough off the surface to which it is attached. Methods of raising the temperature, (some of which have been tried in the field), include the injection of high pressure steam or hot water, and the in situ heating of the water with an electric heating element. As can be imagined large amounts of energy are required to heat up the water column in and around a well screen and this together with the equipment required means that pasteurisation is likely to be limited to developed countries where such resources /facilities may be more readily available.

2.2 Chemical processes

(C1) Acids are commonly used in well development and rehabilitation. In the UK and elsewhere hydrochloric acid is used in the development of newly drilled chalk and other limestone wells. The acid dissolves the calcium carbonate, resulting in reaction products: calcium chloride and carbon dioxide. The latter is controlled in order to build up pressure and force the unspent acid out into the formation via joints and fissures. There has been some discussion as to whether the acidisation process primarily results in dissolution of the rock, thus increasing hole and fissures diameter, or whether it primarily results in the dissolution and removal of the drilling slurry plastered around the hole and within fissures. Some laboratory work by the author shows that the reaction rate of acid with chalk slurry can be up to 15 times greater than that with chalk rock. This suggests that the acidisation process is primarily a cleaning one, with the acid preferentially dissolving and removing the carbonate slurry coating the walls of the hole and clogging joints and fissures, rather than attacking fresh rock. Field studies using pre and post-acidisation caliper logging and analysis of pumping tests, tend to confirm this.

An important fact to note is the recent trend, in view of environmental considerations, to using food-grade acid as opposed to industrial-grade, which was normally used in the past. The use of inhibitors and other additives to the acid is now less common, both because of environmental considerations and because there are no longer considered necessary. There is also a realisation that careful application of the acid can reduce the typical 20 tonne dosage (in UK) by at least 50%. Effective acid concentration need not be greater than the commonly available 36%, but should not be less than 18%.

Acids are also commonly used in the rehabilitation of wells subject to iron fouling, ie where the formation, gravel pack and well screen become clogged by iron based deposits. The acid treatment is equally effective for deposits which are chemically and/or microbially generated. Normal hydrochloric acid is probably as effective as the specially produced pellitised products which are composed of sulphamic acid with an inhibitor, sequestering agent and surfactant added.

...Several problems can occur with acid treatments, in that secondary gel-like products can be generated which may cause further clogging and permeability impairment. In some cases iron hydroxide gels are generated at the acid/ferric oxyhydroxide reaction front when the pH of the reacting acid rises to a value of 3 or more. It is also possible for the protein/carbohydrate material occurring in biofilms to become denatured by the acid to form an organic gel-like substance. Furthermore care should be taken when using acid in cemented or partially cemented sandstones, since the acid may dissolve away some of the cementing material with the consequent release and migration of fines to cause clogging.

(C2) Biocide application is used where biofouling has been identified. It is often part of a sequence of treatments. The biocides commonly used are strong oxidising agents such as chlorine based compounds (liquid sodium hypochlorite, chlorine dioxide gas), and hydrogen peroxide. The latter is particularly attractive because it generates heat which aids the process, and water as the main by-product poses no threat to the environment. On the other-hand, chlorine compounds such as sodium or calcium hypochlorite are much more readily available.

Hydroxyacetic acid is sometimes used to treat biofouling as it acts as both acid and biocide. Quaternary ammonium compounds, such as Hyamine, can also be used. These often have surfactant as well as biocidal properties, which may assist penetration of the clogging material, thus enhancing the biocidal process.

It is important to appreciate the nature of a biofilm when considering this sequence of treatments. It is not particularly effective for instance to a apply biocide directly to a badly biofouled well. The reasons for this are (a) the bacteria are to a large degree protected from bacteriocidal agents by the biofilm itself and (b) there is such a large amount of material for the biocide to work on that even large doses of high concentrations are easily spent. For badly biofouled wells the procedure should usually be: remove the bulk of the clogging material by physical and/or chemical processes; mop up the remaining bacteria with a biocide; remove any secondary materials generated by the first two processes; disinfect the well in order to inhibit (it can rarely be totally eradicated) regrowth. It is important to note that the concentration/dosage of disinfectants for the sessile bacteria involved in biofouling, need to be much greater than those used for disinfection in normal water supply systems, where planktonic (ie free floating, and therefore easy to get at) bacteria are of concern. For instance in badly biofouled wells concentrations of 1000 ppm free chlorine need to be used compared to the 0.1 ppm residuals required for domestic water supply systems. A further point to remember with chlorination is that sodium hypochlorite should be used in preference to the calcium form, since salts of the latter which will be generated as part of the process, are much less soluble and thus present the possibility of precipitation, deposition and secondary clogging.

(C3) Dispersants (or deflocculants, surfactants). These such as polyphosphates (eg Calgon) and detergents (eg drilling foams) are applied in order to help break-up and disperse wall cakes, developed during drilling (eg bentonite skin) and deposits of redistributed clays and fines originating from the formation. The chemical process needs to be used in conjunction with physical agitation processes, with simultaneous pumping, in order to be fully effective. There is little point in chemically breaking down a clogging structure if the material is not removed, but left to be deposited elsewhere.

2.3 Hydrodynamical processes

In simple terms this refers to any process in which involves a washing/agitation action with water.
(H1) Surging in its various forms is probably the most common, and can be accomplished directly by the use of a surge-block or indirectly by intermittent pumping, using an airlift system or a mechanical pump. The principle of the surging process is the inducement of a forward/reverse flow in the well, through the screen slots, gravel pack and formation, in order to break-up and disperse the clogging media (eg drilling mud cake, fines). The bi-directional flow is applied because this is considered to help to prevent the bridging of fines in pore spaces, a process thought likely to occur with uni-directional flow. Observations of these processes using micro-pore models which have fixed 2D pore systems suggests however that bi-directionnal flow is not that effective in reducing bridging. In unconsolidated formations and gravel-packs the process therefore also needs to induce grain agitation in order to be fully effective. On the other hand over-zealous surging can lead to agitation/movement at the gravel pack/formation interface, where. mixing of the two materials may produce a material of reduced permeability.

(H2) Jetting is a process like surging which is often referred to in the literature. Yet there seems to have been a lack of understanding and quantification of the process. Questions such as:" What pressures/flow rates are required to remove the various types of encrustations encountered on clogged well screens?" "How effective is a jet of water after it passes through the different sizes/shapes of screen slots?" "For what distance can a jet of water effectively penetrate a gravel pack?" "At what pressures/flow rates will jetting damage well linings?", are not easily answered. Yet so called high pressure jetting, is in practice applied by anything from a 300 psi (2000 KPa) drilling rig mud pump, to a 30,000 psi pump used in the cutting away of sewer linings.

Recent studies have shown that the effectiveness of a jet of water drops off extremely rapidly with distance from the jet nozzle, something like 85% reduction in pressure within a distance of 10 x the nozzle diameter. With a 1.5 mm jet nozzle this would an effective working distance of only 15 mm. In practice with well jetting it is not easy to achieve a nozzle to screen distance of less than 25 mm. Quite high pressures, ie of the order of 5000 psi, are therefore required to remove encrustations. Properly applied, high pressure jetting has proved to be very effective process for cleaning the internal surfaces and slots of clogged well screens. In jetting through screens to develop the gravel-pack/formation beyond, energy dispersion is even more severe. Screen slot shape and frequency will influence the effectiveness of the jetting process and it is suggested that jetting will be of minimal use for gravel-pack/formation development where torch-cut, louvre, bridge-slotted and geotextile wrapped screens are installed. Too high pressures of course may lead to damage to the well lining. A stationary water jet in air at 1500 psi and in water at 4000 psi using a 1-2 mm nozzle, at a nozzle to surface distance of less than 10 mm, can cut through thermo-plastic well linings. With plastic lined wells high pressure jetting should only be applied below the water table and with pressures not exceeding 3000 psi.

2.4 Operational procedures

(O1) As discussed in the Introduction to Cause Processes, many problems can be alleviated by altering operational procedures, eg sand pumping can often be reduced by reducing pumping rate and/or by adopting a continuous instead of a intermittent operating schedule. Similar measures can also serve to reduce the degree of iron fouling.

(O2) Fouling in pumps may be reduced by selecting pumps with lower inlet and through-flow velocities.

2.5 Structural/mechanical procedures

In the event of structural failure of part(s) of a well there a three principle options: repair, replace or reline. The same procedures apply equally to pumps and distribution.

(S1) Repair. A casing break for instance may be repaired in situ by backfilling the well with sand to a short distance below the break; inserting a cement plug at the break; forcing some of the cement out in to the annulus; drilling at a reduced diameter through the cement plug and then removing the temporary fill, to leave a cement collar seal.

(S2) Replace. Should the condition of the well be suitable an alternative option is to retrieve the casing/screen string and replace it with a new string of appropriate design and materials (eg the use of non-ferrous linings in situations where iron fouling is a problem).

(S3) Reline. Where an insitu repair is not feasible and where the condition of the well lining and/or hole preclude the option of pulling out the casing/screen string, then relining the well may be an appropiate option. This is a relatively straightforward operation and the reduced diameter of the new lining need not restrict pump size if only the screened section is relined. The ability to develop or rehabilitate the old screen/gravel pack should it become clogged in the future, will however be very limited.

Pumps and pipework are relatively easy to retrieve to the surface, to be cleaned and repaired or abandoned and replaced, depending on condition.

3 Discussion

This introductory paper serves to highlight, not only the main 'cure' processes/procedures, but also to stress that to be successful in well rehabilitation it is essential to fully understand the nature of the processes and the consequences of their application; ie Don't just try a technique because it has been used before or someone else says its good. Think it through; find out what the processes actually does and how it works; consider whether it is appropiate for the case in hand; condsider how/when it should be best applied.

16 EFFECT OF ACIDISATION ON CHALK BOREHOLES

D. HARKER
Anglian Water Services Ltd, Cambridge, UK

Abstract

The standard tools of the hydrogeologist for the investigation of the borehole performance of test pumping and downhole geophysical logging are used to quantify the effects of acidisation.

Examples are given of the use of hydrochloric acid to rehabilitate a borehole whose yield had fallen due to encrustation and to improve the yield of an inefficient borehole.

It is concluded that the primary role of acidisation is in the initial development of chalk and limestone boreholes by the removal of soft material from near to the borehole rather than in attacking hard encrustation or in creating or extending fissures.

It is suggested that neither large volumes of acid nor repeated acidisation produces enhanced improvement in borehole performance; also that borehole yield is controlled by the success of intercepting a fissure network and is limited by the properties of the aquifer at a particular locality.

Keywords: Acidisation, Rehabilitation, Source Reliable Output, Test Pumping, Flow Logging.

1 Introduction

Anglian Water Services Ltd (AWSL) provides public water supplies to a population of nearly 4 million people in the Anglian region of the UK, covering an area of some 22,000 square kilometres. About half of the water supplied is groundwater from a variety of sandstone and limestone aquifers; the most important of which is the Chalk, which forms a broad belt sweeping through the region from the Lincolnshire Wolds in the north until it disappears beneath the London Basin in the south.

The Company operates some 150 groundwater sourceworks with between one and ten individual borehole sources at each. In total there are around 500 borehole sources in operational use. The age,condition and operational requirements of these sources varies considerably. Some date back to Victorian engineering with shaft and adit systems designed to meet the needs of steam driven pumps rather than modern design criteria. The information available on them is also variable. One of the main tasks of the Water Resources team at AWSL is to undertake investigations on sourceworks performance as part of the development of new sources or the investigation of source reliable output (SRO) of existing ones. The evaluation of the SRO requires an understanding of the behaviour of the borehole under variable water table conditions in order to consider its optimum level of operational use. The SRO investigation provides a bench mark against which any deterioration due to a loss of well or aquifer performance can be measured, as well as a means to measure the effects of borehole development or rehabilitation.

2 Source reliable output investigation techniques.

The techniques of SRO investigation are the standard tools of the hydrogeologist for the study of borehole performance.

Variable rate step pumping tests are conducted to define the yield characteristics in relation to drawdown. Under the confined conditions found in many of the area's aquifers a single set of tests will define the response for all piezometric conditions. However in unconfined aquifers the level of the highest major inflow is critical and step, tests under high and low water tables are needed to fully define the borehole's performance. Repeated tests under comparable aquifer conditions measure the deterioration of borehole yield and provide the means of quantifying the effects of acidisation.

Continuous rate pumping tests define the time-drawdown relationship in terms of the variation of the rate of drawdown with time. Changes of drawdown rate in time measured in minutes providing information on the behaviour of the borehole; while the changes of drawdown rate after longer periods will normally be associated with changes in the behaviour of the aquifer at some distance from the abstraction point. Simple plots of drawdown against log time showing changes of rate of drawdown as changes of gradient. These gradients can be represented by values of transmissivity using the basic Jacob equation and provide the method for diagnosing the causes of poor borehole performance as well as measuring the improvements due to acidisation.

Flow logging measures the vertical flow in the borehole and identifies the levels of inflow. Operational requirements can restrict the detail of the flow log available. Quantitative information from an impellor flow meter in the pumped borehole is best. However qualitative information from temperature and conductivity measurements in pumped or unpumped boreholes, or even in observation boreholes, can provide useful information provided that care is taken in its interpretation. Flow profiles derived from flow logging provide a means to decide on suitable injection levels for acid and to measure the effects of acidisation in improving existing or developing new inflows into the borehole.

Close circuit television is used to confirm the levels of inflow into the borehole as well as to observe the condition of the borehole wall and lining tubes. CCTV logs taken before and after acidisation will show the changes in physical appearance of these caused by the acid.

In combination these techniques can thus be used to investigate the need for rehabilitation and to study the effects of development by acidisations. Two sets of examples of acidisations are now considered. First where the purpose of the acidisation was to improve the yield of a new or previously unacidised borehole. Second where acidisation was used to rehabilitate a borehole where the reliable output had fallen due to encrustation, or to further improve the yield of a previously acidised borehole.

3 Initial acidisation

Examples are drawn from a study carried out on boreholes in the confined and semi-confined Chalk of Essex and Suffolk in the early 1980's. The study used data from before and after the first acidisation of ten boreholes, eight of which were acidised immediately after construction, two had been previously used for public water supply.

Table 1 summarises the results of the study. Improvements to the boreholes outputs were quantified from the yield - drawdown curves before and after acidisation. The factors given are the increase in output after acidisation, at the maximum drawdown before acidisation; and the reduction in drawdown after acidisation,at the maximum pumping rate achieved before acidisation.

The conclusion drawn from the study was that the degree of improvement was of the order of 2 to 3 times. It also appeared that the improvement was independent of the volume of acid used, indicating that a relatively small volume of acid would suffice in most cases.

Table 1. Yield and drawdown changes following initial acidisation

No	Nominal diameter	Open depth	Acid used	Yield Change			Drawdown Change		
	mm	metres	tonnes	from tcmd	to	factor multiply	from tcmd	to	factor divide
1	300	100	5	5	10	2	22	11	2
2	600	82	5	5	10	2	14	7	2
3	600	105	5	6	12	2	12	4	3
4	600	95	9	5	10	2	24	8	3
5	450	52	8	2	4	2	36	12	3
6	200	80	2.5	4	6	1.5	16	10	1.6
7	600	106	5.3	5	10	2	14	6	2.3
8	600	70	18	6	10	1.6	28	14	2
9	680	19	9	4	8	2	24	8	3
10	680	54	13	4	8	2	30	13	2.3

Cooks Mill is borehole number 10

Typical of the boreholes studied was Cooks Mill to the west of Colchester. Figures 1 to 4 give details in the form of a construction diagram, yield - drawdown curves, time - drawdown graphs, and flow profiles before and after acidisation.

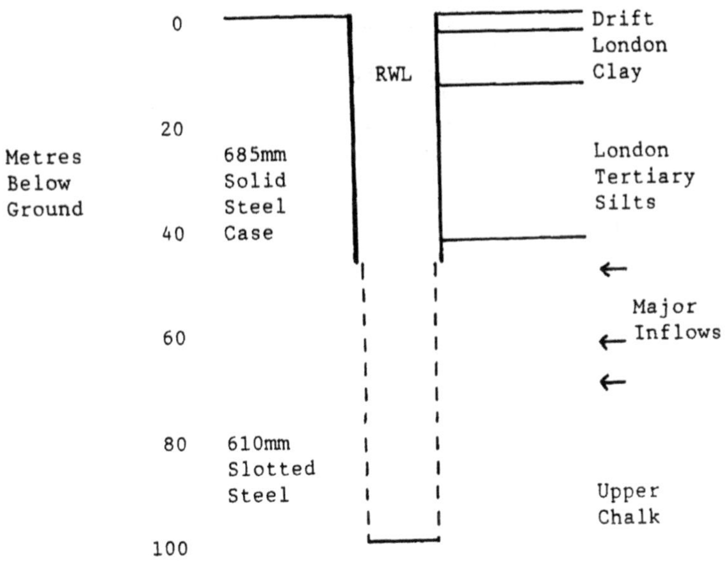

Fig. 1. Borehole Construction for Cooks Mill

The curves of yield versus drawdown for one hour of pumping show improvements at all pumping rates. More detailed analysis of the data shows that the effect of the acidisation has been to reduce the well loss factor [C] to 15% of its previous value.

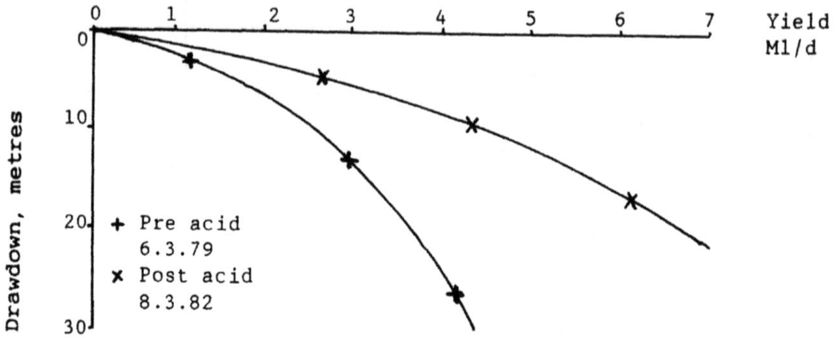

Fig. 2. Yield - drawdown curves for Cooks Mill

The graphs of log time versus drawdown are from step tests at similar pumping rates. They show the typical effect on the first few minutes pumping with a reduction of the rate of drawdown following acidisation. This can be represented by a change of gradient in the form of transmissivity from 44m²/d to 150 m²/d. The time after which the rate of drawdown reduces also changes from around eleven minutes to five minutes. The gradients of the graph after the changes of slope are similar, with a transmissivity of 625m²/d. This is comparable with the value calculated from an observation borehole at a distance of 510 metres and represents the transmissivity of the aquifer. The response of the observation borehole is not affected by the acidisation and does not show the initial steep rate of drawdown.

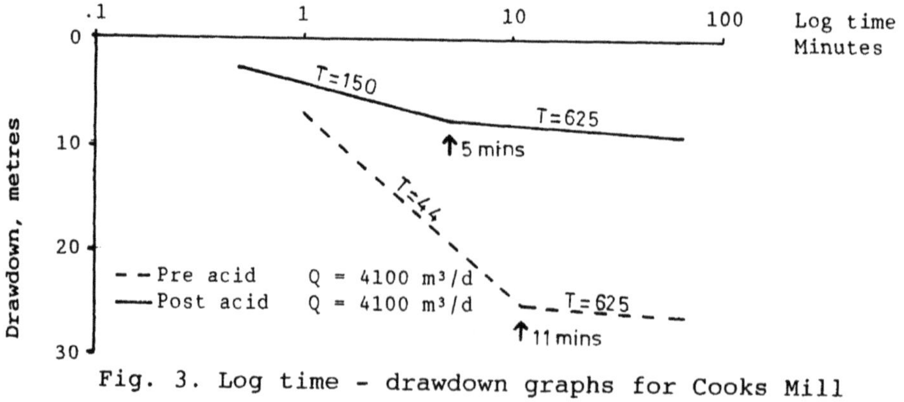

Fig. 3. Log time - drawdown graphs for Cooks Mill

The flow profile is shown as a percentage of the full pumping rate to allow for the different rates of pumping during the two logs. There are a series of inflows from the base of the solid casing at 47 metres to a depth of 70 metres. The nature of the method of measurement of non laminar flow caused by turbulance at each inflow from the side gives some scatter of the discrete measurements. It can be seen though that the general shape of the profile was not altered by the acidisation.

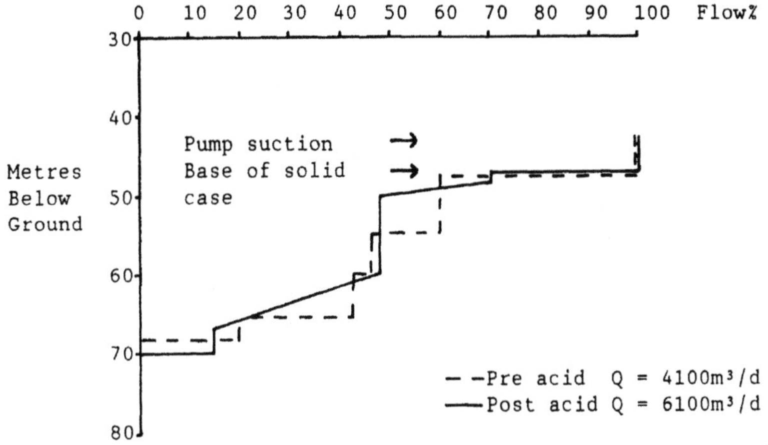

Fig. 4. Flow Profiles for Cooks Mill

4 Repeated acidisations

The reasons for repeating the acidisation of a borehole are usually either to achieve a further improvement in output or to rehabilitate a borehole whose output has deteriorated. An example of each of these cases is given.

4.1 Further improvement

Two boreholes were originally drilled at the Raydon sourceworks to the west of Ipswich. Neither were acidised at the time of construction, but as demand in the area grew it was necessary to increase the sourceworks output. One of the boreholes had a poorer output than the other and following investigation this was attributed to poor connection with the fissure network in the aquifer. The length of solid casing and locally thick drift were also contributing factors.

The poor yielding borehole was acidised twice, initially with 20 tonnes and then with 18 tonnes of acid at two injection points. Figures 5 to 7 show details of the borehole's construction, yield - drawdown curves and time - drawdown graphs at each stage of the work. The curves and graphs show an improvement in yield of around 100% after the first acidisation, resulting from the reduction of the rate and duration of the initial drawdown. The borehole was reacidised as it still did not achieve the output required and had not developed the potential of the aquifer. The curves show that although further improvement was achieved, this was only of the order of another 15%. A replacement borehole was constructed at a distance of 80 metres where the overburden was thinner. The curves, graphs and construction diagram of the new borehole which was acidised with 10 tonnes of acid, successfully developing the potential yield of the aquifer, are shown for comparison.

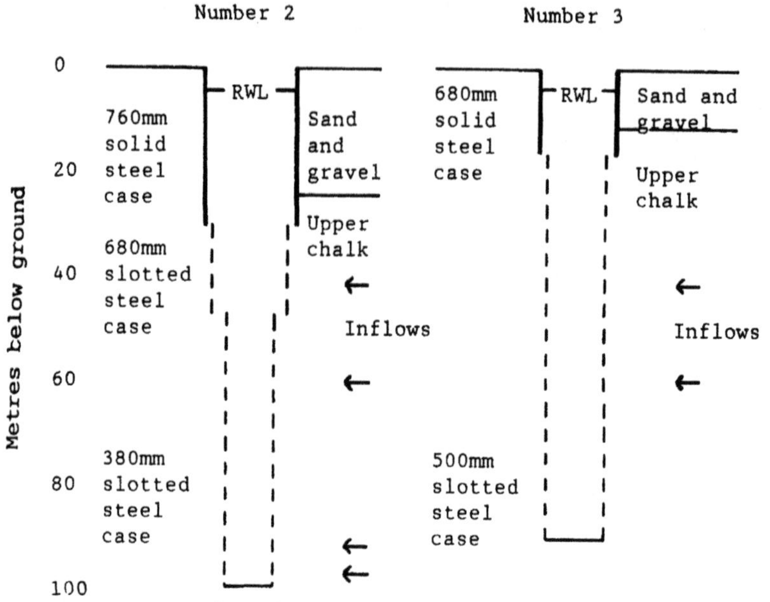

Fig. 5. Borehole construction for Raydon

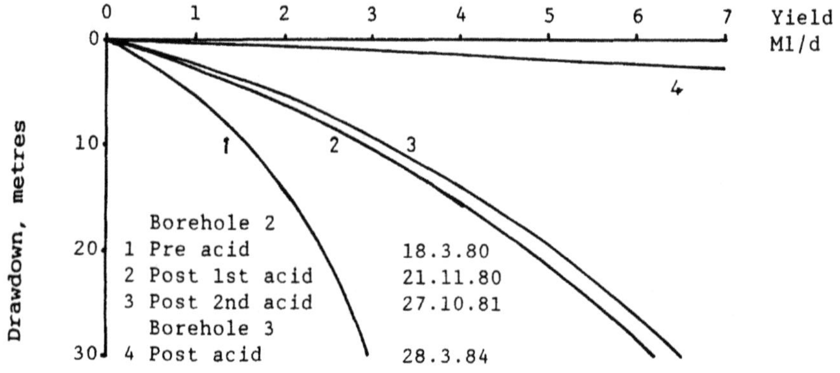

Fig. 6. Yield - drawdown curves for Raydon

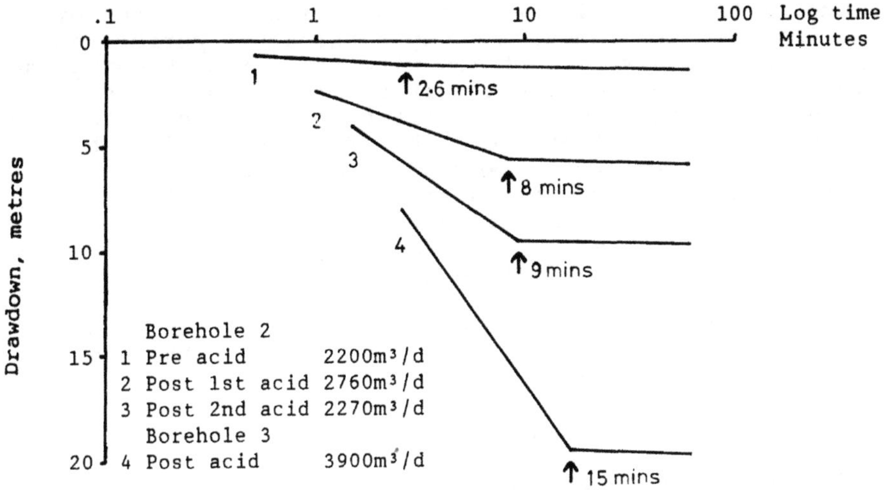

Fig. 7. Log time - drawdown graphs for Raydon

4.2 Rehabilitation

Cases of deterioration of the output of chalk boreholes due to encrustation are luckily rare, but an area where it has been seen to occur in a number of boreholes is the lower Stour valley on the Essex - Suffolk border.

The two boreholes at Bowdens were drilled in 1963 and acidised with the usual improvement in yield. During the 1980's a steady fall in pumping water levels was noted, rest water levels did not fall. Figure 8 shows the deterioration of the boreholes in terms of yield - drawdown curves at successive dates. The boreholes at Bowdens are less than 5 metres apart and act as one, they are both lined with slotted steel casing in the aquifer. CCTV examination showed that these slots had become severely constricted by encrustation.

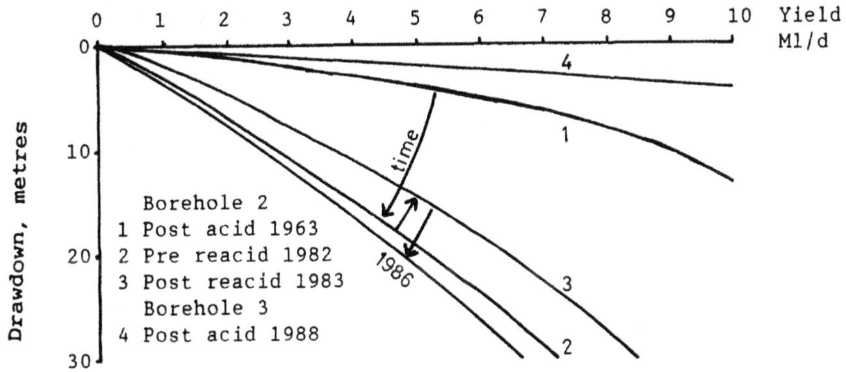

Fig. 8. Yield - drawdown curves for Bowdens

One of the boreholes was acidised with 5.5 tonnes of acid injected throughout its depth through a perforated pipe. Pressures were kept low with the aim of keeping the acid in the borehole to allow it to clean the slotted casing. The effect of the acidisation is shown by the two curves before and after the work. The output of both boreholes was improved by some 25%, although not insignificant it did not restore the original yield. The improvement was also shortlived, figure 8 includes a more recent curve which shows that the boreholes have continued to deteriorate. The solution adopted has been to drill a new borehole 300 metres from the existing ones. The yield - drawdown curve of the new borehole number 3 which has successfully developed the potential yield of the aquifer is shown for comparison.

5 Discussion of the results of acidisation

The chemistry of acidisation is simple:

$$2HCl + CaCO_3 = CaCl_2 + H_2O + CO_2$$

In terms of quantity, 1 tonne of 32% commercial grade hydrochloric acid can dissolve 1.4 tonnes of chalk and will give off 100 cubic metres of carbon dioxide gas at atmospheric pressure.

The way in which the acid acts on the chalk in the borehole is open to conjecture. The following conclusions are put forward from the examples given above. Although the examples are drawn from an area of confined and semi-confined chalk aquifer, the conclusions are borne out by practical experience of the acidisation of limestone and chalk boreholes throughout the Anglian region, carried out using various techniques by most of the UK drilling contractors and direct labour.

1. Acid will preferentially attack material with the largest surface reactive area. This means that it will react with putty chalk, and in particular slurry from drilling left on the borehole wall or in fissures, in preference to hard rock chalk.

2. The volume of acid needed to remove material restricting the inflow of water into the borehole is relatively small. The same degree of improvement has been achieved in boreholes where the volume of acid used has been 25% of the borehole open volume as when it has been 100% or more.

3. The use of the large volumes of acid can be counter productive if high pressures or chemical attack damage the grout seal outside the solid casing or otherwise effect the structure of the borehole.

4. The potential yield of a borehole is defined by the local aquifer characteristics and the distribution of inflows from fissures. Acidisation can clean up a borehole to improve its yield, but is unlikely to develop a new inflow to radically improve the yield and can never improve the yield beyond the local potential of the aquifer.

5. Successive acidisations can provide further improvement in yield but the relative further improvement diminishes with each operation.

6. The success of hydrochloric acid in attacking encrustation on a steel screen by chemical action is only limited. The effectiveness could be improved by physical cleaning methods as well, but if the underlying problem is bacteriological growth then any improvement is likely to be shortlived.

6 Conclusion

Much folklore exists on the methods and objectives behind the acidisation of shallow water supply boreholes. The practical conclusions drawn above have been developed by direct observation of the effects of acidisation using the tools of pumping tests and geophysical logging, before and after acidisation. It is not always possible to carry out a full investigation of each borehole source but any good quality data is invaluable in understanding its behaviour and in monitoring its performance in the future.

7 Acknowledgements

The permission of the Regional Engineering Manager to publish information on Anglian Water Services sources is acknowledged. Thanks are due to the Water Resources staff of AWSL and its predecessors for the collection of the data and the development of ideas. The views expressed are those of the author.

17 WATER WELL REGENERATION – NEW TECHNOLOGY

K.F. PAUL
Charlottenburger Motoren and Gerätebau KG, Berlin, W. Germany

Abstract

Regeneration of water wells became more important with the growing knowledge of the causes of degeneration of water wells. When regeneration is advisable and conditions for it are favourable, the use of a regenerating chemical maybe necessary to remove all blocking deposits from the gravel pack. Great differences can be observed in the efficiency of regeneration by appliances and methods offered in the market today.

The description of a new technique in hydraulic-chemical regeneration of water wells is given. The On-line Well Cleaning System KIESWASHER KW 3G *»with Process Regulation«* demonstrates the present technical standard.

Keywords: Water Wells, Well Degeneration, Well Clogging Problems, Well Rehabilitation, Chemicals for Well Rehabilitation, Environment Protection, New Technology

1 Introduction

Water wells have been regenerated in West Germany since about 1900. The cleaning results varied; mostly they were economically unsatisfactory and very difficult to control. CMG-Berlin has been developing improved cleaning systems for water wells since 1976, and selling them under the name "KIESWASHER" (= Gravel Washer). The system is updated on a continuous basis depending on the state of the technology and practical requirements.

The KIESWASHER is a twin-chamber apparatus for removing waterborne substances precipitated biologically or chemically, and harmless as far as quality is concerned, but reduce well performance. The KIESWASHER combines hydraulic and chemical actions to remove these substances one section of the well at a time:

- from well screen slots
- from the well screen exterior
- from the void volume of the gravel pack as well as
- from the void volume of the transition between the gravel pack and the adjacent aquifer.

At the beginning of 1990 the system has been completed to include the "On-line Process Regulation". For the first time, the cleaning of wells is now systematically controlled, regulated and balanced.

The On-line Process Regulation is the result of extensive development work with intensive testing in cooperation with the Stadtwerke Wiesbaden AG, West Germany.

The On-line Well Cleaning System KIESWASHER KW 3G *»with Process Regulation«* is used:

- for intensive regeneration,
- for intensive disinfection,
- for the removal of well drilling additives and also
- for controlling the quality of ground water.

The system parts are (Fig. 1)

- On-line Process Regulation
- KIESWASHER KW 3G
- cable winch with self-supporting underwater power circuit line
- electrical switch unit.

2 On-line Process Regulation

Over the last few years, environmental protection groups have increasingly demanded a reduction in the use of well regenerating chemicals as well as the supervision of their effect and removal.

In addition, years of experience with water well regeneration posed the question whether there are not more deposits in the well, in the gravel pack, and adjacent formations, than could be dissolved in just a single application of cleaning liquid.

Practical examination showed that, in fact, there are more deposits in a well than can be removed in a single dissolving process, regardless of the current methods. This held true equally for all degenerative processes such as

- incrustations from bacterial precipitation of iron and/or manganese,
- incrustations from chemical precipitation of iron and/or manganese,
- incrustations from precipitation of the carbonates of calcium and magnesium,

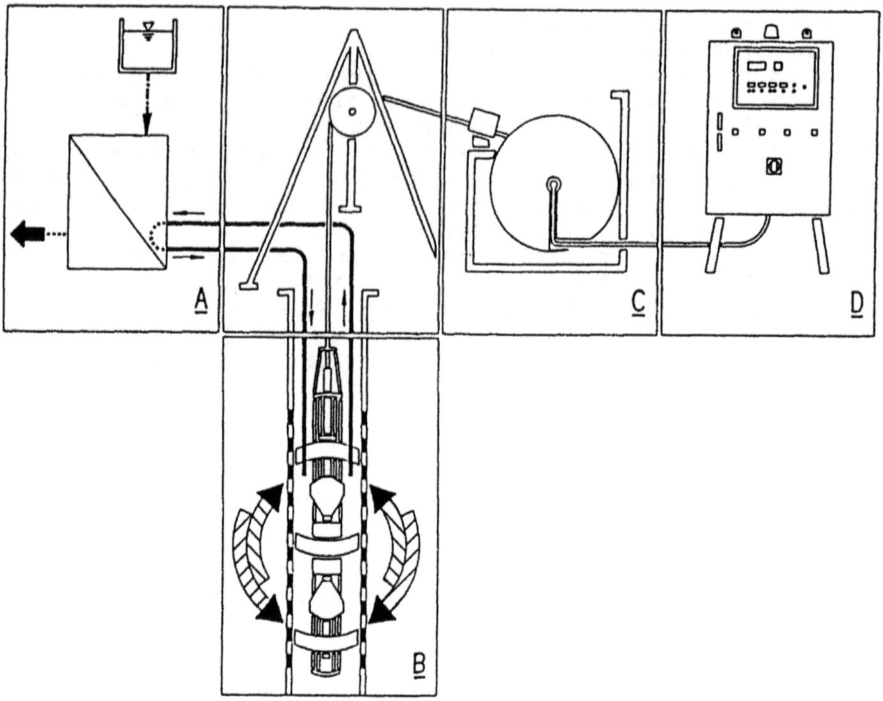

A On-line Process Regulation

B KIESWASHER KW 3G

C Cable winch
with self-supporting
underwater power circuit line

D Electrical switch unit

Fig.1

On-line Well Cleaning System KIESWASHER KW 3G *»with Process Regulation«*

- incrustations from precipitation of slime-forming organisms or
- incrustations from precipitation of aluminum.

Knowing this, led to the development of the "On-line Process Regulation".

The principle of the On-line Process Regulation is that the whole cleaning process is controlled and regulated simultaneously with the measuring of the actually dissolved ion types (in the case of incrustations from bacterial or chemical precipitations: iron, manganese and/or calcium). For analytical purposes some of the active cleaning liquid is continuously pumped out of the section being cleaned. No regenerating agent is lost since the volume being analyzed is returned directly to the cleaning section. The concentration of the regenerating agent is checked continuously in the flow and the dosage, or rather the dosage adjustment, is made process-regulated to maintain the prescribed process parameters.

The continuous on-line monitoring with a connected measurement and graphic printer serves to observe and evaluate the deposit solution directly (Fig.2):

- When a selected ion reaches its maximum concentration in the solution, the section being cleaned is partially pumped out in order to allow fresh cleaning solution access to the performance-reducing deposits that have yet to be dissolved.
- The solution treatments in one section are repeated until the maximum solution concentration in the Nth repetition cannot be reached since all soluble deposits have been removed.

The continuous on-line monitoring of the ions also determines the time actually necessary to dissolve the deposits, and the dissolving capacity of the regenerating agent. Therefore, the On-line ion analysis also serves to evaluate the "effectiveness" of the regenerating agent while the regenerating process is in progress.

By pumping out spent cleaning solution into an extra container which will later serve for neutralization of the solution, and by means of ion analysis and ion balance calculations, the actual amount of dissolved deposits and regenerating agents pumped out of the well can be demonstrated.

CMG-Berlin has applied for international patent protection for the On-line Process Regulation technology with all the features described above, including the balancing calculations.

On-line Process Regulation is applicable with any chemical well-regenerating procedures or systems but the

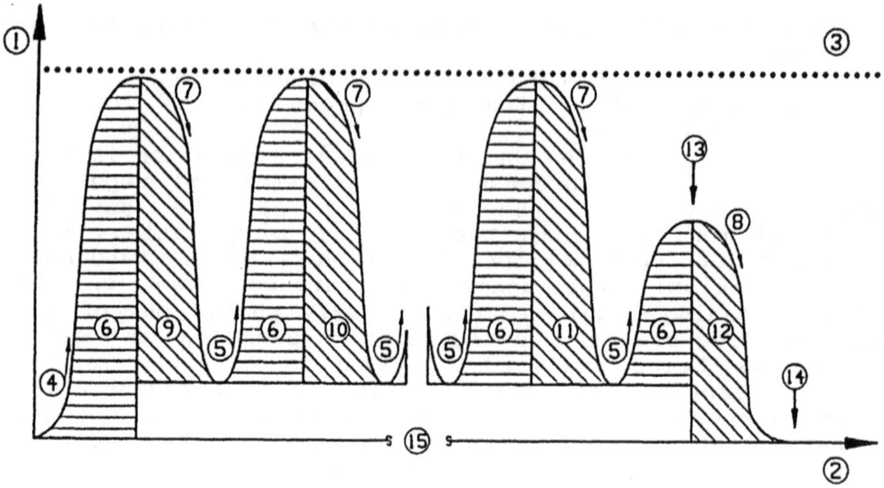

1. specific ion concentration in the KIESWASHER cleaning section, e.g. total iron (Fe^{2+} & Fe^{3+})
2. time
3. maximum absorption capacity of the cleaning fluid for each ion being measured
4. regulated dosage of the regenerating agent
5. regulated post-dosage of the regenerating agent
6. dissolving process
7. intermediate pumping out
8. final pumping out of the working section
9. removed volume, first solubility flushing
10. removed volume, second solubility flushing
11. removed volume, next-to-last flushing
12. removed volume, last solubility flushing
13. all deposits dissolved
14. repositioning of the KIESWASHER for the next working section
15. the time taken for complete dissolving of the actually present deposits, possibly required cleaning cycles at the working section

Fig.2

Schematic presentation of the solution programme for a specific ion to be dissolved from the well.

best results are achieved when the On-line Process Regulation is used with the KIESWASHER KW 3G.

3 KIESWASHER KW 3G (Gravel Washer)

The patented features of the KIESWASHER are the reversible flow direction of the washing solution and the hydrodynamic sealing of the washed section in the screen. The sealing creates two chambers within the screen, between which a pressure difference ensues. This technique causes a flow from the screen into the gravel pack upto the adjacent formation. Outside the screen the current basically flows vertically (into the pore spaces) and the direction is reversed (up/down) at regular intervals (for example every 30 seconds). The KIESWASHER's flow rates outside the screen have been examined and proven in an extensive scientific study.

A regenerating agent, chosen according to the type of deposit, is introduced into this flow to dissolve the deposits. The amount introduced is controlled by the On-line Process Regulation. The study on the KIESWASHER's flow showed that the regenerating agent's ability to reach soluble deposits is optimal because of the constant reversal of direction and the permanent, at least partly, turbulent flow in the pore space of the gravel pack and the adjacent formation. Using a frequency-controlled variable-speed drive for regulating the pressure difference between the KIESWASHER's two chambers also guarantees that the physical integrity of the gravel pack is not destroyed by the KIESWASHER's current.

The removal of deposits from the pore spaces is absolutely necessary because of the accelerated degeneration which would otherwise take place. Most of the deposits are in the pore spaces (Fig.3), and they can only be reached fully by the KIESWASHER's vertical, periodically reversed up and down flow (Fig.4).

Mechanical cleaning techniques reach the interior surfaces of the screen as well as some of the screen slots. Hydraulic cleaning techniques reach the interior surfaces of the screen, the filter slots as well as some of the deposits in the horizontal pore canals. Therefore, mechanical and hydraulic cleaning techniques are only suitable for pre-cleaning as a further possibility to save cleaning agents. The KIESWASHER is a much more efficient cleaning system.

The On-line Well Cleaning System KIESWASHER KW 3G *»with Process Regulation«* can also be used as a packer-sampler for measuring the quality of groundwater. The advantage of the system is that it is not necessary to construct extra observation pipes for testing; instead, measurement is possible directly in existing wells with a screen pipe diameter ≥ 200mm (approx. ≥ 8"). It should be noted that

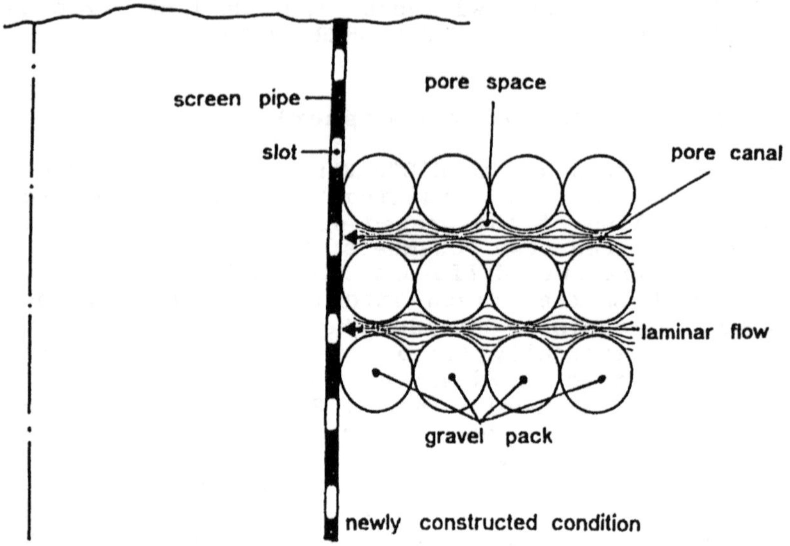

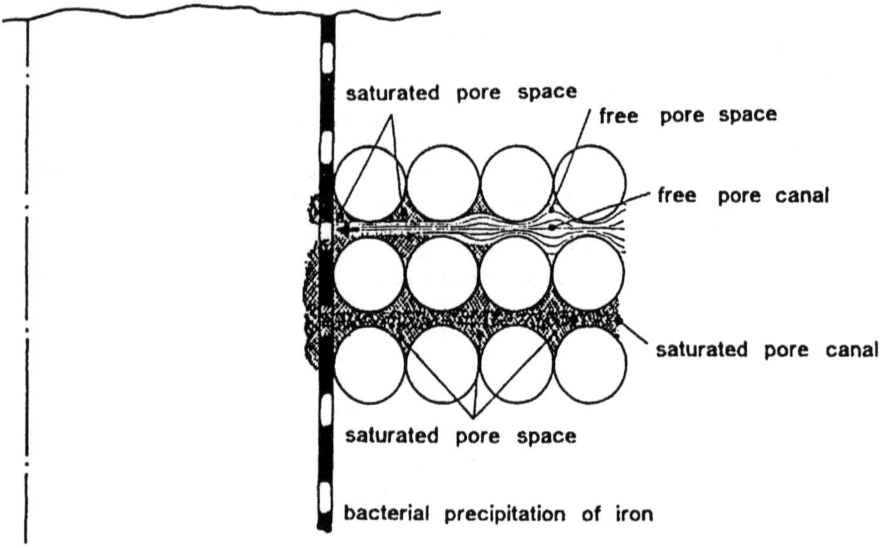

Fig.3

Flow, degeneration and blocking deposit conditions in the gravel pack

7

it is not the quality profile of a single aquifer that is recorded, but rather the qualities of different aquifers are tested selectively. For this purpose the KIESWASHER is used without producing pressure differences between its two chambers.

4 Working Conditions

In its standard version, the On-line Well Cleaning System KIESWASHER KW 3G *»with Process Regulation«* can be used under the following conditions:

- up to 750m well depth,
- up to 80m depth to groundwater,
- up to 30°C groundwater temperature.

The KIESWASHER KW 3G is a working instrument of a plug-in and bolt-together construction, consisting of the KIESWASHER basic apparatus I for a screen inner diameter ≥ 200mm ≤ 250mm (approx. ≥ 8" ≤ 10"), and the KW basic apparatus II for a screen inner diameter ≥ 300mm ≤ 800mm (approx. ≥ 12" ≤ 32"), available in diameter increases of 50mm (approx. 2").

The adjustable pressure difference between the two chambers of the KIESWASHER is designed for a drilling diameter up to 2.000mm (approx. 7ft.) with bonded as well as with multi-layered gravel pack. The effective cleaning height of the KW apparatus allows for working in 1-metre cleaning sections.

5 Operating Experience

The On-line Process Regulation for the KIESWASHER KW 3G was tested intensively and in detail in the development phase by the Stadtwerke Wiesbaden AG, W. Germany, at its Schierstein Water Works. The practical experience has been incorporated in the final design of the equipment.

At Schierstein Water Works there is a gallery of 42 water wells which all have

- screen pipe diameter 300mm (approx. 12")
- material: stoneware or copper
- drilled diameter: 1.000mm (approx. 3,5ft)
- triple gravel pack
- type of degeneration: incrustations from bacterial and chemical precipitation of iron, manganese and calcium

Before and after every well regeneration a diagnostic examination by Colour TV and a pumping test for specific capacity were undertaken.

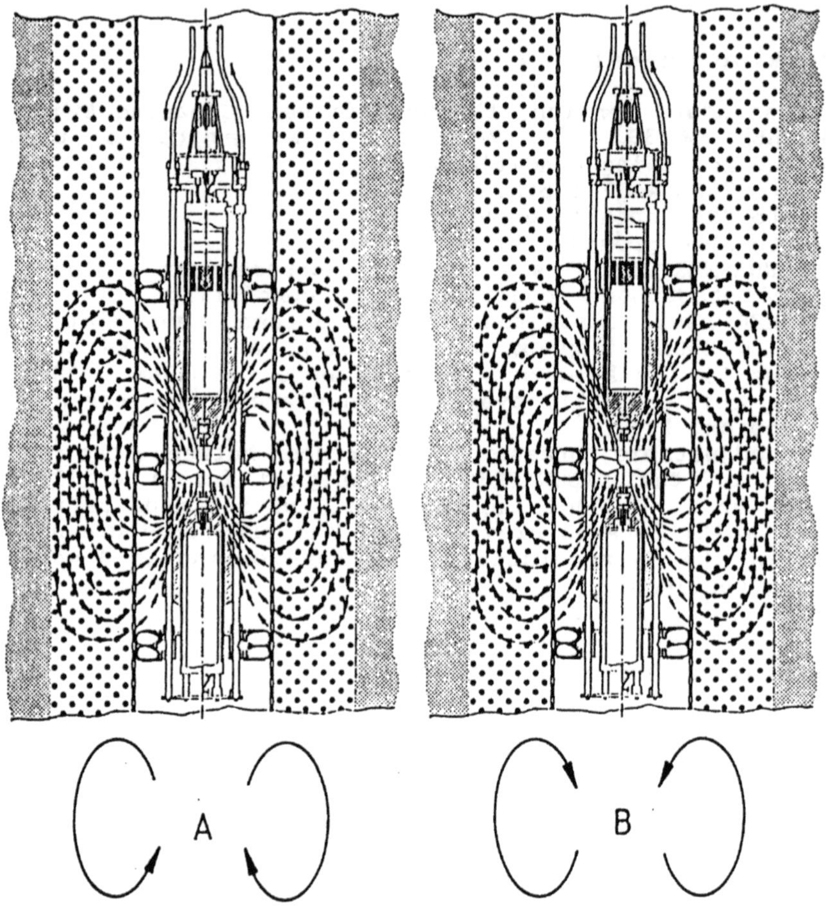

Fig.4

KIESWASHER KW 3G - cleaning section - change of flow direction - The gravel pack is intensively purged with the regenerating agent, which reaches the pore space of the gravel pack via the vertical flow.

The regeneration of these water wells took place in the following sequence:

- mechanical:
  cleaning the interior surfaces of the screen pipe by brushes
- hydraulic:
  partial cleaning of the horizontal pore canals of the gravel pack by pumping out section by section
- hydraulic-chemical:
  Removing the blocking deposits from the void volume of the triple gravel pack and the annulus between the gravel pack and the adjacent formation by On-line Well Cleaning System KIESWASHER KW 3G *»with Process Regulation«*

The first two steps were used for pre-cleaning to remove incoherent deposits as well as to save on regenerating chemicals.

The application of the new hydraulic-chemical regenerating technology gave the following conditions and results:

- regenerating chemicals: special well regenerating type of acid;
- optimum process pH: 0,8 - 0,9
- maximum absorption capacity of the regenerating fluid for dissolved iron, manganese and calcium:
  1.200 mg Fe total/l
  600 mg Mn total/l
  400 mg Ca^{2+}/l
- time taken to reach the maximum absorption capacity for each period of solution: 45 minutes to 60 minutes
- process-regulated regenerating chemical consumption for each period of solution: approx. 2% of the drilling diameter of the working section
- necessary solution periods per working section:
  1 to 12, mostly 2 to 5.
- volume of the cleaning fluid (groundwater + regenerating chemicals + dissolved deposits) per solution period removed by intermediate on-line pumping out: 800 l.
- proportion of regenerating chemicals removed by the intermediate pumping out: 60% to 80%
- maximum quantity of iron pumped out per 1 m screen pipe: 2 kg Fe_{total}/m (dry weight; without quantity of mechanical and hydraulical pre-cleaning)
- a comparison of test results of the application of the KIESWASHER with and without On-line Process Regulation showed at least a 10-fold increase of the amounts of dissolved and removed deposits through the application of the On-line Process Regulation. In comparison to other well-cleaning procedures, the On-line Well Cleaning System KIESWASHER KW 3G *»with Process Regulation«* a still greater proportion of deposits can be removed

- within the limits of measurements, a 4-fold performance increase in specific capacity was proven by comparing well performance before and after the regeneration

After the development and testing stage was successful, basic research in the area of water well regeneration is continuing with the Stadtwerke Wiesbaden AG using their own On-line Well Cleaning System KIESWASHER KW 3G »with Process Regulation«. This research contains extensive laboratory and field tests, statistical calculations, cost effectiveness studies and long-term observations. With these experiments the environmental protection aspect has been given a high value and the results will be published soon by the Stadtwerke Wiesbaden AG.

6 Conclusions

A. The On-line Well Cleaning System KIESWASHER KW 3G *»with Process Regulation«* is a proven and effective technique with the following applications:
   a. hydraulic-chemical water well regeneration by removal of performance-reducing deposits of various types.
   b. hydraulic-chemical well cleaning including the disinfection and removal of well drilling additives
   c. packer sampling from wells with a screen pipe diameter ≥ 200mm (approx. ≥ 8").

B. The advantage of the system when used for water well regeneration are:
   a. minimal consumption of regenerating chemicals thanks to process regulation
   b. short, process-regulated application periods of the regenerating chemical in the well
   c. ability to reach the performance-reducing deposits, especially in the pore space of the gravel pack and upto the adjacent formation.
   d. controlled and complete removal of all soluble, performance-reducing deposits from the pore space area thanks to:
      - On-line ion analysis
      - On-line intermediate pumping out
      - repetition of dissolving processes until the removal of all soluble deposits is complete
   e. ion analysis directly at the regeneration site allows:
      - observation and evaluation of the dissolving process
      - summation of the amounts of deposits removed
      - control of the removal of the regenerating chemical
      - examination of the effectiveness of the regenerating chemical

f. costs of regeneration compared to these of drilling new wells are between about 10% to 20% including incidental expenses; depending on the degree of degeneration and the actual amount of cleaning required.

C. Before <u>every</u> well regeneration a diagnostic examination (at least by Colour TV) should be undertaken. This examination shows the structural condition of the well and determines the type and the extent of the cleaning techniques to be used (mechanical, hydraulic and/or hydraulic-chemical).

D. The new process technique not only optimizes the cleaning results in a so far unachievable manner, but it has also contributed significantly to the process-regulated and minimized use of regenerating chemicals. With the On-line Well Cleaning System KIESWASHER KW 3G *»with Process Regulation«* the service companies and well operators, as well as the environmental and supervisory authorities now have immediate access to a high capacity, environment-friendly well cleaning system with extensive control and balancing possibilities.

7 References:

Krems, G. (1972) **"Studie über die Brunnenalterung"** ("Study of Well Degeneration") Expert opinion on behalf of Federal Ministry of the Interior of the Federal Republic of Germany; in German; source of supply: Bundesamt für Zivilschutz, Deutschhorrenstr. 93-95, D-5300 Bonn 2

Haesselbarth, U. & Lüdemann, D. (1972): **"Biological incrustation of well due to mass development of iron and manganese bacteria"** Institut für Wasser-, Boden- und Lufthygiene des Bundesgesundheitsamtes (Institute of Water, Soil and Air Hygiene, Berlin, W. Germany); Water Treatment and Examination, vol.21, pp.20-29

Paul, K.F. (1985): **"Untersuchung der Strömung in einem Versuchsstand mit Kiesschüttung zur Beurteilung der Wirksamkeit eines Brunnenreinigungsgerätes"** ("Examination of flow in a test stand with gravel pack to prove the effectiveness of a well cleaning apparatus") Institut für Chemieingenieurtechnik (Institute for Chemical Engineering, TU Berlin, W. Germany); in German; source of supply: CMG, Potsdamer Str. 98, D-1000 Berlin 30

Paul, K.F. (1987): **"Anforderungen an Verfahren zur Brunnenregenerierung"** ("Requirements for the procedure of water well regeneration") 12th Water Technology Seminar, TH Darmstadt, W. Germany; WAR Vol. 32 "Neuere Erkenntnisse beim Bau und Betrieb von Vertikalfilterbrunnen" (Latest experiences in construction and maintenance of water wells); in German; pp. 235-306

18 THE USE OF HIGH PRESSURE WATER JETTING AS A REHABILITATION TECHNIQUE

J. FOUNTAIN and P. HOWSAM
Silsoe College, Cranfield Institute of Technology, UK

Abstract.
Jetting is often referred to in the literature as a well development /refurbishment technique and it appears to have been widely applied. Yet there seems to be little understanding of the process. 'Suck-it-and-see' applications should no longer be regarded as good practice. This paper describes a project in which the nature of a high pressure water jet was studied and its behaviour under application situations evaluated. The aim was to improve scientific and practical understanding of how the processes works and how it should best be applied.
Keywords High pressure water jetting, Borehole, Gravel-pack, Well screen, Development, Rehabilitation, Jet characteristics

1 Introduction

In the early days jetting typically involved using the mud-pump to pump water down the drill-string and out via 2 or more small holes drilled into a short sealed drill sub, with nozzle exit pressures of several hundred psi. When observed working at the surface it might have looked an impressive cleaning agent, but down in the hole the behaviour of a water jet in water, though a screen slot and into a gravel-pack was not really understood. One could only imagine what effect the jet might have. From observing the use of high pressure water jetting for cleaning in a number of industrial circumstances, it would not be difficult to imagine jetting when applied at close range, to be an effective surface cleaning technique. This is deceptive however since water jets in air will behave differently from water jets in water. In more recent times proper high pressure systems have been used. Systems of up to 20,000 psi had been developed for sewer renovation. Such systems could be used for cleaning, removing and cutting a variety of materials found in sewers. It was a logical step to carry the application from horizontal subsurface tubes to vertical subsurface tubes.

As far as the authors are aware there has been little systematic evaluation of jetting as a development or rehabilitation technique. Yet with increasing caution from environmental considerations, to the use of chemicals in wells and boreholes, it is anticipated that there will be a increasing interest in the use of environmentally-friendly techniques such as jetting. It was therefore thought appropriate to try and put a little bit more science into the understanding and application of this technique. The 15 month study carried out as part of an ODA funded research project on borehole rehabilitation, was supported by a Department of Trade & Industry (DTI) scholarship and sponsored by MTC Well Systems Ltd, a leading company in the UK offering a patented high pressure water jetting service.

The main objectives of the project were to:

A: examine water jet in water characteristics

B: quantify jet characteristics

C: study the behaviour of a high pressure water jet

- against different well screen materials
- through various screen slot shapes
- through a gravel-pack

D: evaluate the effectiveness of high pressure water jets in removing encrustations from well screen surfaces and slots.

E: relate the laboratory test results to current practices and results on site

2 Description of the jetting process

The equipment required for jetting is relatively simple: a high pressure pump and hosing and a jetting head. The head is normally made from solid steel and is designed to carry four nozzles, placed at 90 to each other. A more recent design involves the actual jetting head rotating, but normally the jetting head and the high pressure hosing from the surface, is rotated.

The procedure consists of slowly raising (about 6 m/hr) the rotating jetting head up inside the well screen/casing, with the nozzle positions adjusted so as to keep the nozzle to screen distance as short as possible. In practice this is often not less than 25 mm. To avoid damage to the screen itself, stationary operation of the jetting is usually avoided. However because the jetting energy can only be focused on a small area at any one time, the rate of travel must be such as to allow enough time for proper cleaning.

3 Principles of jet behaviour

A jet, in the terms considered here, may be defined as a discharge of fluid from an orifice into a large body of the same or similar fluid. As with pipe flow jets may have either laminar or turbulent flow patterns. However in the case of high pressure jetting, flow will invariably be turbulent.

The jet behaviour will depend on three types of parameter:

1: Jet parameters

- the initial jet velocity distribution and level of turbulence
- the flux of the jet mass, momentum, and of any jet tracer material

2: Geometric
   parameters - jet shape and orientation
   - proximity to solid boundaries/free surfaces
   - whether jet is submerged or not

3: Environmental
   parameters - level of turbulence
   - currents
   - density stratifications

These factors usually begin to influence jet behaviour at some distance from the actual jet orifice. In this study the area of interest was that close to (ie within 100mm) of the jet orifice and therefore these parameters were not considered.

When a submerged jet emerges from an orifice there is at that point a significant difference between the velocity of the jet fluid and that into which it is passing. This will create a pronounced degree of instability, with the kinetic energy of the jet fluid steadily decaying through viscous shear. This tangential shear causes a reduction in kinetic energy leading to deceleration of the jet, ie a decrease in the velocity of flow with an associated increase in the area of the flow section. The same shear force at the same time results in the acceleration of the surrounding fluid. In the shear zone at the edge of the jet, instability occurs in the form of small eddies which will develop inward and in the direction of flow. The point at which this mixing reaches the centre line of the jet is defined as the end of the flow establishment region (see Figure 1). Beyond this point the flow may be considered as fully established and the diffusion (mixing) process will continue without any essential change in character. Further entrainment of the surrounding fluid by the expanding eddies will now be balanced inertially by a continuous reduction in velocity of the central region of the jet.

The extent of the establishment zone and the established flow limit is theoretically not possible to define. Experimental work on air in air jets between 1920 and 1950 suggested that the zone of flow establishment was of a length, anything from 5 to 40 times the nozzle diameter (Fountain 1990).

4 Laboratory Testing: methods and apparatus

A high pressure (nominally 3000 psi/20,700 KPa) CAT pump powered by a 14 HP electric motor, together a range of pencil jet nozzles (1 to 3 mm diameter) were used in the tests. Checks on the nozzles were made and the effect of nozzle diameter and nozzle wear on discharge in relation to pressure was evaluated. The results produced nozzle discharge coefficients of 0.6 to 0.9 and indicated that the discharge from a nozzle at a given exit pressure will vary as the nozzle becomes worn.

To examine water jet in water characteristics, a 3 metre long, 0.5 by 0.5 metre steel tank was built, incorporating a 1 x 1 metre glass observation panel at one end. The dimensions of the tank were chosen so as to minimise any edge, surface and end effects on the behaviour of a jet of up to 3 mm original diameter. Special rigs were designed and fabricated in order to fix/adjust the testing system within the tank.

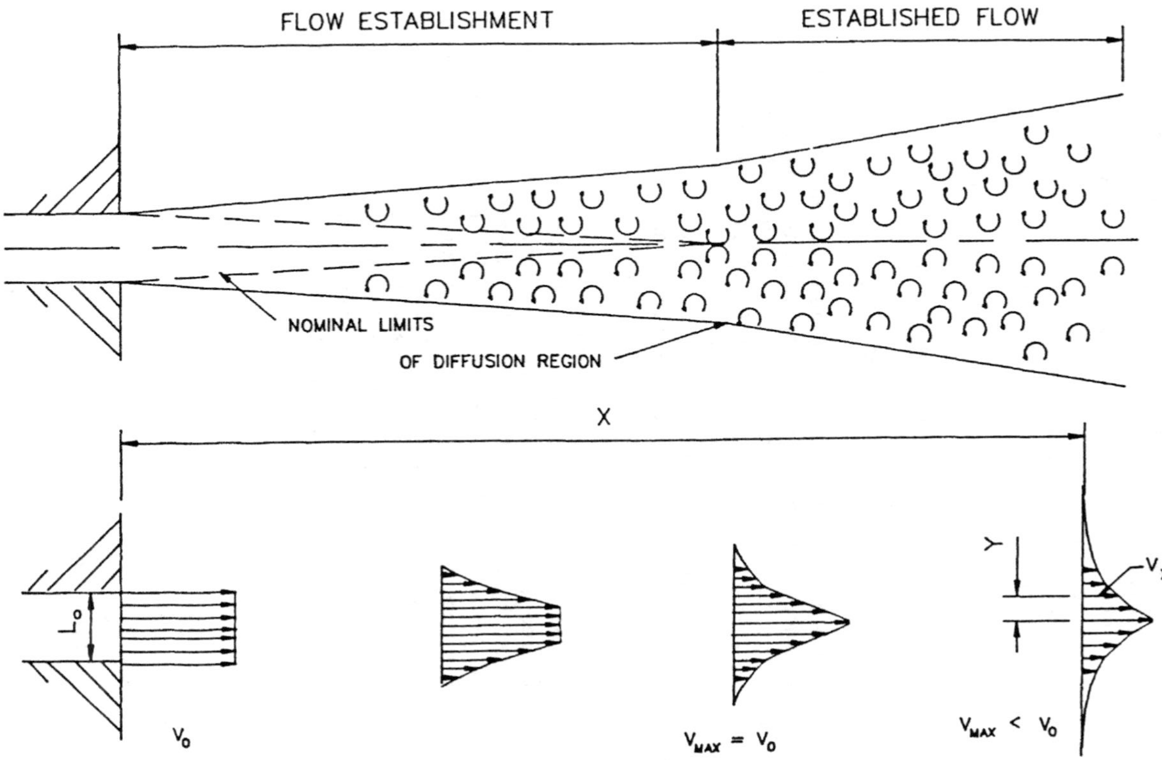

Figure 1. Schematic representation of jet diffusion

Pressure measurements throughout jet profiles were made using a calibrated digital voltmeter with a 25,000 KPa rated pressure transducer, fitted with a specially fabricated cone-shaped head for point measurements. The measurements were taken at 1mm intervals across the width of the jet, and at different distances from the nozzle, for different diameter nozzles and at different nozzle exit pressures.

For tests for observing jet behaviour in passing through and cleaning borehole screens, a variety of types/materials were used, ie stainless steel wedge-wire-wrapped with 1mm slots, plastic with machine-cut 2 mm slots, geotextile-fibre-wrapped slotted plastic, mild steel with 2 mm bridge slots, mild steel with 2 mm louvre slots

In order to evaluate the effectiveness of jetting in removing encrustations from borehole screens the following procedure was adopted:

1: A variety of encrusting materials, obtained from borehole rehabilitation field trials were tested to evaluate their strength characteristics. Samples of the weaker materials could not be tested but those of adequate strength (3 types) were tested in modified shear box and uniaxial compressive strength test rigs. Twenty samples of each type of material were tested.

2: Experiments were conducted on plaster/cement mixtures in order to create simulated encrustations with similar strength characteristics to those obtained from the field. Three different simulated materials of strengths (cement/plaster ratios 25/75, 35/65, 55,45) up to the maximum measured of the field samples were prepared.

3: The artificial encrustations were applied to different screens and subject to water jets, using different nozzle sizes and pressures at a fixed distance of 25 mm. The jet aimed directly at a screen slot was applied for 1 minute.

For tests on the use of jetting for gravel pack/formation development, a model well-rig, incorporating full scale 150 mm screens was used. Controlled jetting could be applied inside the screen which could be surrounded by different types and thicknesses of gravel. Behaviour of the jet within the gravel pack could be observed directly during the test by means of a fibre-optic borescope. Jet reaction was defined in four categories of grain agitation

ie very strong: grain movement very fast/complete blur
strong : almost able to identify individual grains
medium : movement of individual grains can be observed
weak : slow, intermittent movement of individual grains

5 Jet behaviour evaluation

Since it was considered that in applying high pressure jets to cleaning, the most effective part of the jet would be the flow establishment region, it was decided to evaluate jet establishment zones experimentally in the specially designed tank, using dyed jets and photography. From the photographs it was possible to measure the different jet angles for different size nozzles (1.10, 1.70, 2.34, 3.00mm) at different exit pressures (300 - 1500 psi, ie 2070 - 10350 KPa), and then from these calculate the length of the flow establishment region. The data is summarised in Table 1:

Table 1. Flow establishment zone evaluations

Nozzle diameter. (mm)	Pressure (KPa)	Top length	Bottom length	Av.length
		(lengths are in terms of nozzle diameters)		
1.10	2070	12.14	13.40	12.8
	3450	14.76	7.16	11.0
	6900	5.72	16.71	11.2
	10350	9.55	8.49	9.0
			Average =	**11.0**
1.70	2070	6.60	7.90	7.3
	3450	7.18	7.91	7.6
	6900	12.70	8.95	10.8
	10350	15.22	13.47	14.3
			Average =	**10.0**
2.34	2070	8.41	8.67	8.5
	3450	11.87	12.33	12.1
	6100	12.30	7.91	10.1
			Average =	**10.2**
3.00	2070	8.14	9.17	8.7
	2760	11.22	7.57	9.4
			Average =	**9.1**

Mean length of establishment zone for all nozzles = 10 x nozzle diameter
(Average of three measurements for each pressure)

This mean value fits within previous experimental findings but meant that in field applications, where nozzle to surface distances cannot usually be practically set at less than 10 nozzle diameters, it is the established flow zone which must be considered as the part of the jet that will have to do most of the work.

Measurements were also made on radial and axial jet pressure profiles. Typical results are plotted in Figures 2 & 3. As can be seen from Figure 2, the rate of pressure fall-off with distance is very rapid within the established flow region, reducing to 5% of the nozzle exit pressure within a distance of 40 nozzle diameters away from the nozzle. This indicates that for the jetting process to be efficient the nozzle to surface distance must be kept as short as possible. Reference to current field practice, where nozzle diameters are usually 1 to 1.5mm and achievable nozzle to screen surface distances are usually not less than 25 mm, shows that only a small proportion of the applied pressure is being used to clean the surface with the rest being lost through fluid friction.

Figure 3 shows the effect that different nozzle sizes has on the radial jet profile. This indicates that at a certain pressure the larger the nozzle size, with an associated larger flow rate, the greater the distance the nozzle can be from the target material, and useful work is still being done.

From theoretical considerations and from these experimental results it has been possible to determine equations which define the radial and axial diffusion of a water jet in water (See Figure 4). By combining the two equations a single equation has been derived (Figure 4) which describes the 3-dimensional diffusion of a water jet in water. Therefore by knowing the nozzle diameter and exit pressure the pressure

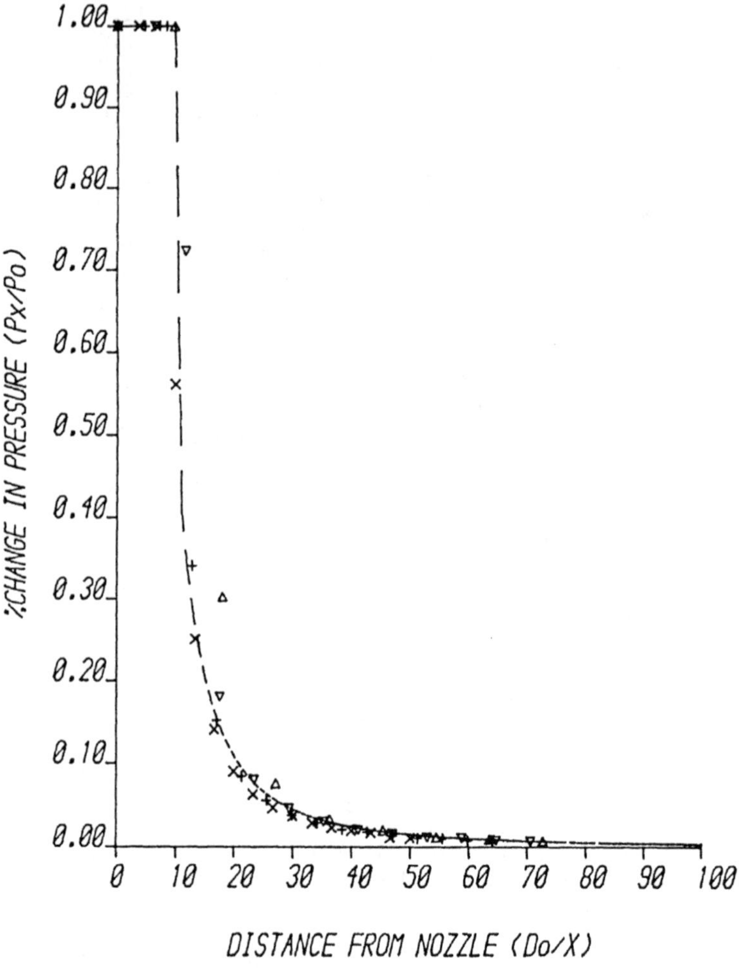

Figure 2. Pressure loss in a water jet with distance
(Po = pressure at jet nozzle exit
Px = pressure at distance x from nozzle
Do = nozzle diameter
X = distance x)

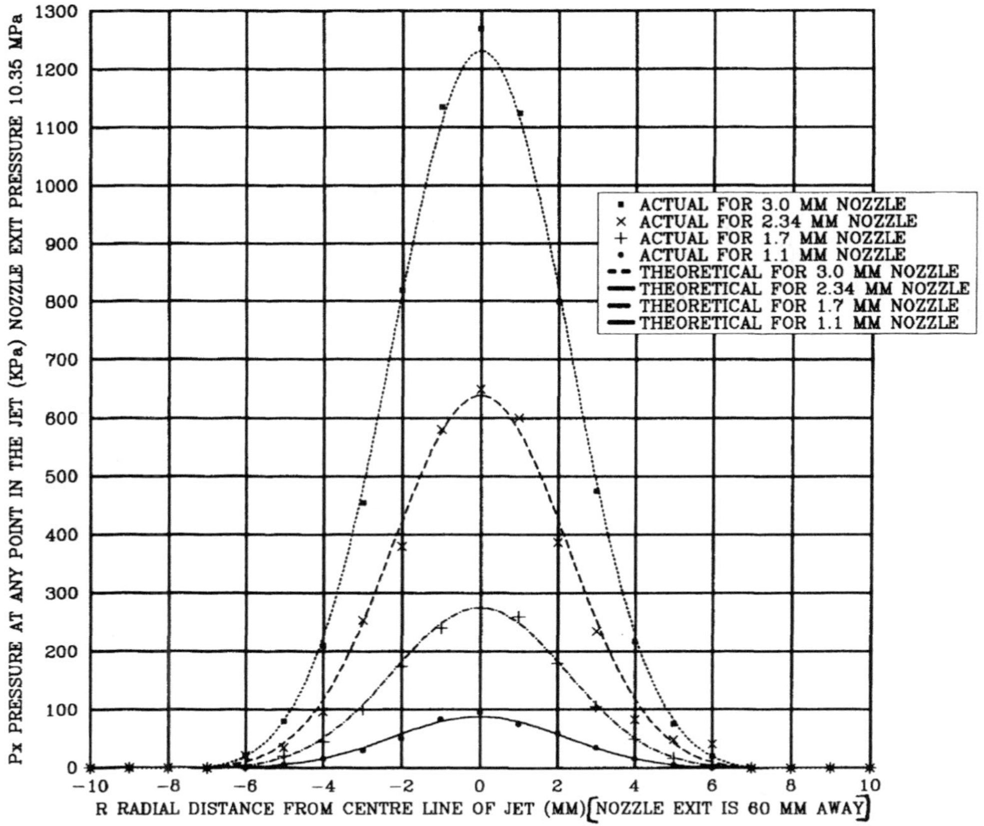

Figure 3.
An example of radial distance/pressure profiles for different nozzle sizes

[nozzle exit pressure = 10350 KPa
nozzle exit axial distance = 60 mm]

for 2D axial diffusion:

$$\frac{P_{max}}{P_o} = 332 \left[\frac{D_o}{X}\right]^{2.64}$$

Do = nozzle diameter

Po = nozzle exit pressure

for 2D radial diffusion:

$$\frac{P_x}{P_{max}} = \left[\exp^{\frac{-y^2}{2(0.05x^2)}}\right]^2$$

Px = pressure at point x

Pmax = maximum pressure

Write above Equations in Terms of P_{max}:-

$$P_{max} = P_o\, 332 \left[\frac{D_o}{X}\right]^{2.64}$$

$$P_{max} = \frac{P_x}{\left[\exp^{\frac{-y^2}{2(0.05x^2)}}\right]^2}$$

$$\therefore\ P_o\, 332 \left[\frac{D_o}{X}\right]^{2.64} = \frac{P_x}{\left[\exp^{\frac{-y^2}{2(0.05x^2)}}\right]^2}$$

for 3D diffusion:

$$\frac{P_x}{P_o} = 332 \left[\frac{D_o}{X}\right]^{2.64} \left[\exp^{\frac{-y^2}{2(0.05x^2)}}\right]^2$$

Figure 4 Equations to describe the axial/radial diffusion of a water jet in water

at any point within the jet can be determined. Or in practice, if the appropriate pressure to produce effective rehabilitation is known then, for any nozzle diameter, the required nozzle exit pressure to be applied can be determined.

6 Encrustation removal from well screens

A summary of the results of the strength tests on field encrustation samples is given below in Table 2

Table 2 . Strength characteristics of borehole screen encrustations

	Keresley	Meppershall	Slough Est.
Shear Strength			
mean value	535	656	1940
range	335 - 1890	446 - 1381	808 - 3022
Compressive Strength			
mean value	20	155	-
range	13 - 31	100 - 250	-

(values in KPa)

To be conservative simulated encrustation material with a shear strength of up to 3000 KPa and a compressive strength of up to 250 KPa, (from a 55/45% cement/plaster mix) was used in the encrustation removal trials. The results of some trials are summarised in Table 3

Table 3 . Results of laboratory encrustation removal trials on simulated encrustation (55/45 cement/plaster ratio)

Pressure: / **Screen type**	1000	1500	2000 KPa
Wedge-wire-wrapped	12.7	15.0	19.7
Louvre slotted	30.0	33.6	33.5
Bridge slotted	10.9	13.8	15.1
Machine-cut slot (plastic)	26.4	37.2	44.0

NOTES:
Nozzle-screen distance = 25 mm
Nozzle diameter = 1.7mm
Results in diameter (mm) of area of encrustation removed (average of three readings)

The cleaning width to nozzle diameter ratio, at 2000 KPa ranged from 9:1 to 26:1. The differences in encrustation removal for the different screen slot designs reflects factors such as slot width and angle (see Figure 5). It may also reflect encrustation to screen material bonding. The straight machine cut slot of 2mm width in the plastic provided minimal restriction on the 90 degree jet impact to the encrustation layer. In the case of the louvre slot, the jet was deflected downwards by the slot surface, so that it impacted the encrustation layer at what appears to be an optimum angle for breaking the encrustation to screen adhesion. The 90 degree, left and right, deflection of the jet by the bridge slot was in comparison, obviously an impediment to effective removal of the encrustation layer.

7 Slot design and gravel pack/formation development

A series of tests were conducted to examine the effectiveness of high pressure water jetting in the post construction development (cleaning) of the formation and installed gravel pack.

General experience suggests that to achieve effective gravel pack development, grain agitation needs to be induced. Observations made during this and previous studies show that jet pressure reduction with distance into the gravel is extremely rapid (Figure 6). When in these tests the equivalent of a 10 metre overburden pressure was applied to the model-well rig, the region of grain agitation was reduced dramatically (Figure 6).

Dye tests were also conducted in the jet diffusion study tank in order to evaluate the effect of screen slot design (Figure 5) on jet efficiency. The best screens for jetting through were observed to be those with high open area and continuous slots (ie the wedge-wire-wrapped screen) These provided the least resistance to jet flow. On the other-hand those designs, eg the bridge and louvre, which caused deflection of jet flow caused significant distortion and dampening of the jet. Severe dissipation of the jet was also observed when passing through geotextile screen coverings.

8 Well screen damage potential

Earlier work using a water jet in air conditions, had shown the thermo-plastic to be susceptible to erosion/damage at pressures exceeding 10,000 KPa, applied for 1 minute, at nozzle-surface distances of about 5 mm.

For this project, conditions more related to most field applications ie water jet in water, were used. Tests at nozzle exit pressures up to 17,000 KPa (maximum attainable by laboratory system) were conducted on a variety of screen materials, ie stainless and mild steel, glass reinforced plastic and thermo-plastic. In no case was any damage observed after the 1 minute period of application at a nozzle to surface distance of 5 mm.

Simple on-site tests/ field observations by MTC staff have shown that water jet pressures exceeding 30,000 KPa can cut through plastic casing. In practice the recommended maximum applied pressure for plastic lined boreholes has therefore been set at 20,000 KPa. In steel lined boreholes pressures of up to 55,000 KPa have been observed to cause no damage.

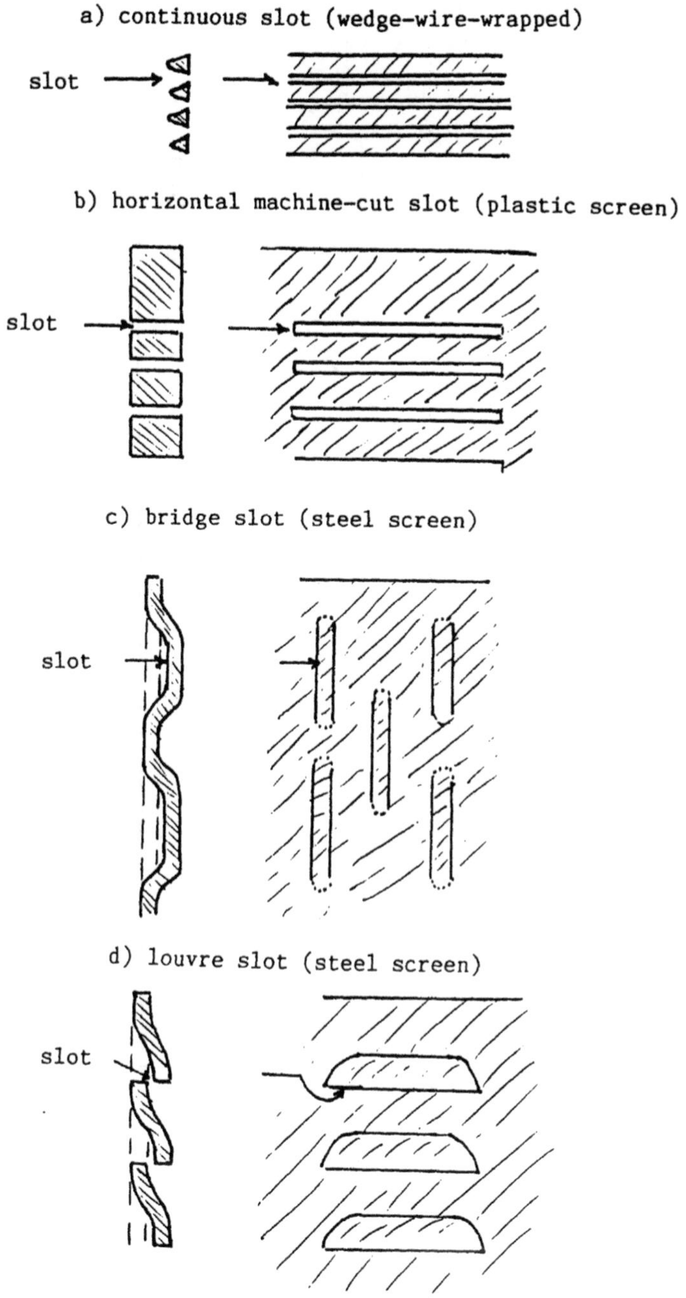

Figure 5, Screen slot types

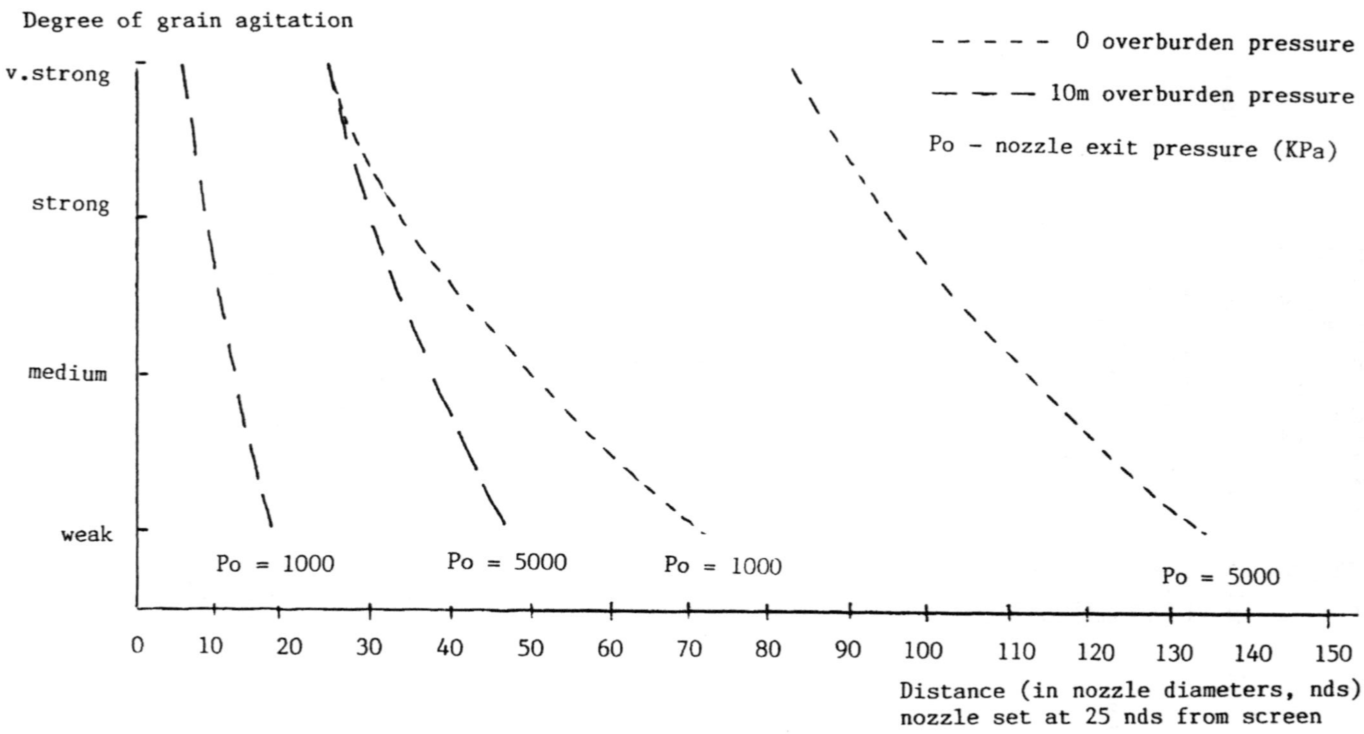

Figure 6. Gravel-pack jetting tests (using 1.7mm nozzle)

9 Comments on field trials

Information was sought on 40 borehole high pressure water jetting contracts carried out for 13 different Water Authorities/Companies in the UK, since 1980 by MTC Well Systems. It is a sad reflection of past attitudes to borehole rehabilitation that of these projects only in a relatively few cases were proper assessments of the rehabilitation carried out as part of the work. The records also show that in many cases even details of the borehole design/condition were not available. Whilst there are verbal reports that the jetting was successful or unsuccessful, there are hardly any cases where proper pre- and post-rehabilitation monitoring and testing was conducted. It is not therefore possible to quantify the technical and/or economic benefit of borehole rehabilitation by high pressure water jetting, from this data. Of those case histories, with closer project involvement, and/or where CCTV inspections were carried out, the general conclusion was that the jetting was very effective in cleaning of encrusted/clogged internal screen surfaces/slots, but that this did not always result in any improvement in borehole performance.

10 Conclusions and recommendations

Laboratory and field evidence show that high pressure water jetting can be an effective method for cleaning the internal surfaces and slots of clogged well screens when applied in a technically and scientifically correct manner.

The laboratory tests show that nozzle exit pressures of 17,000 KPa, applied at nozzle-surface distances of 25 mm, with a normal (ie 1.5 - 2.0 mm diameter) nozzle will be effective in cleaning all types of the materials likely to be encountered, from the internal surfaces and slots of clogged screens. In practice pressures of up to 40,000 KPa are used because attaining nozzle-surface distances of 25 mm or less has not always proved to be practical.

To avoid the risk of damage from the jetting process it is recommended that pressures applied in plastic lined boreholes should not exceed 20,000 KPa. (assuming a working distance of 25 mm and a standard 1.5 - 2.0 mm nozzle)

Similarly in open, unlined holes, current practice restricts application pressures to less than 40,000 KPa, in order to minimise unnecessary erosion/instability of the formation.

With increasing environmental concern over the use of chemicals in groundwater systems, the technique will be viewed as having advantages for many situations over chemical rehabilitation techniques.

The project results show that units of about the 100,000 KPa size commonly used in the UK and Europe for sewer and borehole rehabilitation, are too big for many borehole situations and therefore smaller and cheaper 40,000 KPa units would be appropriate to cover most conditions in developing countries.

The equipment required is not yet readily available in developing countries, but there is no reason why the technique cannot be developed by many groundwater organisations in developing countries which already have borehole/tubewell drilling/construction resources.

Reference and selected bibliography:

Fountain, J. (1990) Unpublished MPhil thesis, (in preparation). Silsoe College, CIT, UK.

Albertson,M.L. (1950) Diffusion of submerged jets. **Am. Soc. Civil Eng.** Paper 2409, 639-664.

Corrsin, S. (1943) Investigation of flow in an axially symetrical heat jet of air. **Nat. Advisory Committee for Aeronautics, War time Report 3L23,** Washington.

Dean, R.B. (1976) A single formula for the complete velocity profile in a turbulent boundary water. **.Trans. A. M. E., J. Fluid Engineering.** 723-727.

Driscoll, F.G. (1986) Groundwater and Wells. St Paul, Johnson Division

Erdmann-Jesnitzer, F. Material behaviour, material stressing - Principle aspects in the application of high speed water jets. **4th Int. Sym. on Jet Cutting Technology.** Paper E3, 29-44

Gronauer, R.W. (1972) Cleaning and descaling of equipment with high pressure water. **1st Int. Sym. on Jet Cutting Technology.** Paper D3, 25-28.

Shavlovsky D.S. (1972) Hydrodynamics of high pressure fine continuous jets. **1st Int. Sym. on Jet Cutting Technology.** paper A6, 81-92.

Tollmein, W. (1926) Calculation of turbulent expansion process **Zeitschrift fur angewandt Mathematik und Mechanik,** 6, 1-12.

Turner, J.S. (1973) **Buoyancy effects in fluids.** Cambridge University Press

Vickers, G.W. (1974) Water jet impact damage at convex, concave and flat-inclined surfaces. **Trans. A.M.E.,** Dec., 907-911.

Watson, A.J. (1982) Impact pressure characteristics of a water jet. **6th Int. Sym. on Jet Cutting Technology.** Paper C2, 93-106.

Acknowledgements

Department of Trade and Industry
Engineering Division of the Overseas Development Administration
MTC Well Systems Ltd, UK
Technical/Computer staff at Silsoe College

19 BOREHOLE YIELD DEVELOPMENT TECHNIQUES IN AN IRON-RICH AQUIFER

M.J. PACKMAN
Southern Science Ltd, Working, UK

Abstract
Over a period of some 8 years various yield development techniques were tested on recently constructed production boreholes in the Lower Greensand aquifer on the Isle of Wight. The techniques can be divided into physical, which included air-lift pumping, surging and high-pressure jetting, and chemical involving chlorination and polyphosphate dispersants. Any improvements in yields were quantified by carrying out short step-test pumpings before and after each technique was used. The type of screen installed in the borehole was an important factor in determining the choice of physical yield development tool used and its success. For this type of tool the best results were achieved by jetting, although it is very dependent on the quality, volume and velocity of the water used and the type and open area of the screen. The main factor with chemical yield development was contact time, rather than degree of agitation. A combination of physical and chemical techniques was the most effective, since well clean up after construction was best achieved by chemical dispersants, whereas the development of yields by physical means were felt to be more long lasting.
Keywords: Yield Development, Iron-rich, Air-lift, Surging, Jetting, Chlorination, Dispersants, Step-test.

1 Introduction

1.1 Water supply and demand

The Isle of Wight is situated just off the south coast of England, being some 30 kms. long and up to 25 kms. wide, with a resident population of nearly 130,000. However the Island is a popular holiday centre, which causes the population to nearly double over the summer months and significantly increases the demand for water during the peak period. Being an island with a relatively small surface area, most obvious water resources were eventually developed, so that by the 1970s there was a shortfall between demand and supply. This culminated in the drought of 1976, when there were severe restrictions imposed on demand.

1.2 Lower Greensand aquifer

Southern Water Authority, the public water supply undertaking at

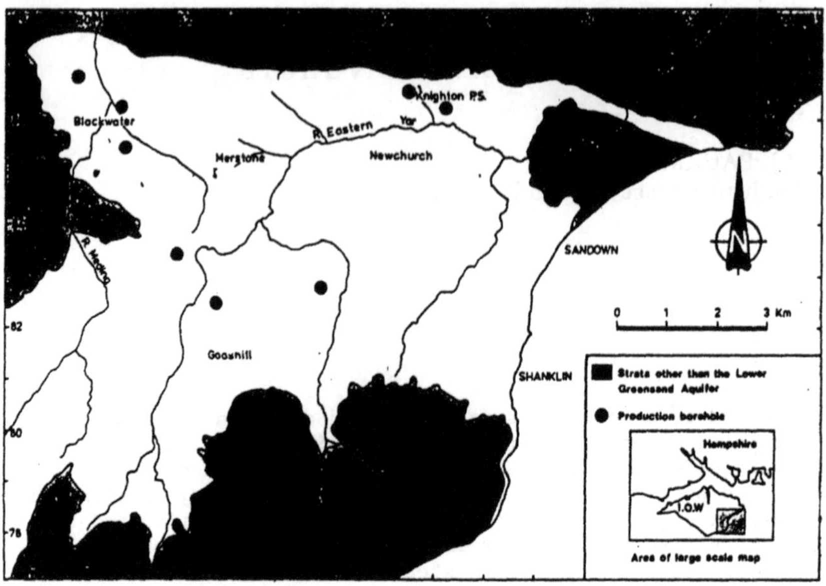

Fig.1. Location of the Lower Greensand aquifer on the Isle of Wight

the time, therefore carried out studies into all possible sources of water. This included an investigation of the groundwater resources of the only permeable strata on the Isle of Wight not fully developed, the marginal Lower Greensand aquifer (figure 1). Although yields of up to 2,000 m^3/day had been obtained from the aquifer, it was often short-lived and at the expense of sand ingress or eventual clogging of the screen. The grain size of the Lower Greensand on the island differs from that on the mainland, being much finer overall with anything from clay to medium sand and even pebble beds. As well as vertical variations in lithology, there are also lateral facies changes, which makes accurate correlation between sites virtually impossible, However the fence diagram illustrated (Figure 2) will hopefully convey some of the aquifer configuration. The predominant lithology is a glauconitic or limonitic silty fine sand/sandy silt.

1.3 Hydrogeology

Past borehole failures can be classified into three types:-

- Those where the yield was insufficient from the start.
- Those that failed due to corrosion of the screen.
- Those that gradually silted up due to incorrect borehole design.

Although the strata covers a relatively small area, there is a tremendous variation in the hydrogeology, from unconfined through semi-confined to totally confined aquifer conditions with overflowing artesian heads in excess of eight metres above ground level. Where the strata reaches a reasonable thickness, it becomes a multi-layered aquifer, often with leakage between layers.

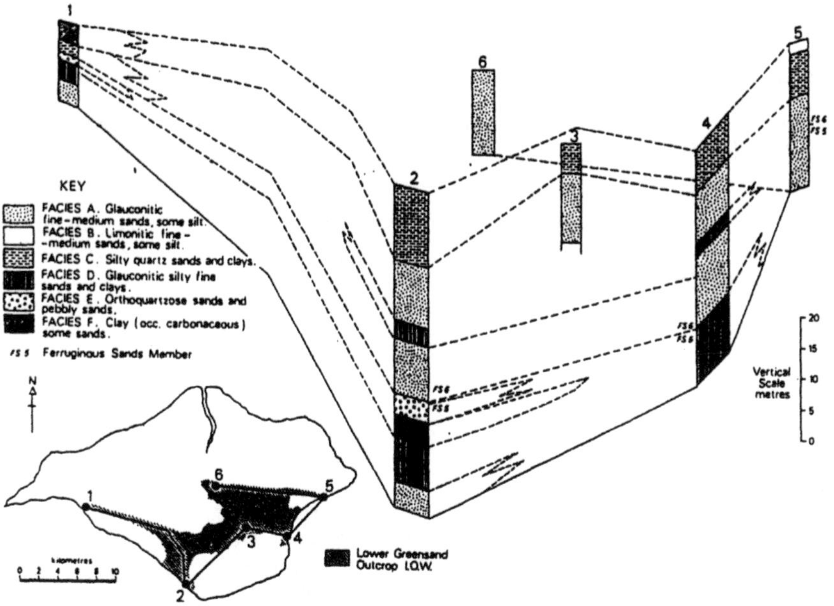

Fig.2. Fence diagram of the upper Ferruginous Sands formation.

1.4 Hydrochemistry

Carbon 14 dating of the groundwater from the confined aquifer gives ages in excess of 4,000 years. A reflection of the various conditions is shown by the range encounted in the groundwater chemistry (Table 1).
Hydrogen sulphide is often encountered in small quantities at abstraction boreholes in the confined or partially confined aquifer. The groundwater is moderately to severely aggressive in character and will quickly corrode unprotected low carbon steel pipework. Iron bacteria has been detected within a number of boreholes.

2 Aquifer development

2.1 Production borehole design

After initial investigations were promising, an area of maximum groundwater development potential was delineated and eight production boreholes sunk between 1981 and 1988. To minimize the risk of failure due to inadequate design, a fully cored borehole was sunk at each production site to provide undisturbed samples upon which to base the criteria for the selection of screen and pack. All the sites were located where the various sandy facies reached their maximum thickness and therefore, in theory, all three types of aquifer conditions could have been screened in a borehole. However, in practice, sufficient plain casing was often installed from the surface downwards to preclude the unconfined strata. A typical borehole construction is illustrated schematically in Figure 3.

Table 1. Groundwater chemistry of the Lower Greensand aquifer, I.O.W.

Chemical parameter (mg/l)	Unconfined aquifer	Semi-confined aquifer	Confined aquifer
pH	5.5-6.5	6.5-7.2	7.0-8.2
Eh (mV)	>300	+0-250	<+100
Dissolved Oxygen	>3.0	2.0-5.0	<3.0
Calcium	<30	50-85	20-50
Magnesium	4.0-7.0	4.0-7.0	6.0-9.0
Sodium	15-30	20-40	30-50
Potassium	1.0-6.0	2.0-6.0	3.0-5.0
Total Iron	3.0->10	2.0-6.0	<2.0
Total Manganese	>0.2	0.1-0.9	<0.2
Bicarbonate	<100	>200	100-200
Chloride	<30	25-35	>30
Sulphate	<20	10-40	<10
Nitrate as N	<10	<0.05-0.5	<0.05
Conductivity (uS/cm)	<300	300-500	<350
Silica	<10	10-20	>20

2.2 Screen and pack selection

The often rapid changes of lithology with depth and the requirement to maximize borehole yields, meant that as much of the aquifer as possible was screened. The two trial production boreholes sunk in 1981 were located at either end of the aquifer's grading spectrum within the area of maximum development potential. Each had a custom-graded sand pack and screen slot size to suit the aquifer lithology.

The large open-area afforded by wire-wound screen was chosen for the following reasons:-

To keep screen-entrance velocities to a minimum (<0.01 m/sec).

To provide as large a "safety margin" as possible for any reduction in open-area due to iron-encrustation.

To facilitate the removal of up to 35% of the pack in the vicinity of the screen during yield development, to produce the required sorted structure which gives optimum stability and permeability and minimizes further sand ingress.

To maximize access to the pack for any future yield rehabilitation.

Type 304 stainless steel was chosen as the screen and casing material for corrosion protection. The dimensions of the production boreholes are given in Table 2.

2.3 Borehole construction

Production boreholes were drilled using reverse mud-flush rotary rigs. This technique was preferred to percussion drilling as the

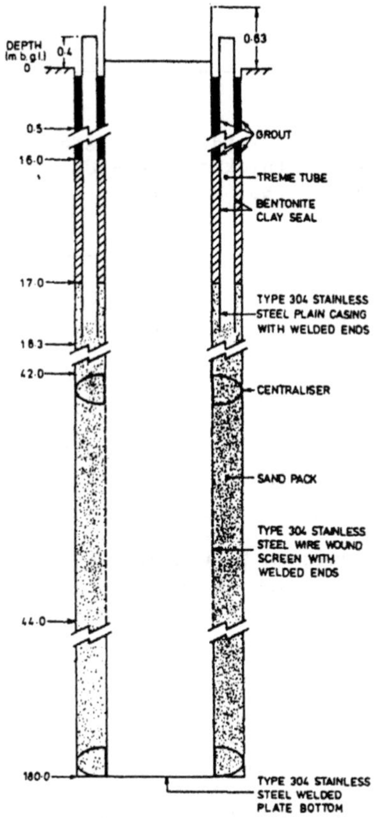

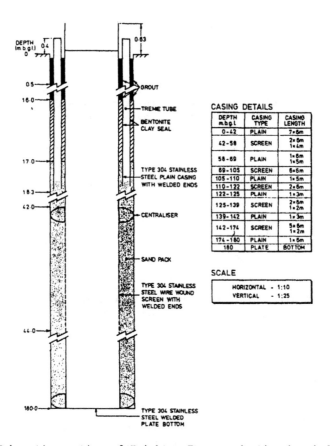

CASING DETAILS

DEPTH m.b.g.l.	CASING TYPE	CASING LENGTH
0-42	PLAIN	7×6m
42-58	SCREEN	2×6m 1×4m
58-69	PLAIN	1×6m 1×5m
69-105	SCREEN	6×6m
105-110	PLAIN	1×5m
110-122	SCREEN	2×6m
122-125	PLAIN	1×3m
125-139	SCREEN	2×6m 1×2m
139-142	PLAIN	1×3m
142-174	SCREEN	5×6m 1×2m
174-180	PLAIN	1×6m
180	PLATE	BOTTOM

SCALE

HORIZONTAL - 1:10
VERTICAL - 1:25

Fig.3. Schematic section of Knighton Farm production borehole.

uncemented and unconsolidated nature of the strata often lead to problems of running sands and clay squeeze which precluded the extensive use of temporary casing. Only polymer drilling fluids were allowed with great care being taken in the breaking of the "muds", the subsequent flushing of the boreholes and disposal of the effluent. A technique was developed with the contractors to install the graded pack by means of compressed air, whilst ensuring that only the viscosity of the polymer mud at the point of emplacement was broken. This achieved the two objectives of making sure the pack stayed well graded with no settling out and also maintaining the viscosity of the mud in the borehole thus preventing collapse.

3 Yield development

3.1 Rationale

Past failures suggested that regular borehole rehabilitation would be required to prevent screens and packs clogging with iron encrustation and the migration of fines. Developing the yield of a borehole after

Table 2. Production borehole Dimensions.

Site	Depth (m)	Drilled dia. (mm)	Liner dia. (mm)	Av. pack thickness (mm)	Pack grading 90%-10% finer (mm)	Screen slot width (mm)
Godshill	93	500	270	115	1.18-0.3	0.50
Blackwater	95	500	270	115	2.00-0.425	0.75
Lessland Lane	103	550	305	120	2.00-0.425	0.75
Lower Yard	136	550	305	120	2.00-0.425	0.75
Birchmore	99	550	305	120	2.00-0.425	0.75
Marvel	144	550	305	120	2.00-0.425	0.75
Knighton Farm	180	600	355	120	2.00-0.425	0.75
Knighton Old Mill	180	600	355	120	2.00-0.425	0.75

construction involves similar techniques to rehabilitation, especially in a fine-grained aquifer such as the Lower Greensand. The opportunity was therefore taken to test various methods appropriate to the aquifer conditions.

3.2 Programme

Following completion of construction, a small amount of air-lift clearance pumping was carried out to remove any debris that had accumulated in the bottom of each borehole and might interfere with subsequent tests. The yield-drawdown characteristics of each borehole were then determined by air-lift step-drawdown test pumping. Three or preferably four 90 minute pumping stages were undertaken with the discharge rate measured by a 90° V-notch weir tank equipped with a water level chart recorder.

Air-lifting was used initially, as the amount of material drawn through the screen precluded the use of submersible pumps. This provided the base data upon which any improvements in yield-drawdown could be assessed.

All but two of the boreholes were developed by chemical methods first, as it was felt that well "clean-up" to remove drilling mud cake should precede the physical methods, which tend to have a more localized effect within the screen and pack.

The level of sand in the pack was constantly monitored, throughout the yield development work and topped up when necessary via the tremie tubes.

At two of the sites intermediate step-test were undertaken between the chemical and physical techniques to try and assess the relative contribution of each method to any improvement in borehole performance. At some of the boreholes it was possible to carry out a detailed C.C.T.V. inspection after each stage of the yield development to give a visual impression of the cleaned screen.

Finally the overall affect of trying to develop the yield of each borehole was then quantified by a series of step-tests involving both air-lift and submersible pumps. The former to provide comparison with the earlier tests and the latter to extend discharges beyond the limit of 1,500 m^3/day imposed by the air-lift submergence ratios.

The opportunity was taken in 1988 to lift the submersible pump from the Knighton Farm borehole that had been operational for three years, carry out a C.C.T.V. inspection and any rehabilitation necessary. Step test pumpings were carried out to assess any improvements.

3.3 Chemical methods

Two types of chemicals were used; chlorine and polyphosphates. Chlorine has the dual use of a sterilizer and oxidizer. The first cleans the well and kills any bacteria either accidentally introduced during construction or indigenous to the aquifer. Secondly it was found after various laboratory tests to be the cheapest, safest and most effective method of breaking all polymer drilling muds commonly available to the water well drilling industry.

Chlorine in the form of hypochlorite was found to be the easiest way of handling the chemical. It is commonly available in two forms, Calcium and Sodium. Of these the powdered calcium hypochlorite with 35% free chlorine was found to be by far the easiest to handle on a drilling site. Liquid sodium hypochlorite with either 10 or 14% free chlorine could be splashed onto bare skin, especially if the container becomes pressurized after being left out in the sun.

A minimum concentration of 1,000 mg/l free chlorine was found to be required to break polymer drilling muds quickly and disinfect the well effectively. The easiest method of introducing the chlorine into the borehole was by mixing the hypochlorite with water in a sealed paddle mixer and pumping it down using the piston pump and drill rods attached to the rig.

Two types of polyphosphates were used; a sodium polyphosphate glass containing 67% phosphate and sodium hexametaphosphate. Concentrations of up to 40 Kg/m^3 were tried, where the volume calculated included both the casing and the pack. This is because as well as cleaning up the screen, the main purpose of using dispersants was to remove any drilling mud cake on the walls of the borehole and to remove any fine material which had either invaded the pack or the aquifer in the vicinity.

The polyphosphates were first dissolved in a holding tank using compressed air driven paddle mixers, before being pumped down the boreholes. Two basic methods of dispersing the fines were tried:-

Maximizing contact time by injecting it and then leaving it for varying periods from 12 to 48 hours.

Maximizing agitation by jetting and air-lift pumping at the same time to recirculate it.

3.4 Physical Methods

The following techniques were used:-

Air-lift clearance pumping

Air-lift surge pumping

Surge blocks

High pressure jetting

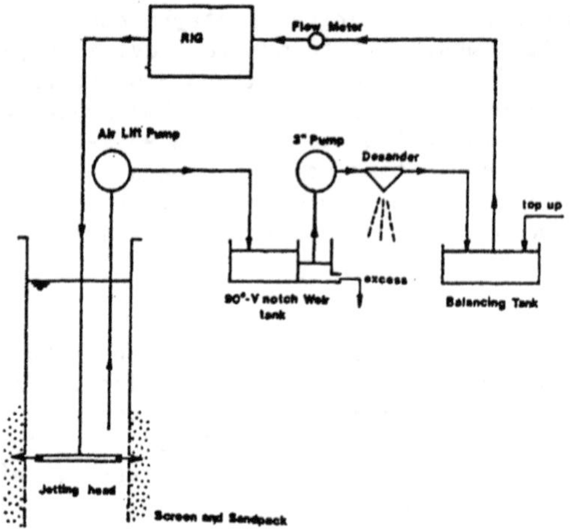

Fig.4. Schematic diagram of jetting recirculation system.

The first method was straight forward air-lifting to remove as much sand and fine material until the discharge was reasonably clear. Rising mains of up to 150 mm diameter were used and lowered in steps of one metre throughout the screened portion of the boreholes using the winch system on a drilling rig.

Surging involved using the same equipment as the first method, but by lowering air lines below the rising main, compressed air could be used to surge the borehole. The water in the rising main was also used to surge, by raising and dropping the water column using the valve on the air compressor. The surge head could be increased by lifting the rising main to the top of the rig.

Surge blocks were also used on some of the earlier boreholes before an efficient high pressure jetting system had been successfully tested. Various methods were tried to combine the use of surge blocks with air-lift pumping, in an attempt to remove any mobilized fines before they could settle back down into inaccessible parts of the borehole.

The jetting system utilized as much of the equipment found on a reasonably large rotary drilling rig as possible. Initially jetting on its own was tried, but very quickly the facility to pump simultaneously on the rig was taken advantage of. Therefore an air-lift pump followed close behind the jetting head to "hoover up" the fines. The jetting head was mounted on a drill stem, which allowed it to rotate at speeds as low as 3 to 4 r.p.m. and also utilized the rigs modified piston pump. The configuration of the head could be varied, as it allowed up to four nozzles of between 5 and 9.5 mm diameter to be mounted within 20 mm of the inner face of the screen. The head was lowered in short steps of about one screen diameter over each 6 metre length of screen during a period of an hour or so.

The rates of injection through the jetting head were varied between 3 and 6 l/s. As the jetting was primarily for yield development, the exit velocity of each jet was kept in the range 15 to 40 m/s, which is somewhat lower than the minimum 50 m/s often recommended for rehabilitation of encrusted screens. Attempts were made to measure the flow accurately using various types of flow meter, in order to control exit velocity. However the combination of the pressures involved and the pulsing nature of the pump made this difficult, especially with the higher pressures. The rate of fall of the water level in a balancing tank of known volume was used when high pressures were involved.

Early on during the trials it was found that the cleaner the water used the more effective the jetting. It was especially important not to jet with water containing any significant particles, as at high velocities this could damage the screen by sand blasting. Where possible mains water was used, either directly by pipeline from a hydrant or tankered in at 2,000 gallons at a time. However many of the sites were relatively remote and inaccessible, and therefore a recirculation system was developed (Figure 4).

The key element was the hydrocyclone type desander, which works on the vortex principle and is used on many drilling mud recirculation systems. The one used for this work was composed of four 50 mm diameter cones, in parallel, with an inlet manifold and was designed to remove all particles above 15 micron in size. In practice some problems were experienced in achieving this degree of removal due to the pulsing nature of most standard surface pumps. In addition any small public water supply that was available was used to top up the system.

4 Results

4.1 Chlorination

In most cases it was not possible to reliably quantify the results of chlorination alone, as it was an integral part of the borehole construction, being the method used to break the drilling mud. However when the pump at Knighton Farm was lifted there was evidence of some iron bacterial slime on the rising main pump and screen. The opportunity was therefore taken to flash chlorinate the borehole and assess the results by step test pumping. Since construction there had been up to a 4% reduction in specific capacity, that is yield per unit drawdown, at the highest discharges. After chlorination much of the lost specific capacity was restored.

4.2 Dispersants

The results from the use of chemical dispersants are shown in Table 3. It is rarely possible to exactly match the discharges of the step tests carried out before and after development, especially with air-lift pumping. Therefore the results were plotted graphically and a best-fit line drawn through them, to allow comparisons of specific capacities to be made for the same discharge. Although improvements of up to 45% were measured, it was felt that the air-lift clearance pumping, which was necessary to remove material fixed in suspension, before the step-test, contributed significantly to the method.

Table 3. Chemical dispersant yield development results

Pre-development yield/unit drawdown ($m^3/d/m$)	Post-development yield/unit drawdown ($m^3/d/m$)	Difference (%)
128	128	0
112	120	+7
83	117	+41
200	222	+11
167	205	+23
135	184	+36
102	148	+45

There was little to chose between the two types of polyphosphates used. However it did appear that the dispersant obtained from a drilling fluid company was easier to mix on site and produced an "effluent" with higher suspended solids, than that obtained from a water treatment company. If there are various grades of polyphosphates available, then the higher the phosphate content the better the dispersing properties.

One of the main problems encountered with the use of dispersants was the disposal of the effluent subsequently pumped from the boreholes. It is unlikely that the National Rivers Authority will give permission for it to be discharged directly to a water course of reasonable quality. The pollutants involved are twofold; phosphates are a nutrient and therefore affect dissolved oxygen levels and the very high suspended solids produces problems associated with turbidity. Soakaways also soon blind with the suspended solid load.

On site treatment, in the form of settlement, was found to be not feasible, as the fines are fixed in suspension. Dilution before discharge is only possible where a suitably large pit or tank is present. Spreading it over land may be a possibility if a suitable site is nearby.

The two methods of disposal that were resorted to when all else failed, were discharge to a sewer with the agreement of the relevant authority and tankering it away.

4.3 Air-lift pumping

Air-lift pumping, on its own as a development technique without any chemicals, produced improvements in yield drawdowns of up to 17% (Table 4.) However the greatest improvements tended to be confined to the better boreholes with the higher specific capacities to start with, whereas a more general increase of some 5-10% was the norm.

There was an overall lower level of improvement compared to the results with the chemical dispersants. The other main difference was that the greatest degree of improvement occurred at the lowest discharges and decreased with each incremental increase in pumping

Table 4. Air-lift pumping yield development results

Pre-development yield/unit drawdown ($m^3/d/m$)	Post-development yield/unit drawdown ($m^3/d/m$)	Difference (%)
96	106	+10
101	107	+6
98	103	+5
94	99	+5
308	360	+17
290	328	+13
272	299	+10
234	253	+8

rate, whereas the opposite was the case with the chemical dispersants.

4.4 Surging

Of the two methods of surging tried, the one involving air-lift produced the greatest improvement in well performance (Table 5.). A conventional surge block has to be used with some caution in a newly constructed borehole, as the localized pressure differentials set up by the "piston type" action can cause screens to collapse, if the pack surrounding them was not installed properly or had time to consolidate. The somewhat careful use of the surge block may partly account for the results obtained, although the velocities involved were gradually increased as development progressed.

Various techniques were tried to combine what was felt to be the high localized velocities generated around a surge block with the backwashing and removal capability of air-lifting. It involved either the use of the air-lift capability of reverse circulation drilling pipe with a type of block tool on the end, or conventional air-lift apparatus with a perforated main between two blocks/packers. Work on this is continuing as the early results were difficult to quantify due to the yield of the boreholes already being partially developed. However there were definite indications that the combination was a substantial improvement on the use of surge blocks by themselves and may be superior to surging by air-lift, if the same degree of fines removal can be achieved.

4.5 Jetting

Definite results were obtained from the jetting trials and these are shown in Table 6. The jetting only was carried out on the Lessland Lane and Godshill boreholes that had responded well to the use of dispersants, to see if additional fines could be flushed out. However jetting appears to have made matters worse, as the yield drawdown characteristics deteriorated. This was felt to be due to

Table 5. Surging yield development results after chemical treatment.

Surging technique	Pre-development yield/unit drawdown ($m^3/d/m$)	Post-development yield/unit drawdown ($m^3/d/m$)	Difference (%)
Air-lift surging	85	113	+33
	83	101	+22
	78	89	+14
Surge block	96	110	+15
	101	109	+8
	105	109	+4
	100	102	+2

either fines mobilized by the dispersant or by the jetting being forced back into the pack/aquifer. Subsequent air-lift clearance pumping did not fully restore the lost yield, which was finally achieved by another treatment with dispersants. This experience clearly demonstrated the importance of removing fines as soon as they are mobilized, before they have time to settle back down again.

Table 6. Jetting yield development results.

Development technique	Pre-development yield/unit drawdown ($m^3/d/m$)	Post-development yield/unit drawdown ($m^3/d/m$)	Difference (%)
Jetting only	222	210	-5
	205	186	-9
	184	162	-12
	148	123	-17
Jetting and air-lift pumping	128	139	+11
	120	135	+13
	117	120	+3
Chemical treatment followed by jetting and air-lift pumping	158	178	+13
	138	178	+29
	130	173	+33
	118	166	+41
	27	82	+204
	29	78	+170
	31	84	+171
	30	79	+163

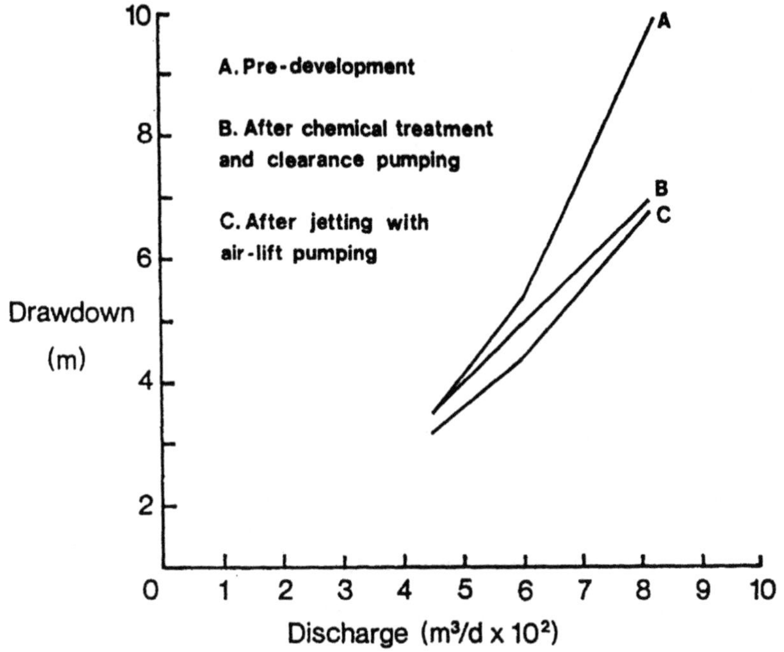

Fig.5. The affects of different development techniques at one site.

Simultaneous jetting and air-lift pumping was then tried on the Lessland Lane borehole with the result that improvements due to the dispersants were not only consolidated, but actually improved upon (Figure 5.)

The final development technique that evolved was a combination of chemical and physical means, involving first the use of dispersants, followed by air-lift clearance pumping and finally jetting with simultaneous air-lifting. Initial results gave average improvements of some 30%, which increased to a maximum of some 200%, once the method had been perfected (Figure 6.)

5 Conclusions

C.C.T.V. inspection after flash chlorination of a borehole shows it to be reasonably effective in restoring yields affected by iron bacteria, so long as it is treated early enough. Current work in Kent indicates that where bacteria is suspected of having been active for a long time, jetting at very high pressures is required to disrupt iron encrustation before chlorination.

Chemical dispersants are very effective at entrapping fines if a long contact time is allowed. Yield development is dependent on efficient removal of the fines from the borehole, with the greatest improvements occurring at the highest discharges/entrance velocities.

With physical methods of development, the converse was generally true, as most improvement in yield occurred at the lower discharges. It is felt that this could be due to the migration of fine material

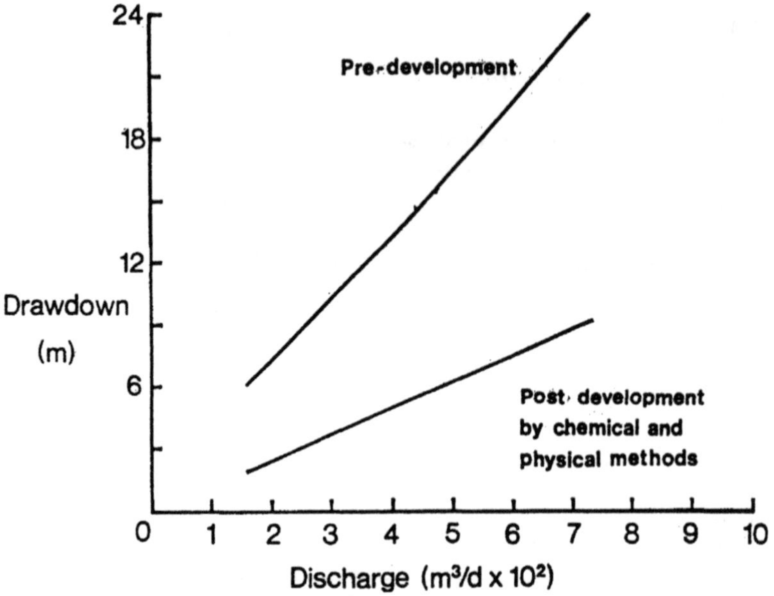

Fig.6. Yield development of Knighton Farm borehole.

towards the borehole with higher groundwater velocities, thus limiting improvements in permeability within the aquifer immediately around the well and in the pack.

The single most effective physical method was air-lift surge pumping, with jetting on it's own actually having a detrimental affect. The most successful combination of development tools for this design of borehole, in this type of aquifer, was found to be chemical dispersants to mobilize the fines, air-lift pumping to remove those easily accessible and jetting with simultaneous removal to "blast out" those not.

Subsequent step-test pumping using submersible pumps confirmed that the yield drawdown improvements could be sustained at higher discharges. In fact at some boreholes, where there was a large discrepancy between the maximum discharges with the different methods of pumping, further development of yields occurred during the submersible pump tests. Five out of the eight boreholes now have yields in excess of 3,000 m^3/day.

Acknowledgements

The author wishes to thank Southern Water for permission to write this paper. The views expressed are the authors' own and not necessarily to be regarded as being those of Southern Water.

20 EUCASTREAM SUCTION FLOW CONTROL DEVICE: AN ELEMENT FOR OPTIMIZATION OF FLOW CONDITIONS IN WELLS

R. PELZER
Belgium
S.A. SMITH
Kabelwerk Eupen AG, Eupen, Ada, Ohio, USA

Abstract
Hydrodynamic testing by German researchers from the 1950s onward showed that pumping wells are typically hydrodynamically heavily loaded in the upper screen interval. Hydrodynamic loading decreases exponentially downward along the screened interval. Further, there is a strong vertical velocity component in the near-well environment. The results are:

(1) for practical purposes, water passes though only the upper 5 to 15 % of the screen;
(2) excavation occurs in the wall of the borehole, followed by
(3) local collapse of the gravel pack, leading to
(4) sloughing and subsequent sand infiltration into the well.

The suction flow control device (SFCD) was developed to counteract this flow pattern, forcing an equalization of inflow over the total length of the well intake. The Eucastream-SFCD element is a long slotted pipe (slot interval increasing upward) inserted into the screen area, through which all pumped flow is conducted. The effect of the slot pattern is to force greater inflow through the lower section of the SFCD and hence the screen, resulting in the more cylindical inflow pattern.

The practical result in installations of Eucastream devices and similar devices in the United States has been a dramatic reduction or elimination of sand infiltration as well as improvements in other aspects of well performance, including increased specific capacity and total yield.
Keywords: Suction Flow Control Device, Well Hydrodynamics, Well Development

1 Introduction

The effects of inflow velocity on a pumping well have been variously described in the literature (e.g., Driscoll, 1986; Williams, 1981) with differing views among proponents of various types of well screens. Excessive velocity at the borehole/aquifer interface has been commonly associated with sand infiltration and accelerated encrustation. In most situations, the only real opportunity to limit inflow velocity usually occurs during well design and installation.

If poor design or well deterioration cause problems, remedial measures such as redevelopment, restriction of yield, or installing sand-removal devices have generally been required. This paper describes the theory and application of modifying well inflow velocity, using an insert element known as a suction flow control device (SFCD).

2 Inflow Velocity

The velocity of flow into a pumping well can be described using hydrodynamic principles as applied to flow through porous media in the narrow confines of the near-well environment (generally defined as well hydraulics). In hydraulic calculations, velocity is proportional to head differential as well as the cross sectional area of flow. In hydraulics, velocity (v) is calculated using the formula $v = Q/A$, where Q is the rate of flow and A is the cross-sectional area. Darcy's Law ($Q = KA \{dh/dl\}$) is used to describe flow in porous media. Combining these two equations, following Heath (1983), with the hydraulic equation rendered $Q = Av$, the result is

$$Av = KA\ (dh/dl)$$

Cancelling the area terms, the result is

$$v = K\ (dh/dl)$$

This equation is valid for a stratum with 100 % porosity. Returning to $v = Q/A$, approach velocity is calculated

$$v = Q/A \quad = Q/\pi\, dh \quad \longrightarrow \quad v \approx 1/d$$

Where: d = specific diameter within stratum; h = thickness of stratum. The approximation $v \approx 1/d$ is useful for calculating approach velocities through porous media. Near-well flow for fractured and fissured rock is more appropriately described directly by the "pipe flow" equation, $v = Q/A$, with A being the total cross-sectional area of fractures intercepted by the open borehole. Values for A can be estimated.

Mogg (1959) used several laboratory tests to demonstrate empirically that velocity increases as groundwater converges toward a well. Mogg's model, as widely disseminated in U.S. groundwater industry literature, is that inflow passes into the gravel pack and screen in an essentially horizontal direction in fully penetrating wells. Using this model of flow to a well, flow velocity is controlled by proper screen slot and gravel pack selection, and suitable screen length and placement.

German researchers identified a vertical component in this inflow velocity, showing that inflow to the well did not pass through the near-screen gravel pack in a horizontal direction. Flow measurement tests (early 1950s by Truelsen (1988) and more recently, University of Darmstadt (Kirschmer, 1977) and Technische Hochschule Aachen (Ehrhardt, 1986)), were conducted in conventionally constructed screened wells and models. In all cases, where the pump was positioned above the screen the bulk of the flow into the well occurred through the top of the screen.

The Aachen tests (Ehrhardt, 1986) detected vertical flows in the gravel pack that accelerated upwards to 30-to-50 times the inflow velocity at the borehole wall. Albrecht (1987) demonstrated that for his experimental conditions, a maximum velocity of 0.11 m/s and inclination of 88 degrees would result.

Figure 1 illustrates water flow into a conventional well and the inflow velocitycharacteristics. The inflow velocity profile (as illustrated in Figure 1) is more triangular in a highly permeable formation, and more rectangular in a poorly permeable formation, with a sharp vertical velocity change at the aquifer-gravel pack interface in the latter case.

Thus, the greatest part of the inflowing water passes through the upper 5 to 15 % of the screen and then flows upward to the pump. Consequently, only a small upper part of the screen is strongly acted upon hydrodynamically, while 75 % of the zone is acted upon weakly or not at all.

Two practical well maintenance problems result from this unequal hydrodynamic loading and vertical flow through the near well gravel pack or developed formation: (1) increased minerological incrustation due to velocity effects in the high-velocity zone, and (2) excavation of the wall of the borehole, sloughing in washed-out caverns in the gravel pack and subsequent sand infiltration (Figure 1). Biologically generated iron and manganese deposits also will tend to collect in these high flow zones, while the biofilms develop where redox and flow conditions are optimal.

3 Suction flow control devices (SFCD)

3.1 SFCD function and development

SFCD were developed as a result of the observation that a variably perforated insert element in the well would counteract the vertical flow velocity in the gravel pack by transferring the vertical velocity component to the inside of the insert element. An early concept, the SFCD-I (in our parlance), consisted of two concentric, identically slotted tubes, with the annular space between them filled with a granular material. Unpublished field tests conducted separately and independently in Germany and the United States showed that this design did not function as expected. While only a small part of the vertical flow component was transferred out of the gravel pack, upward flow resulted in (1) the annular space between the well screen and the SFCD-I and (2) the gravel-filled space between the coaxially arranged tubes of the SFCD-I element. This annular flow within the SFCD-I resulted in washing out of the filter pack from the element.

A second design, much simpler in concept, was developed by Pelzer for Kabelwerk Eupen. This design, the SFCD-II (or by commercial designation, Eucastream-SFCD), consists of a single pipe, closed at the bottom and slotted. The Eucastream product is plastic, reducing weight and cost and providing improved corrosion resistance compared with metal fabrication.

Distances between the slots are small at the bottom (away from the pump) and increase upwards. The slot pattern is calculated using software which utilizes basic hydrodynamic principles and empirically determined hydrodynamic coefficients. Using this program, the slot pattern is calculated according to the characteristic features of each well. Decisive parameters are the maximum delivery of the well, length (L_E) of the filter zone, and diameters of the screen and borehole.

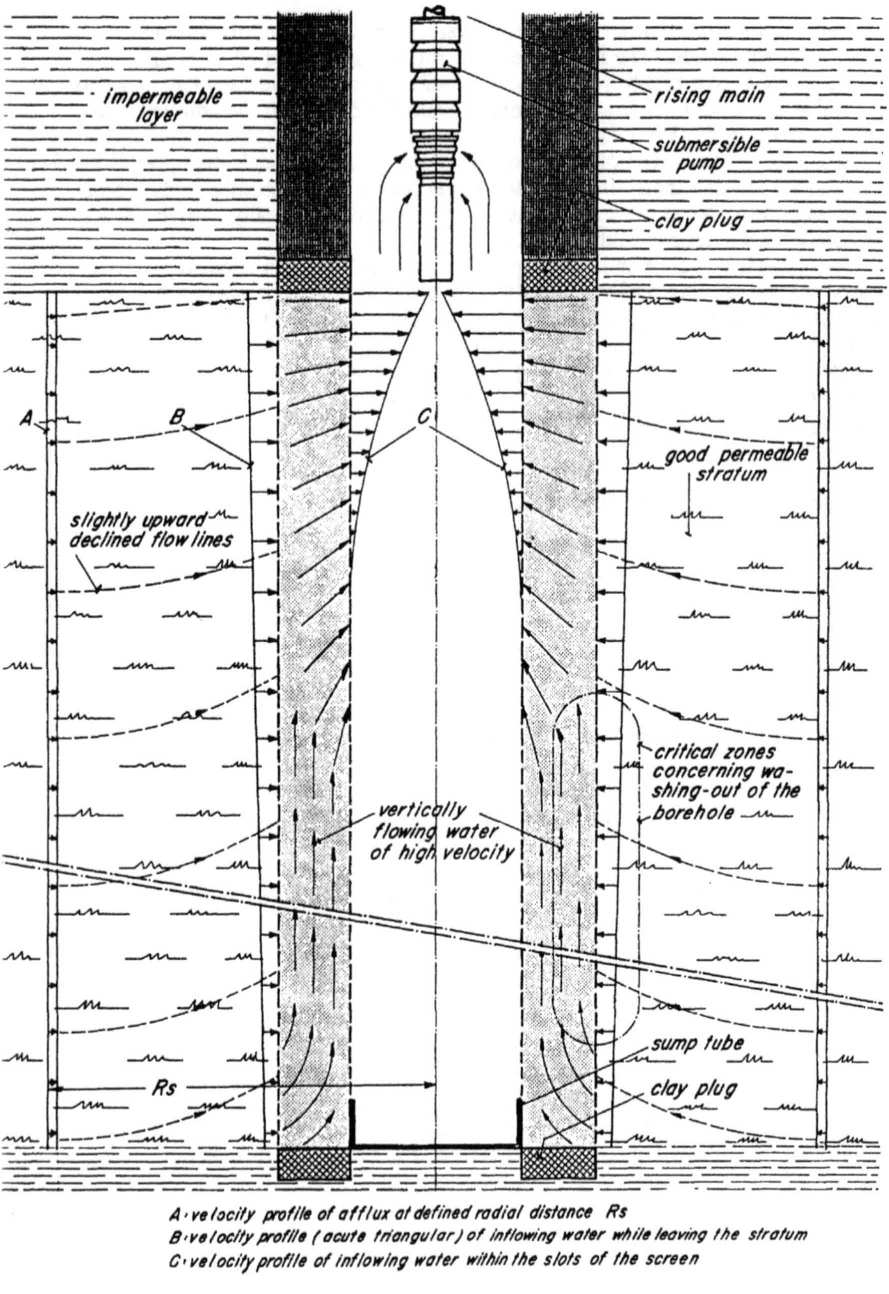

Figure 1. Flow pattern in a conventionally constructed gravel-packed well.

3.2 Eucastream-SFCD effects in the pumping well

With the Eucastream-SFCD design, the vertical velocity component is entirely transferred to the inside of the SFCD, and radial flow to the well cylindrical in profile (Figure 2). Water is flowing inward horizontally over the total length L_E within the surrounding near-well aquifer, filter pack, and across the screen slots. Velocity is strongly reduced in the critical upper zone.

Over 100 SFCD-II-type installations are on record, either Eucastream-SCFD, (installed with the consultation of Kabelwerk Eupen AG engineers) or a similar device developed by Aquastream, Inc., Dallas, Texas, USA, (installed by affiliates of Layne-Western Co., Shawnee Mission, Kansas). Results reported to the authors show that sand pumping slowed or stopped immediately and has not recurred or increased in any properly installed settings, saving a number of expensive, high-capacity wells from abandonment. Specific capacity (Q/s) increases have also been reported for the USA installations, probably through natural development of the formerly underutilized and underdeveloped screen sections. With increased efficiency (increased Q/s), reduced drawdown per unit yield and reduced power consumption have resulted for these wells. Some examples are presented in Table 1.

While there is only a few years' field experience with SFCD-II installations, the following additional advantages to SFCD installation are expected:

(a) Reduced incrustation of the screened interval.
(b) Permanent elimination of sand pumping and consequent increased lifespan for well, screen, pump and their components.
(c) Possibility of increased total delivery to the well due to increased hydrodynamic loading of the full aquifer interval.
(d) Reduced filter pack thickness and consequently a smaller borehole diameter can be designed. Larger-diameter filter pack material can also be used, increasing screen slot size and reducing well loss.
(e) Development time and intensity for removing fine particles from the near-well aquifer can be reduced, reducing well construction costs.
(f) Increased time until redevelopment/rehabilitation are needed and reduced flow-restriction impacts of incrustation and biofouling.

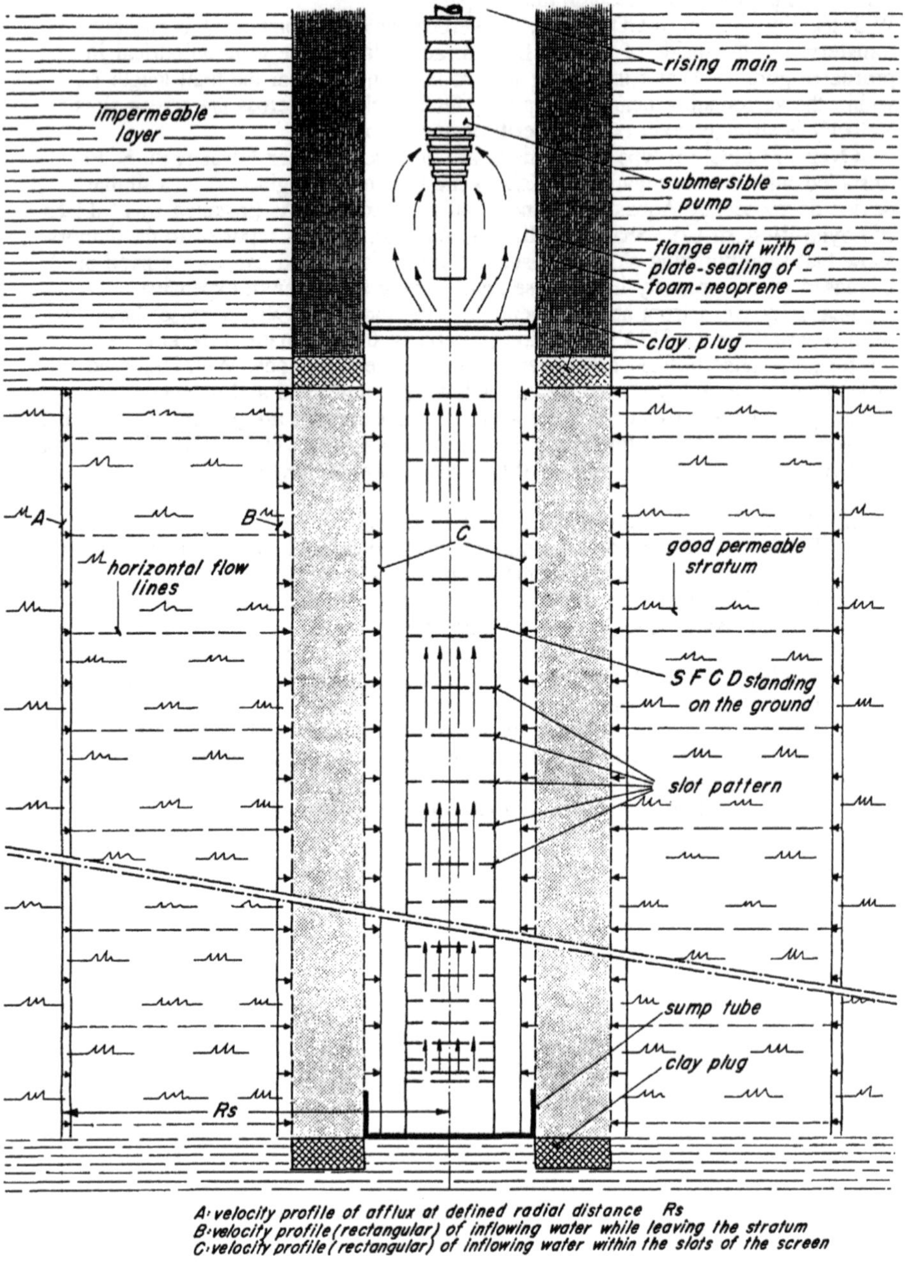

Figure 2. Flow pattern in a well equipped with Eucastream-SFCD.

Table 1. Case Histories of SFCD Installations and Results

Tilberg, Netherlands (1987). Device: Eucastream. Well characteristics: Before SFCD: Q = 33 l/s, sand 20 cm^3/m^3 water. After: Sand production 0.5 cm^3/m^3.

Mettenheim, Bavaria, FRG (1988). Device: Eucastream. Wells: 3 with Q = 70 l/s. Sand before SFCD: 100 cm^3/m^3. Immediately after (1 hr): 1 cm^3/m^3. One year later (same conditions), sand production was zero.

City of Madison, Indiana, USA (1988). Device: Aquastream. Before SFCD: Well no. 3 producing sand at 100 gpm (approx. 7.5 l/s). With SFCD: After well regeneration and SFCD installation, well produced sand-free water at 1200 gpm (approx. 90 l/s). Specific capacity (Q/s) increased from 54 to 112 gpm/ft, with 10-15 % reduction in electrical input per unit water pumped.

Trendelberg/Kassel, FRG (1989). Device: Eucastream. Well: Q = 22 l/s. Before SFCD: Water was brown-red with fine clay/silt. With SFCD: After some hours, water was clear, free of particles.

City of Paramount, California, USA (1989). Device: Aquastream. Well: 5-year-old, 1200-ft (400 m), 4000-gpm (approx. 299 l/s). Before SFCD: Well intermittant sander, wearing pump bowls. With SFCD: Sand alleviated, yield increased to 4300 gpm (approx. 322 l/s).

City of Vinton, Iowa, USA (1989). Device: Aquastream. Before SFCD: Well disconnected for unspecified reasons and regeneration attempted in 2/88, with some improvement, but well lost production and began pumping sand later in 1988. SFCD was then installed. With SFCD: No sand pumping and Q/s improved to 61 gpm/ft (at 610 gpm, approx. 46 l/s) from 48 gpm/ft (at 384 gpm, approx. 29 l/s) at end of 2/88 regeneration pump test.

Einsburen/Rheine, FRG (1990). Device: Eucastream. Well: 11 l/s. After 1.5 year, biofouling completely plugged screen and partially plugged gravel pack. After regeneration, drawdown (s) = 1 m, s increasing to 1.5 m within 2 months due to return of fouling. Installation of SFCD reduced s to 1.1 m, and has remained constant for 3 months.

---

4 Conclusions

The current SFCD design corrects the undesirable inflow patterns around pumping wells completed in porous-media aquifers. The hydrodynamic result of the installation of a Eucastream-SFCD is the transferral of the vertical flow velocity component from the near-well environment to the interior of the SFCD. The practical result is elimination of sand pumping, increased specific capacity in many instances, lower-cost and more efficient wells, and reduced impact of incrustation and biofouling.

5 References

Albrecht, B. (1987) Saugstromstauerung der zweiten Generation beim Bau und Betrieb von Vertikalfilterbrunnen (SFCD of the second generation for the construction and operation of vertical bore wells). **12th Wassertechnischen Seminar**, Instituts für Wasserversorgung, Abwasserbeseitigung und Raumplanung der Technischen Hochschule Darmstadt, West Germany.

Driscoll, F.G. (1986) **Ground Water and Wells**. Johnson Division, St. Paul, MN.

Ehrhardt, G. (1986) **Theoretische und experimentelle Untersuchungen an Sammelzylindern für Brunnen**. Instituts für Wasserversorgung, Abwasserbeseitigung und Raumplanung der Technischen Hochschule Darmstadt, West Germany.

Heath, R.C. (1983) **Basic Ground-Water Hydrology**. Water Supply Paper 2220, U.S. Geological Survey, Reston, VA.

Kirschmer, K. (1977) **Nold-Brunnenfilterbuch**. 5th ed., Bieske and Wandt, eds., J.F. Nold & Co., Stockstadt am Rhein, West Germany, pp. 112-115, S. Annex.

Mogg, J.L. (1959) The effect of aquifer turbulence on well drawdown. **Proc. ASCE Jrnl. Hydr. Div.**, HY 11, pp. 99-112.

Truelsen, C. (1988) Groesste Brunnenleistung. **Die Wassererschliessung: Grundlagen der Erkundung, Bewirtschaftung und Erschliessung von Grundwasservorkommen in Theorie und Praxis**. 3rd ed., H. Schneider, ed., Vulkan-Verlag, Essen, West Germany.

Williams, E.B. (1981) Fundamental concepts of well design. **Ground Water**, 19, 527-542.

PART FIVE

REHABILITATION IN PRACTICE

21 APPLICATION OF PHYSICO-CHEMICAL TREATMENT TECHNIQUES TO A SEVERELY BIOFOULED COMMUNITY WELL IN ONTARIO, CANADA

J. GEHRELS
Ontario Ministry of Environment, Thunder Bay, Ontario, Canada
G. ALFORD
George Alford Inc., Daytona Beach, Florida, USA

Abstract
After one year of operation, the growth of nuisance bacteria caused a community water supply well in Northwestern Ontario to steadily lose its efficiency. During a 15 month period, the well yield with an original capacity of 17 l/sec had decreased 65% (from 13.85 l/sec to 6.06 l/sec).

Remedial work, using an iterative process of applying heat and adding chemical disinfectants, increased the well yield by 92% above the minimal production (to 11.5 l/sec). Subsequently, however, the yield decreased by 7.9%/week until pre-treatment yields were reached. Following extended physical surging and various chemical treatments the well recovered to its original design capacity. A maintenance program has been implemented to provide on-going control of nuisance bacteria.

This experience highlights that there is no one method which can be consistently prescribed as a formula to effectively rehabilitate all biofouled wells. Once a biomass is firmly entrenched, several treatments may be required to restore well yields to acceptable levels. Rehabilitation efforts must be followed by on-going maintenance programs. Further research into the bacterial and inorganic composition of waste water during treatment is warranted. Finally, additional consideration must be given to public perception of bio-remediation work done on community water supply wells.
Key Words: Biofouling, Nuisance Bacteria, Remediation, Maintenance, BCHT, Well, Water, Supply, Perception.

1 Introduction

The term "nuisance bacteria" is frequently used to refer to the diverse group of non-health related bacteria which include iron related, sulphate or sulphur reducing, and slime forming bacteria. These organisms usually form masses of gelatinous and filamentous organic matter which can act as a host environment for health related bacteria (such as opportunistic species of pseudomonas), affect the aesthetic quality of drinking water and completely plug and corrode pipes and water supply equipment.

Nuisance bacteria are present in most soils and surface waters. Since nuisance bacteria are mainly sessile, aerobic organisms, they were historically not considered to be of concern in groundwaters. Symptoms of their growth were widely misdiagnosed as being physical or chemical in nature.

Research, however, is beginning to discover that nuisance bacteria are much more prevalent in ground water than was thought possible. It is also becoming apparent that ultra microcells are able migrate considerable distances through the groundwater regime (Costerton, 1990). These ultra microcells are 10% of normal diameter, are effectively in a state of suspended animation and are non-attachable (Cullimore, Alford and Morrell, 1990). Since the scope of the problem of nuisance bacteria in water supply wells has only been recently identified, there is a lack of understanding on how to prevent their establishment and growth, identify their presence and treat established biomasses.

This paper describes a remediation case study in which a variety of techniques were used to treat a severely biofouled water supply well. It is hoped that lessons drawn from this experience will help those in the water well industry to move towards developing a systematic approach to treat biofouled wells.

2 Background

The well reported on in this study is the main supply well for a small community located in northwestern Ontario. The town is located within a large coarse sand and gravel outwash deposit. These materials are underlain by Precambrian age intrusive granite. Due to the poor water transmitting capability of the bedrock, the well was completed within the unconfined sand and gravel aquifer which has a estimated Transmissivity of 680 m^2/day.

The well, installed in 1985, consists of a 13.7 meter long, 50.8 cm dia. outer casing and a 25 cm inner casing which extends to a depth of 19.8 metres. The well terminates in a layer of coarse gravels, sands and boulders and is screened with a 3 meter long Johnson 50 slot stainless steel screen. A 7.5 hp submersible pump was installed 16 meters below grade, 12.6 meters below the pitless adapter. The capacity of the well was estimated to be 17 l/sec.

In January, 1987, following the construction of a community water distribution system, the well was put into use. After one year of operation, the well began to steadily lose its efficiency. During the 15 month period January, 1988 to April, 1989, the well yield decreased 65 % (Figure 1). At this point the community began to experience water shortages and the problem with the well was noticed.

When the drop in yield was reported, it was suspected that the declining yield was due to severe biofouling. Water samples were analyzed and were found to contain Iron Related Bacteria (Gallionella) and Sulphate Reducing Bacteria. Subsequent bacteriological testing using the patented Bacterial Activity and Reaction Test system (BART™) confirmed that the well was severely biofouled.

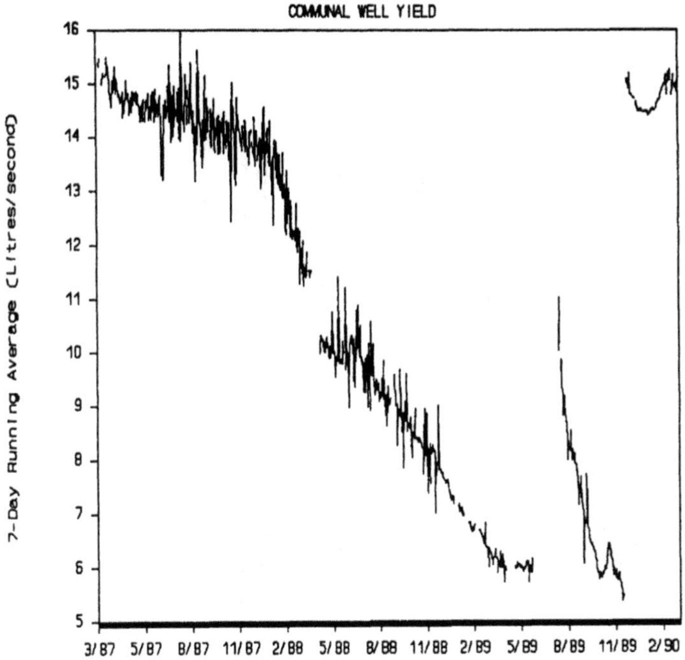

Fig.1. Communal Well Yield during primary plugging event (3/87, mm/yr), after BCHT in 7/89, and Carela treatment (12/89).

Particle counts and indicators of bacterial population in the well were found to increase with depth (Table 1). The highest levels were found in the upper region of the well screen and in the area adjacent to the pump intake. This pattern of bacterial fouling is related to high flow velocities in these areas enabling bacteria to accumulate maximum amounts of nutrients from the water column.

3 Initial well treatment

Treatment of the well was initiated in July, 1989 using a modified version of the patented "Blended Chemical Heat Treatment" (BCHT) process. This iterative procedure, applied here for the first time in Canada, utilizes heat in conjunction with a blend of chemicals. As discussed in Alford et al. (1989), this technique has been demonstrated to be effective in restoring biofouled wells throughout the U.S.

The treatment of the well consisted of the following sequential steps: shock and dispersion, partial disruption and dispersion, full disruption, and dispersion. The BCHT process usually has one disruptive stage which combines application of heat and chemicals to dissolve biochemical encrustations. In this case, however, applications of heat were limited due to the boiler being experimental rather than industrial in nature. The disruptive stage was therefore split into two shorter phases: partial and full disruption. Each treatment

TABLE 1

Bacterial Activity Information*

Sample Description	Particles: Counts (#/mL)	Particles: % of Volume	BART Reaction Times (Days): IRB	BART Reaction Times (Days): SRB	BART Reaction Times (Days): SLYM	Bacteria Counts/ml: Total Aerob.	Bacteria Counts/ml: Pseud.	Bacteria Counts/ml: IRB
Prior to pump removal	172,100	0.0050	6	6	4	76	14	36
From pump	2,010,000	0.6080	2	1	2	560	280	350
Bottom of well	154,100	0.0129	2	2	2	5,600	2,390	1,810
6 feet from bottom	158,600	0.0109	6	7	2	6,350	2,410	2,450
12 feet from bottom	179,700	0.0310	2	2	3	320	130	160
18 feet from bottom	175,000	0.0068	2	5	3	240	200	120
12 hrs after first BCHT	116,300	0.0036	14	10	6	80	45	67
12 hrs after second BCHT	7,200	0.0001	20	20	9	10	0	0
1 day after treatment	65,500	0.0006	20	20	10	0	0	0
2 days after treatment	132,300	0.0005	15	20	10	4	0	0

*Acquired before, during and after the treatment of the communal water well by BCHT.

phase was designed to address a specific aspect of the biomass problem and to enhance subsequent treatment phases.

The primary purpose of the initial "shock" phase was to soften biochemical encrustations and to place the biofouling microorganisms in a state of shock, thus reducing the chemical demand in subsequent treatment phases. This phase consisted of a disinfectant (sodium hypochlorite) and a wetting agent (similar to Arc One) being added to the well water. Sodium hypochlorite was used in preference to the calcium form since the latter can result in the production of low solubility calcium minerals (see Howsam, 1988). The wetting agent was used to reduce the surface tension between water molecules, thus allowing the disinfectant to better penetrate the biomass (see Hackett, 1987). Sufficient sodium hypochlorite was added to approximately displace the volume of water in the well casing. This procedure was designed to ensure that a high free chlorine residual would be maintained in the well and in the surrounding aquifer over the entire contact period.

The chemicals were mixed in the well and forced into the aquifer by repeated surging. This process is essential to the success of the treatment since it loosens the ferric hydrate and organic slime deposits and results in a thorough contact between the bacteria and the chlorine residual (Hackett, 1987). Since a surge block was not available, surging was completed using air lift.

Chemical contact times generally vary from 5-48 hours (Smith, 1983). In this situation, the chemicals had been neutralized after 12 hours and the well was purged in order to expose fresh contact surfaces for subsequent treatments. The initial discharge water had a bright reddish colour. Analysis of this initial purge water showed that levels of metals were extremely elevated with concentrations of aluminum being surprisingly high (Table 2). Of greatest interest, however, is that the sample was found to consist of 1 percent iron (10,000 ppm).

These analytical results appear to indicate that the bacterial population had been well entrenched and that much of the biomass had been loosened from within the well. This interpretation was confirmed by the BART™ reaction times and the bacterial counts (Table 1) which showed a dramatic decrease in the level of bacterial activity following this first treatment procedure.

Partial disruption was accomplished by adding a mix of sodium hypochlorite, wetting agent and sulfamic acid to the well. Sulfamic acid chemically dissolves ferric hydrate deposits, keeps metals in solution, and disperses and loosens extracellular slime material (Hackett, 1987). The corrosive effect of the sulfamic acid, which is less than that of hydrochloric (Muriatic) acid, was reduced by stabilizing it with sodium hypochlorite. The addition of the acid brought the pH of the water down to 0.9. This enhanced the effect of the sodium hypochlorite by keeping the chlorine from dissociating from the hypochlorous form into hydrogen and hypochlorite ions. This is important to the disinfection process because the hypochlorite ion is not as effective a disinfectant agent as hypochlorous acid (Hackett, 1987).

TABLE 2

WASTEWATER CHEMICAL ANALYSIS *
RESULTS

(All values reported in mg/L)
(ICP Analytical Method)

	Start First Acid Surge 22/9	End of Pump Test #1 13/12 16:30	Block Surging Waste Water 14/12 10:00	14/12 10:20	14/12 10:40	14/12 11:00	14/12 12:00	14/12 12:45	14/12 13:40	End of Pump Test #2 14/12 15:50	Start Carela Purge 15/12 9:30	Start NaOCl Purge 16/12 9:30	End NaOCl Purge 16/12 13:00	Start Acid Purge 16/12 17:35	End Acid Purge 16/12 21:30	End of Pump Test #3 17/12 15:40
	1	2	3	4	5	6	7	8	9	10	11	12	13	14	15	16
Iron	10000	3.3	100.0	200.0	200.0	200.0	53.0	16.0	25.0	1.2	74.0	200.0	3.8	800.0	13.0	3.4
Aluminum	55	0.3	1.8	12.0	16.0	13.0	13.0	4.5	6.5	0.2	15.0	14.0	1.0	200.0	5.2	1.7
Manganese	24	0.06	0.35	0.65	1.00	1.20	0.86	0.31	0.54	0.09	3.40	0.94	0.09	16.00	0.52	0.37
Copper	9.7	0.01	0.11	0.17	0.24	0.18	0.11	0.05	0.07	0.01	0.37	0.11	0.02	3.50	0.35	0.12
Zinc	9.7	0.06	0.90	0.20	0.18	0.32	0.06	0.02	0.04	0.01	0.33	0.44	0.01	4.60	0.07	0.02
Titanium	3.9	0.02	0.14	0.78	1.10	0.85	0.94	0.32	0.51	0.02	0.69	1.10	0.08	10.00	0.10	0.01
Barium	3.8	0.02	0.09	0.20	0.25	0.22	0.14	0.06	0.07	0.04	0.49	0.16	0.05	3.70	0.28	0.16
Lead	2.0	0.004	0.046	0.070	0.080	0.065	0.032	0.020	0.020	0.020	0.025	0.070	0.020	0.460	0.020	0.020
Molbdenum	1.0	0.001	0.027	0.022	0.031	0.024	0.009	0.005	0.008	0.005	0.019	0.030	0.005	0.130	0.005	0.005
Vandium	0.8	0.002	0.034	0.130	0.160	0.110	0.054	0.020	0.030	0.020	0.066	0.084	0.020	0.760	0.020	0.020
Cobalt	0.8	0.001	0.014	0.022	0.030	0.025	0.017	0.010	0.012	0.010	0.038	0.030	0.010	0.300	0.011	0.010
Chromium	0.5	0.002	0.041	0.130	0.140	0.170	0.054	0.020	0.031	0.010	0.370	0.160	0.027	2.800	0.025	0.010
Nickel	0.45	0.001	0.015	0.050	0.080	0.060	0.047	0.016	0.028	0.010	0.280	0.092	0.010	1.900	0.024	0.010
Strontium	0.40	0.029	0.043	0.055	0.061	0.054	0.048	0.035	0.039	0.031	0.170	0.120	0.040	0.860	0.057	0.046
Cadmium	0.26	0.000	0.002	0.000	0.004	0.004	0.002	0.002	0.003	0.003	0.003	0.004	0.004	0.012	0.003	0.002
Beryllium	0.06	0.001	0.010	0.010	0.010	0.010	0.010	0.010	0.010	0.010	0.010	0.010	0.010	0.020	0.010	0.010

*Before, during and after Carela, hypochlorite and acid purges.

The combining of sodium hypochlorite and sulfamic acid could have resulted in free chlorine being given off as chlorine gas which would penetrate into the aquifer and disinfect the vadose zone.

The well water was then surged in order to disrupt the already softened encrustations and biomass and ensure a more uniform mixture of the acid which is heavier than water. Following a 12 hour contact time, the spent chemicals and detached biomass components were air lifted from the well. The full disruption phase involved heating the well water and then adding chemicals to dissolve remaining biochemical encrustations. The purpose of adding heat was to soften the biomass, increase the efficiency of the chemicals and to increase the metabolism rate of the microorganisms so that they would take-up the chemicals more rapidly. The temperature of the well water was raised by directly injecting steam from a boiler into the well. Alternative techniques for the application of heat include recycling of hot water and implantation of electrical immersion heaters directly at the screen level (Cullimore, 1981). The well water was heated until it reached a temperature of approximately 50°C. Field experiments have shown that dispersion of iron precipitate/organic slime begins when temperatures rise above 25°C and accelerates rapidly above 35°C; lethal temperatures occur at approximately 45°C (Cullimore and Mansuy, 1986). Sodium hydrosulphite, commonly used as the active ingredient in iron and manganese descalers, was added to the well to dissolve remaining biochemical encrustations, and the well was physically agitated followed by a 12 hour contact time.

At the start of the final treatment phase (dispersion), the plugging material should have become disrupted to the point of no longer being attached to the surface matrices presented by the well screen, gravel pack and aquifer strata (Alford et al., 1989). This phase consisted of a prolonged sequence of surging and purging in order to break-up and remove suspended material from the well and to ensure that remaining chemicals were pumped to waste.

Over the 6 hours of surging and purging, the discharge water gradually changed from turbid with visible tubercles to clear with no suspended material. At the end of this treatment phase, particle counts, BART™ reaction times and bacteria counts all indicated that the biomass had been significantly reduced (Table 1). Following the modified BCHT treatment, the well yield increased to 11.5 l/sec, a 92% improvement in flows. However, this yield is only 76% of the 15.13 l/sec original yield from the well when placed on line. In addition, increasing particle and bacteria counts and an increase in BART™ reaction times for iron related bacteria (Table 1) indicated that either the biomass was sloughing-off the aquifer material and/or that the biomass was re-establishing itself in the treated portion of the aquifer.

In the month following the renovation work, the yield decreased by 3.5 l/sec. This decline of approximately 7.9%/week was higher than any previously reported plugging rate and is near maximum rates obtained experimentally (Cullimore, personal communication). In comparison, a small diameter well treated with pasteurization in Saskatchewan plugged at a rate of approximately 0.77%/ week (Cullimore and Mansuy, 1986).

4 Follow-up treatment

In December, 1989, follow-up rehabilitation work was carried out. Despite adverse weather conditions (air temperatures ranged from -25 to -42°C), a treatment program consisting of the following 4 steps was completed: physical development, chlorination, acidification, and extended purging. In order to determine the relative success of these treatments, pump tests were conducted prior to any remedial work taking place, after the physical development stage, and after the chemical treatment stage. Water samples were periodically collected for heavy metal and BART™ analysis.

The first pump test was conducted using the existing submersible. The well was pumped at a constant rate of 5.7 l/s. After 45 minutes, the water level had stabilized at 12.46 meters (a drawdown of 4.3 meters). The rate of flow was then increased to 8.8 l/s and the well responded by rapidly drawing down to the level of the pump intake. Analysis of this data indicated that the well's yield was approximately 6.7 l/s and the Transmissivity of the aquifer was estimated to be 210 m^2/day. When the drop pipe and pump were removed from the well, there was a small amount of slime formation on the inside of the drop pipe and on the external portion of the pipe which had been below the water table. The pump screen exhibited no blockage.

The first step in the rehabilitation program was carried out by physically developing the well for 5 hours using a 10 inch surge block. The purpose of this work was to loosen biological growths and to physically remove sand, silt, and biochemical blockages from the screen and adjacent area of the aquifer. An airlift hose was used on top of the surge block in order to purge particulate matter from the well as it was loosened by the surge block.

Using the surge mechanism on the Cable Tool Drill rig, the surge block was gradually moved up and down the well screen. For approximately 3 hours the discharge water was highly turbid and had a strong red colour. As shown on Table 2, metal concentrations in the discharge water remained elevated during most of this procedure.

Initially, the upper 1 meter portion of the screen was tightly blocked and yielded no measurable volumes of water. The blockage of the screen appeared to be less severe with increasing depth. When second pump test was conducted, the rate of discharge was increased in steps from 8.8 l/s up to 25.2 l/s. When pumping was terminated, the well had only drawn down 4.8 meters to 12.98 meters. The well recovered to its original static level within 3 minutes. The wells yield was estimated to have dramatically increased to 40.54 l/s and the Transmissivity was calculated to have quadrupled to 846 m2/day. This indicates that much of the decline in well yield which occurred following the initial rehabilitation work was related to loosened biochemical deposits in the aquifer gradually accumulating around the well screen as the well was pumped.

The second rehabilitation procedure, chlorination, was then started by adding 90 litres of Carela to the well. Carela is a proprietary mixture containing chlorine, a wetting agent and hydrogen peroxide that has recently been introduced in Canada. This procedure was designed to remove most of the biomass which was attached to aquifer material adjacent to the well screen. The Carela was surged into the formation for a period of 2 hours and then provided with a 14 hour contact period. The well was purged for half an hour to remove dissolved and particulate biochemical deposits from the well. Small particles of iron, sand and silica were noted to be coming up in the discharge water and building-up around the well head. This indicates that the well was being hydraulically developed. Superchlorinated was initiated by adding over 800 litres of sodium hypochlorite (10 percent chlorine equivalent) and 9 litres of Liquid Organic Cleaner (LOC) to the well. The sodium hypochlorite (NaOCl) acts as a biocide and the LOC acts as a wetting agent to maximize penetration of the NaOCl into the biomass. These chemicals were surged into the formation for 8 hours. Following a 12 hour contact period, an additional 4.5 litres of LOC was surged into the formation for 2 hours and the well was purged for 4 hours. The acid was not left in the subsurface for a longer period of time in order to avoid having the acid settle to the bottom of the aquifer and form a stratified layer which could not be subsequently be pumped out of the aquifer. A very large volume of foam was pumped from the well along with highly coloured water.

The third step in the treatment, acidification, was carried out in order to loosen biochemical deposits. Approximately 600 litres of hydrochloric acid was pumped into the well and gently surged into the aquifer for a period of 4 hours. A drop pipe was next inserted into the acidified water to allow removal of the accumulated biofilm.

As shown on Table 2, the highest metal concentrations recorded during the second treatment were recorded at the start of the initial acid purge step. Peak concentrations recorded for most metals in the first treatment were 0 to 4 times higher than peak concentrations in the second treatment, exceptions being molybdenum (8 times higher), iron (13 times higher) and cadmium (22 times higher). The treatment was completed by surging and purging the well for 4 hours using the air lift/surge block technique. This process was designed to remove most of the suspended material from the well and surrounding aquifer. Remaining waste chemicals were removed by conducting a 6 hour pump test. Following 4 hours of pumping at a rate of 25 l/s, the water no longer had a detectable chlorine residual, its conductivity was 245 umhos and its pH was 5.9 (background values were 120 umhos and 6.6 respectively).

After 6 hours of pumping, the water level was 11.9 meters below grade (3.7 meter drawdown) and the pump test was terminated. Within 10 minutes the well had recovered back to its original static level. The well was calculated to have a theoretical yield of 52 l/s and the Transmissivity of the aquifer was calculated to be 892 m^2/day. The well was brought back on-line and the pumping rate measured in the pumphouse was 15 l/s.

5 Maintenance program

It appears that there is still active biological activity in the subsurface. A BART™ sample, obtained at the end of the final pump test, showed a detectable reaction for iron bacteria within 4 days. Laboratory analysis identified iron related (*Gallionella* and *Leptothrix*) and sulphate reducing bacteria in the water. In addition, following the second treatment, the well yield did not hold steady, first dropping, then rising and then showing increasing fluctuations (Figure 1). The rate at which the biofouling will continue cannot be predicted. However, based on the above information and on the rapid decline in yield which followed the initial work, it is assumed that the biomass could rapidly re-establish itself.

A maintenance program of periodic *in-situ* chlorine shocks and acidification is therefore being implemented. On a short-term basis, the well will be chemically treated on a bi-monthly interval. Chemicals were pumped into the well using a tremie pipe which extended down past the pump into the upper screen area. The length of time between treatments will gradually be increased when it is demonstrated that the well is retaining its yield. This maintenance program consists of three stages. First, 12 litres of sodium hypochlorite (NaOCl, equivalent to 2,000 ppm Cl) was pumped into the well in order to kill biological growths in the well and on the well screen. Following a 4 hour contact period the well is purged for half an hour in order to remove waste chemicals and loosen biochemical deposits.

The well was acidified by adding 100 litres of stabilized hydrochloric acid (HCL) in order to soften biochemical deposits in the aquifer adjacent to the screen. This quantity of HCl provides for 20% solution by bore volume. The acid stabilized by adding 75 grams of tri-sodium polyphosphate or 1.5 litres of dry gelatin powder (this quantity may be mixed with water in order that it may be poured into the tremie pipe. Following a 5 hour contact time, the well is purged for 1 hour. Finally, the well was superchlorinated by adding 190 litres NaOCL and a wetting agent. The wetting agent is either 2 litres of Arc One or 1 litre of Liquid Organic Cleaner. This quantity of NaOCl provides for a theoretical penetration distance of 50 cm into the aquifer at a chlorine concentration of 2,000 ppm. The well was finally allowed to sit overnight in order to enable the chlorine to disperse into the aquifer and to penetrate biological slimes. The treatment was completed by purging the well for at least 6 hours.

Within the community, however, there developed a widespread negative perception towards the repeated addition of chemicals to the water supply well. Attention is therefore being focused on the installation of heating devices which can be used to periodically pasteurize the well. Pasteurization would be easier for the local pump house operator to implement and the addition of heat, rather than chemicals, is much more socially acceptable.

If attempts to control the biomass using the above techniques are not successful, a permanent chemical drip feeder may have to be added to the heat treatment system. Consideration is also being given to installing a downhole seal above the pump. The purpose of this seal would be to act as an anoxic block by restricting the supply of oxygen to the system. Potential problems with such a device include (Smith, 1981): (1) Can a complete seal be achieved? (2) How would the seal affect pumping? and (3) How would the seal affect well maintenance?

6 Discussion

Based on this experience, several points regarding the identification and treatment of nuisance bacteria can be made. The first obvious question is "why did the well plug initially?". It would appear that this well was located in an environment which had a high biofouling potential. There is increasing evidence to suggest that nuisance bacteria (possibly ultra microcells), which were known to be present elsewhere within the aquifer, could readily migrate through the permeable substrata until arriving at a favourable environment (e.g., the producing well).

General environmental conditions in the aquifer were conducive to bacterial growth. Iron levels in the local groundwater range from 0.3 to 3.0 mg/L, well within the 0.1 to 10 mg/l range reported by Cullimore (1986) to be critical for the occurrence of iron related bacterial biofouling. The pH of the groundwater in this area was measured to be 6.1, well within the pH range of 5.4 to 7.2 over which most iron bacteria are reported to grow (Hasselbarth and Ludemann, 1972).

The well/aquifer interface presented an ideal location for ultra microcells to form a biomass. It has now become generally recognized that sessile bacteria tend to predominate at the redox fringe between oxidative and reductive states or where there are sudden changes in hydraulic flows (Cullimore, Alford and Morrell, 1990). The bulk of flow into wells occurs in the upper 5-15 percent of the screen and vertical flows in the gravel pack can accelerate up to 30-to-50 times the inflow velocity at the borehole wall (Pelzer and Smith, 1990). This enables sessile bacteria to bioaccumulate maximum amounts of nutrients from the passing water. Although not yet reported in the literature, the stratified pattern of concentrated blockage in the area where water primarily enters the well has been noted in most biofouled wells examined by Dr. R. Cullimore and S. Smith (personal communications). Concentrations of oxygen, critical to the growth of aerobic bacteria, are generally higher in wells than in surrounding geological deposits. Finally, the restoration of the well effectively increased the concentration of dissolved oxygen in the water and provided a mechanism for supplying large quantities of nutrients to the bacteria.

Within this context, the establishment of the thick biomass was related to the declining well yield although this was not noticed until the community began to experience water shortages. This highlights the first, and perhaps most crucial, phase in dealing with nuisance bacteria: recognizing their presence! Symptoms of their growth in a well often appear slowly and are frequently misdiagnosed as being of a physical or chemical nature. By the time the problem was identified here, the biomass was firmly entrenched within the system.

To avoid such problems, water system operators must be fully aware of the symptoms of nuisance bacteria. Some items to watch for include iron concentrations above 0.2 ppm and/or manganese concentrations above 0.05 ppm; water which has an unpleasant, brackish taste and/or a yellow, orange or brown discoloration; occasional high Standard Plate

Counts (SPC), unexplained high coliform readings or periodic slugs of "rusty coloured", turbid water passing through the system; a white or reddish brown gelatinous slime in the water system; a gradual decrease in well yield; corroded water supply equipment. Benchmark physical and biochemical conditions should be established, and well efficiency and the bacterial and chemical quality of municipal water supplies should be regularly monitored. Water supply records can be readily used to monitor well yield on an on-going basis. However, since the "true" well yield may be greater than the capability of the pumping system, a significant amount of biological related blockage may occur before a drop in pumping rate is noticed. It is, therefore, important to periodically record water levels under constant pumping conditions. Pumped well samples rarely contain representative numbers of nuisance bacteria since they are largely sessile and/or anaerobic (Smith, 1983). Indicator parameters, such as iron, manganese, phosphorus, total organic carbon, sulphate, turbidity and colour, must therefore also be monitored. Early warning of bacterial slime build-up can also be achieved using in-situ samplers (Smith, 1983). No case studies were encountered where such a proactive program was in-place prior to the initial well rehabilitation being required. This is highly significant since when the symptoms of biofouling become clear, it is beyond the timeframe when treatment would have been most effective.

The fact that the first treatment was not able to reduce the established biomass to the point that the original well yield was restored may be related to it not being possible to apply heat at each phase of the treatment. In addition, the following three points should be emphasized:

1) Nuisance bacteria are insidious colonizers. Most known techniques for treating biofouled wells are based on the use of chemicals. Physical treatment methods used for treating nuisance bacteria infected wells include pasteurization, irradiation, ultrasonics, VYREDOX (injection of oxygenated water around the producing well), anoxic blocks (exclusion of oxygen from the well) and cathodic protection. The effectiveness of chemicals, however, is reduced by protective slime layers, the presence of organics, low temperatures and by any accumulated deposits of iron and manganese (White, 1972; Smith, 1981). In addition, while an effective chemical concentration can be obtained within the well, chemicals reaching the surrounding aquifer will be rapidly diluted, thus minimizing their effectiveness (Smith, 1981). Very heavy doses and long contact times are therefore required if reduction in the infestation is to be achieved. For example, concentrations of chlorine based biocides need to be several orders of magnitude greater than those which would be applied in normal potable water treatment (Howsam, 1988). The chlorine dosage must exceed the chlorine demand in order to produce a free chlorine residual of at least 1,000 mg/L in the well water (Hackett, 1987).

2) Good well development techniques are crucial to the effectiveness of the treatment. A large recovery in yield following extended block surging at the start of the second treatment

indicates that the air lift well development technique was not able to remove much of the biochemical material which may have been loosened within the aquifer.

Similarly, the air lift technique likely would have limited capability to drive chemicals past this conglomerate of materials to further out into the aquifer where the biological "matt" may also extended. Consequently, the combination of bacterial reinfestation of the cleaned aquifer and consolidation of loosened material around the well screen resulted in the dramatic decline in well yield.

3) Well design can accelerate bioslime formation and can affect well renovation efforts. A relatively narrow screen slot size may have minimized the ability of the materials to pass into the aquifer and limit the amount of biomass which could be pulled back into the well. In addition, bioslime formation was likely accelerated by large quantities of nutrients being filtered from the water as it rapidly passed through the narrow slotted well screen (Pelzer and Smith, 1990; A. Michael and G. Alford, personal communication). Therefore, where wells are constructed in areas where nuisance bacteria are known or suspected to be present, the largest possible screen slot size should be installed.

In addition, a pump installation fault enabled water to cascade down into the well whenever the pump was operated. This injection of heavily oxygenated water into the water is very significant since highly aerated systems can produce up to five times the plugging rate of anaerobic regimes (Cullimore, 1986). Massive growths of iron bacteria have been reported in wells containing less than 5 mg/l 02 (Hasselbarth and Ludemann, 1972).

Finally, the pump was installed with a pitless adapter 12 feet below grade. Two spring operated portals on the pitless act as one way valves, ensuring that water cannot flow from the pumphouse down into the well. However, this configuration makes it impossible to use the pump to surge the well during in situ maintenance treatments.

The rapid "rebound" of the biomass following the initial treatment highlights the recurring nature of biofouling problems. Once a biomass is firmly entrenched, it is virtually impossible to eradicate (Smith, 1983). Several treatments may be required to restore well yield. While bacteria that cause clogging problems in wells which can respond to biocide treatment, growth restarts once the chemical is removed (Howsam, 1988). Nuisance bacteria can also lay dormant in the aquifer as ultra microcells, out of reach of even the most effective development tools. Following treatment, these bacteria and pulled back into the well to start the biomass growth again (Smith, 1981).

It is therefore essential that well rehabilitation works be followed by an on-going maintenance program to retard the inevitable biofouling processes and to extend the operating cycle of the well. Reports on rehabilitation of biofouled wells often reference the need to repeat treatment at intervals of two months (Smith, 1983).

In this case study, public concern over the addition of large volumes of chemicals to the community well began to develop after the rehabilitation work was completed. Reported problems were usually unspecific and generally involved changes in water quality following the well treatments. Some changes in water quality appear to have actually occurred. It is widely known that many nuisance bacteria bioaccumulate metallic ions such as iron and manganese from the water column. Following treatment, therefore, the reduced biomass would have less capability to "filter-out" heavy metals and water pumped from the well would therefore tend to have higher metal concentrations. This explains why concentrations of iron and manganese in the water were elevated following treatment (Table 2) and could account for some of the taste complaints. Within a period of several months, concentrations of iron and manganese in the water had dropped to pre-treatment levels (usually less than 1 mg/L Fe).

In addition, the biomass itself must be considered. This case study would appear to be one of the first reported instances of detailed chemical analysis being conducted on water discharged from a well during treatment. As shown in Table 2, metal concentrations in this waste water were extremely elevated. Continued sloughing of biomass from the aquifer formation following treatment would cause highly concentrated "slugs" of metals to periodically pass into the water distribution system. Increased flow velocities could have also caused sloughing of biomass from the insides of the pipes connecting the well to the pumphouse. This could account for some of the taste and colour (turbidity) complaints.

Some of the complaints, however, are likely related to perceived rather than actual changes. It is understandable that rumours of "unknown chemicals" being added to the well, of "bacteria" being present in the aquifer, and of "red" and "soapy" water being "blown" out of the well would lead to a fear of the unknown. Close examination of tap water would then take place with previously unnoticed characteristics being identified. The existing literature focuses on technical aspects of biofouling and has not addressed the issue of public acceptance. Ultimately, however, this may be the most important aspect of well treatment. Even if the yield of a well is restored and a technically effective maintenance program is established, this work will be useless if it is not accepted by the consumers of that water. If consumers experience a continuing fear and distrust of the water quality, an alternative water supply will eventually be required.

Public consultation and education is therefore required throughout the treatment process. The predominately nuisance nature of the bacteria must be emphasized and it must be demonstrated to the public's satisfaction that all chemicals which will be added to the well can be removed from the aquifer prior to water being pumped into the water distribution system. The community must be prepared to expect transient changes in water quality and must understand that maintenance work will be required on an on-going basis in order to maintain the well yield.

Finally, it is clear that great care must be taken to minimize the potential bio-contamination of wells. When drilling or working with an open well, it is possible for iron related bacteria for the soil and/or from surface waters to be injected into the well, thereby contaminating groundwater which may have been free of nuisance bacteria before that time. Drillers and pump installers must therefore change long established protocols and spend time and energy cleaning equipment between each well.

Everything that goes into the well (including gravel packs, pumps and drop pipes) must be disinfected. Any water used to mix with drilling mud should be taken from a clean groundwater source and should be chlorinated. Use a circulatory tank or lined pit instead of a mud pit to avoid contamination from the soil (Smith, 1980). When wells are newly constructed or are suspected of being contaminated with either nuisance or hygiene risk bacteria, they should be shock chlorinated using a heavy chlorine dose.

7 Conclusions

In order to develop a scientific approach to the treatment of biofouled wells, efforts need to be directed to determine pretreatment biophysical conditions, carefully document treatment techniques and accurately measure the effectiveness of each technique. This case study indicates that additional efforts need to be made to identify hydrogeological environments which have a high potential for biological fouling. Within these areas, careful attention must be paid to the design and construction of water supply wells. Above all, it is clear that every effort must be made to minimize and prevent the spread of nuisance bacteria. The potential value of establishing baseline physical and biochemical conditions and conducting on-going monitoring was demonstrated by describing the difficulty of treating established biomasses. The need to train water system operators to be able to provide early identification of developing biofouling problems was explicitly shown.

This study demonstrates that extensive, effective well development techniques must be applied throughout each phase of the treatment process. The combining of heat and chemical applications has been shown to have much merit and requires further field applications in colder, northern climates using industrial scale equipment. It is clear from this, and other field experiences, that there is not yet one set method which can be consistently prescribed as a formula to effectively rehabilitate biofouled wells. A technique which works in one location may not work in the next well. Even wells located as little as 15 meters apart may require different amounts of acids and the effectiveness of the treatments may vary dramatically (Alford et al, 1989).

The value of conducting step drawdown pumping tests after each treatment phase was shown in this study. The potential usefulness of monitoring changes in the bacterial, inorganic and heavy metal composition of waste water purged from the well during treatment was also shown. Further study into the biochemical composition of waste water appears to be warranted. Improvements in apparent water quality may be early signs of biomass increase and may indicate impending quantity problems. The impossibility of eradicating a firmly entrenched biomass and the rapidity with which biological reinfestation and fouling can occur was documented. This emphasizes the need for on-going maintenance programs to retard the inevitable biofouling processes and to extend the operating cycle of the well. Finally, the need for public consultation and education, beginning as soon as nuisance bacteria problems with a well are identified, was highlighted. Without public understanding and support, the effort and money spent to rehabilitate and maintain a biofouled well may be wasted.

8 References

Alford, G., Mansuy, N. and Cullimore, D.R. (1989) "The Utilization of the blended chemical heat treatment (BCHT)", in Third National Outdoor Action Conference, Association of Ground Water Scientists and Engineers, Orlando, Florida.

Costerton, W.J. (1990) "Transport of vegetative bacteria and ultramicrobacteria through sand and sandstone under reservoir conditions", in 1st Inter. Symposium on Microbiology of the Deep Subsurface, Orlando, Florida.

Cullimore, D.R. (1981) "The Bulyea experiment - controlling iron bacterial plugging by recycling hot water". Cdn. Water Well, 7, 18-21.

Cullimore, D.R. (1987) "Physico-chemical factors in influencing the biofouling of groundwater", in Inter. Symposium on Biofouled Aquifers: Prevention and Restoration (ed D.R. Cullimore), American Water Resources Association, Bethesda, pp. 23-36.

Cullimore, D.R., and Mansuy, N. (1986) "The control of iron bacterial plugging of a well by tyndallization using hot water recycling". Water Poll. Res. J. Canada, 21(1), 50-57.

Cullimore, D.R., Alford, G. and Morrell, R. (1990) "Bacteriology of groundwater sources and implication for drinking water epidemiology", in N.E.H.A. Midyear Conference, Orlando, Florida.

Hackett, G. (1987) "A review of chemical treatment strategies for iron bacteria in wells". Water Well J., February, 37-42.

Hasselbarth, U. and D. Ludemann (1972) "Biological encrustation of wells due to mass development of iron and manganese bacteria", Water Treatment and Examination, 21, 220-29.

Howsam, P. (1988) "Biofouling in wells and aquifers". J. of the Inst. of Water and Environmental Management, 2(2), 209-215.

Pelzer, R. and S. Smith (1990) "Eucastream suction flow control device: A revolutionary additional element for optimization of flow conditions in bore wells", to be in The Monitoring, Maintenance and Rehabilitation of Water Supply Boreholes and Irrigation Tubewells conference, Silsoe College, U.K.

Smith, S. (1980) "A layman's guide to iron bacteria problems in wells". Water Well J., June, 40-42.

Smith, S. (1981) "Physical control of iron bacteria". Water Well J., June, 60-62.

Smith, S. (1983) "The 'Iron Bacteria' problem: Its causes, prevention and treatment in water wells", in Winter Meeting of the American Society of Agricultural Engineers.

White G. (1972) Handbook of Chlorination. Van Nostraad Reinhold Company, New York.

22 A CASE STUDY – REHABILITATION OF AN IRON BIOFOULED PUBLIC SUPPLY BOREHOLE, OTTER VALLEY, SOUTH DEVON, UK

G.G. BOWEN
Marcus Hodges Environment, Exeter, UK

Abstract
Using both physical and chemical techniques, the rehabilitation of one of South West Water's major public supply boreholes in the Triassic Sandstone aquifer of the Otter Valley was undertaken. A gradual reduction in yield had occurred since 1980 which can be attributed to iron biofouling of the screen, gravel pack and aquifer immediately surrounding the borehole. A programme of airlift, acidisation, oxidisation and finally development pumping was carried out, interspaced with regular performance testing and monitoring. The close proximity of the River Otter and a downstream nature reserve demanded strict safety procedures including the careful disposal of waste chemicals and continuous pollution monitoring. An improvement in yield was obtained and the success of each stage of the project examined.
Keywords: Rehabilitation, Iron Biofouling, Sandstone, Performance Tests, Acidisation, Chlorination.

1 Introduction

A gradual reduction in yield had been identified in two public supply boreholes at Dotton in the Otter Valley in East Devon. Performance testing carried out in 1988 indicated a reduction in yield at both Dotton No. 4 and 5 boreholes, when compared with 1980 levels, with No. 5 showing considerably greater deterioration.

The boreholes were drilled in 1967 into the Otter Sandstone Formation which is part of the Sherwood Sandstone Group (Bunter) of Triassic age.

The Otter Sandstone Formation is the main aquifer within the Otter Valley where it is typically over 100 m thick, comprising fine to medium grained, red-brown, micaceous, variably cemented, ferruginous sands and sandstones. This is unconformably underlain by the Budleigh Salterton Pebble Beds Formation which is generally a loosely cemented meta-quartzite comglomerate.

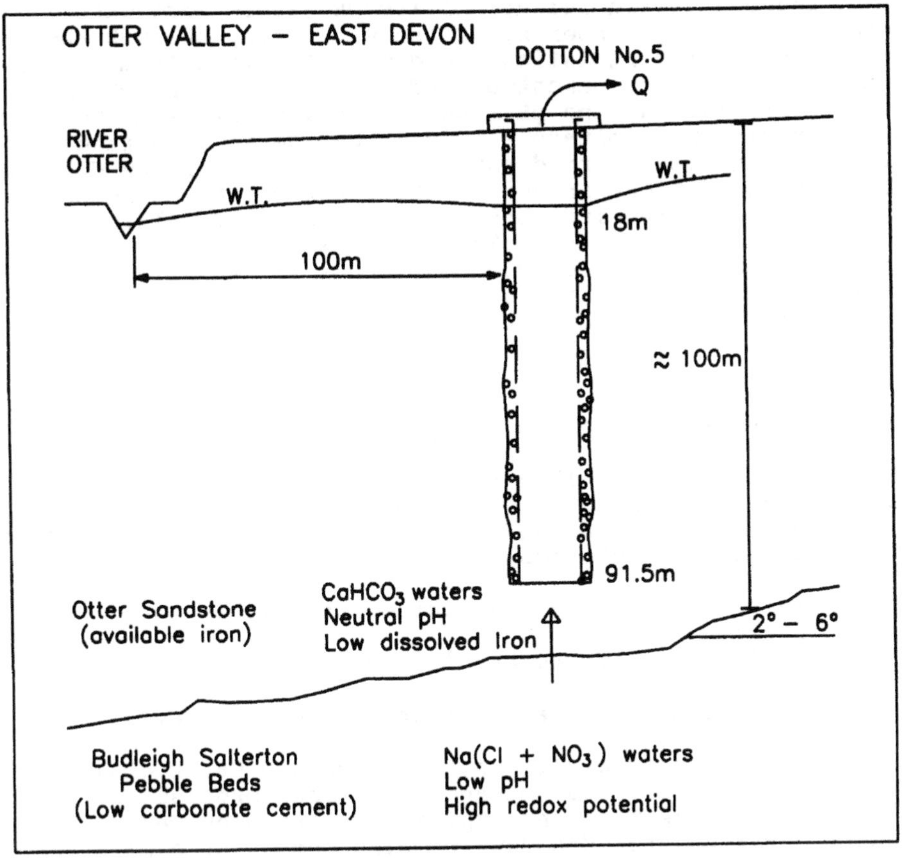

Fig.1. Dotton hydrogeology and geochemistry.

The sandstone has a high porosity and around Dotton also shows high transmissivity typically in the order of 100 - 300 m^2/d and is extensively used for public supply.

The hydraulic continuity between different levels in the sandstone and the underlying pebble beds is unclear and appears to be highly variable throughout the Otter Valley, although generally in the Dotton area the formations are thought to be in continuity. The formation waters are quite different. The Sandstone water reflects the hydrogeological origin, yielding a CaHCO3 water with moderate total dissolved solids (TDS), near neutral pH and low total iron (<20 mg/l) whilst Pebble Bed water has characteristic Na (Cl + NO_3) water with low TDS, low pH and high redox potentials.

One possible cause for the reduction in yield from the borehole is clogging of the the screen, pack and aquifer immediately surrounding the borehole by precipitation of an iron oxide. This may occur where pumping of the borehole induces upward leakage of low pH Pebble Beds water which is grossly under-saturated with iron, through the sandstone, which satisfies the demand for iron from the Fe-coated and calcite cemented sandstone grains, and which then reoxidises as the pH rises to form a colloidal ferric oxyhydroxide precipitate. This mechanism is supported by an investigation carried out by Dr Howsam of Silsoe College to match conditions in the borehole and which was successful in identifying iron bacteria of encrusted Gallionella Ferruginea forming the background to a large iron biofilm mass. A CCTV survey carried out prior to rehabilitation also showed encrustation on the inside of the screen with particularly heavy deposits on the section adjacent to the pump.

2 Refurbishment Techniques

During April and May 1989 South West Water carried out a programme of rehabilitation at Dotton No 5. borehole to improve the yield available for supply. The works included both chemical and physical processes:

An initial airlift at various depths within the borehole.
Injection of acid and pumping to waste in two stages.
Oxidisation using hypochlorite agent.
Pumping at high rates to clear clogged pathways.

The borehole condition, performance and water chemistry were tested before, during and after each process and from this an assessment of the improvement has been made. Both short duration step drawdown pumping tests and simple quick slug tests were used to monitor improvement.

2.1 Initial Airlift

The airlift was carried out at different levels during a 24 hour period. A total of 40 m^3 of brown iron-rich waste was removed to a lagoon. Much iron sludge was removed by this process and is likely to represent most of the encrustation visible from the video survey on the inside of the screen.

2.2 Acidisation

Health and Safety procedures dictated a strictly-applied system of work since extra caution is necessary when carrying out either acidisation or oxidisation processes.

Two separate injections of 5 m^3 and then 8 m^3 of 34% hydrochloric acid were completed over a nine day period. In each case the acid was injected into the borehole quickly at 65 m depth and the borehole sealed. The reaction of the acid with carbonate cement and precipitate caused a rapid build up in pressure within the borehole to a maximum of 35 psi and 55 psi respectively. Venting of the borehole was carried out prior to vigorous agitation by airlift and the acid water mixture was then pumped to waste. Monitoring of the quality of the water removed indicated that the acid in the first injection remained close to the borehole and was quickly spent. However much of the acid from the second injection remained largely unspent reaching some distance from the borehole. Waste acid was neutralised easily within the lagoon and taken to a waste treatment works in Exeter.

2.3 Oxidation - Chlorine Dosing

The chlorination stage was designed to oxidise and to kill any iron bacteria still present after the acid stages. A larger than expected volume of 75 litres of 10% chloros solution was required to maintain free chlorine concentration of greater than 120 ppm for 1 hour and above 1 ppm for 10 hours. The fast fall off in free chlorine indicated that bacteria were still present at this stage.

2.4 Development Pumping

The performance test conducted after chlorination indicated that permeability within the aquifer immediately surrounding the borehole had reduced. It was believed this was caused by residue and fines remaining after acidisation and chlorination which may have blocked fissures and fractures. To remove these and recover permeability the borehole was repeatedly pumped and backwashed at rates of up to 55 l/s over a period of four days. Water discoloration was seen during the first two days, but recovery rates monitored throughout the period showed no significant change.

3 Discussion

A Step test carried out in 1988, prior to the rehabilitation, had shown a significant worsening in yield-drawdown compared with tests done in 1980, although these were poor and difficult to analyse.

The initial airlift removed significant quantities of iron precipitate and almost certainly improved the effectiveness of the subsequent acidisation. It is difficult to assess whether the airlift actually improved the borehole performance or whether reduced cascading and improved techniques provided more accurate data.

From the yield-drawdown analysis the airlift resulted in an overall improvement from 41.0 m to 36.1 m of drawdown when compared with the 1980 figures of 30.8 m for a yield of 2500 m^3/d. Longterm yield drawdown curves are shown in figure 2.

Monitoring of the first injection stage showed the acid to be largely spent and to remain within the vicinity of the borehole. However, the second injection was largely unspent, being diluted as it moved much further from

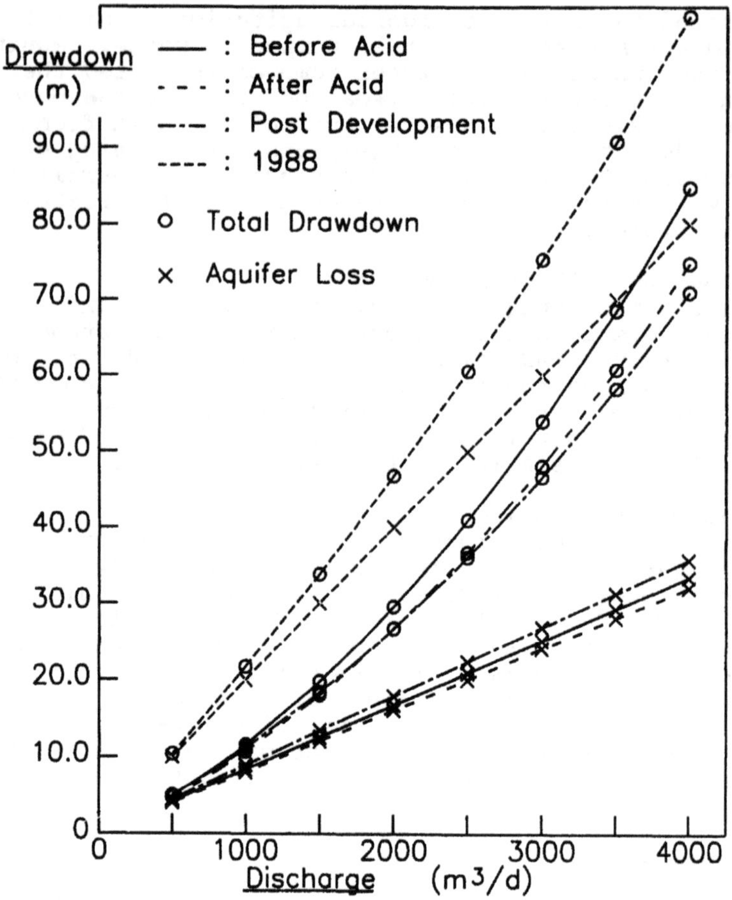

Fig.2. Yield drawdown curves.

the borehole. The movement of the acid further into the aquifer was unlikely to be consistent throughout the length of the screened section and greater preferential movement probably occurred within fissured zones which are common within the Sandstone aquifer. At each stage, the pressure resulting from the reaction of the acid within the sealed borehole was low, rising to a maximum of approximately 55 psi at the top of the borehole. Most of the gas released was carbon dioxide, from the reaction of the acid with both carbonate cement from the sandstone and iron encrustation from the gravel pack and well screen. The low pH and small amount of solid material discharged after the second acid stage indicates that much of the inorganic carbonate precipitate had been removed during the first stage.

The addition of significantly more oxidising agent (sodium hypochlorite) than was theoretically calculated and the quick reduction in free chlorine within the discharge, may be partially due to dilution in the borehole, but may also indicate iron bacteria still present after the acid stages.

The yield drawdown of Dotton 5 was improved at all stages of refurbishment as shown by the long term curves of Figure 2. The total drawdown can be subdivided into losses within the aquifer (BQ) and in the well (CQ^2) and it is therefore possible to identify whether the improvement was in the permeability of the aquifer immediately surrounding the well, or through the cleaning of screen and gravel pack so as to reduce turbulent losses.

The **well loss** was reduced considerably during both the chemical and mechanical refurbishment stages to the extent that losses were similar to 1980 levels. The combination of chemically cleaning the screen and pack of iron precipitate and then the heavy agitation to remove debris, successfully recovered well properties back to 1980 levels.

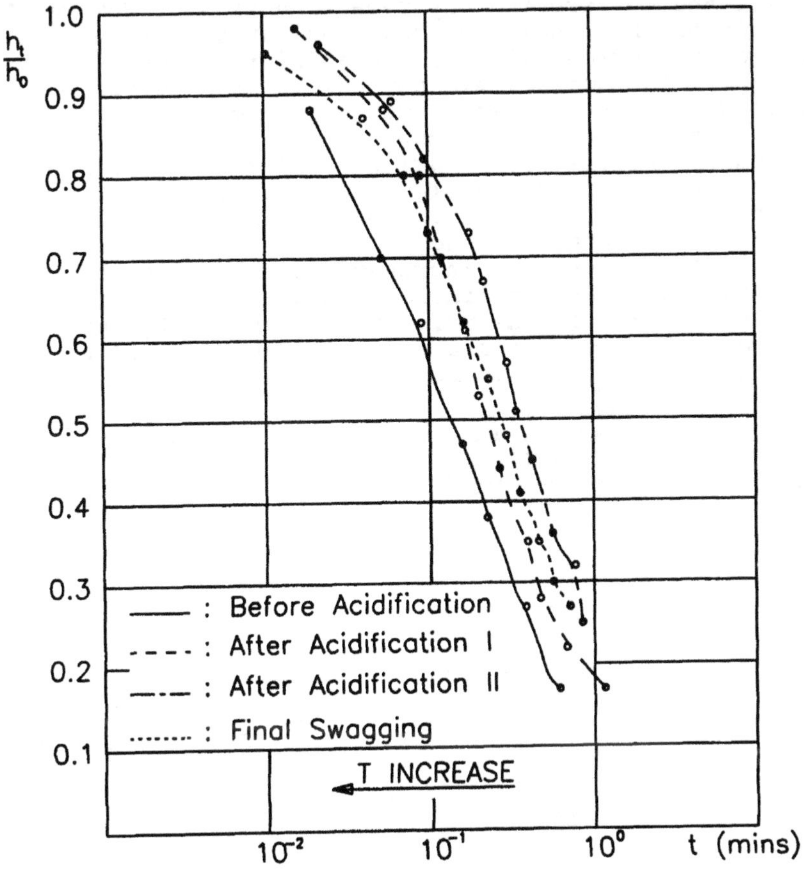

Fig.3. Slug test curves.

Aquifer losses did not improve and the results indicate a slight increase. Information from the chemical refurbishment stages showed that aquifer losses gradually rose, indicating that the transmissivity immediately surrounding the well was reduced (long term effects indicate the reverse). This is supported by the slug test results of Figure 3, which shows the transmissivity to have reduced after each acid injection. This may be a simplification since the test is only accurate to within 10 to 25%.

Discoloration of the discharge from the acid stages and the chlorination stage to dark becoming light brown occurred as iron precipitate and other fines were removed. Perhaps some debris (sludge and fines) may have been left in the aquifer causing a reduction in permeability close to the well. This is supported by results for the final slug test which indicated a small increase in the transmissivity after a period of surging.

The final stage of heavy pumping was designed to pull water further into the aquifer than was possible using the airlift and so removing any remaining fines. However the analysis of the final step test indicates that pumping reduced the permeability of the aquifer surrounding the well still further and did not remove the "debris" as expected, even though water discharged during pumping was discoloured for several days.

Results obtained from various different methods of step test analysis agreed closely, but care is needed in analysing the results for aquifer losses since the trends lie within experimental accuracy.

Well efficiency also increased steadly at each stage (assuming the 1988 unreliable test data is ignored), as the turbulent losses in the well, screen and in the gravel pack were reduced. Results are shown at figure 4 below.

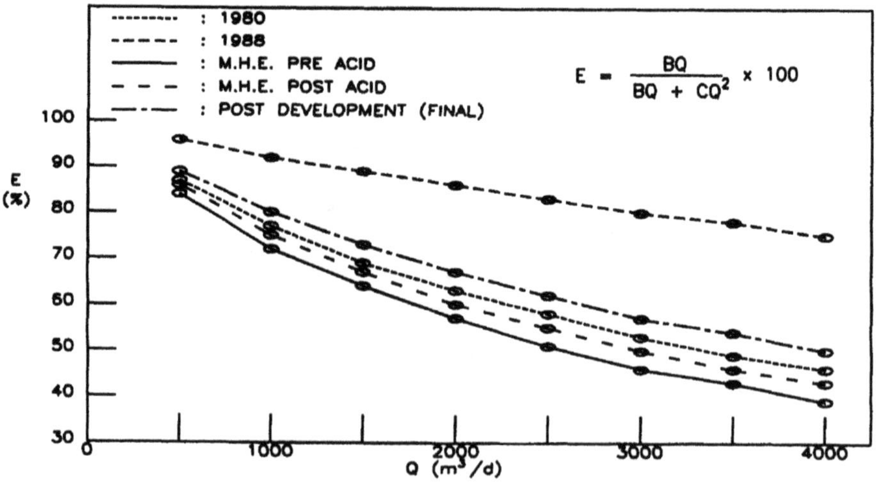

Fig.4. Well efficiencies.

4 Conclusions

1. Dotton No. 5 borehole had become extensively clogged with iron biofouling (Gallionella Ferruginea). This reduced the efficiency of the well. However due to poor test data the original condition was unreliably recorded
2. The initial airlift was successful in removing large amounts of iron precipitate.
The first acid injection reacted fully with the precipi tate and further injections may not have been necessary.
Large quantities of an oxidising agent (chloros) were required to maintain the concentration indicating the presence of bacteria after the acid stage.
Some form of agitation development was required after the chemical stages. However swagging could be more successful than over pumping.
3. Improvement in well performance was due to a decrease in well losses by the cleaning of screen and gravel pack. No improvements were obtained in aquifer properties and indeed the rehabilitation may have reduced permeability immediately surrounding the well.
4. Comprehensive well performance tests are essential to understanding and assessing the improvement. Step tests were successful in identifying trends in well and aquifer performance. Slug tests, although of low accuracy, were quick and easy to perform.
5. Strict safety procedures and extensive monitoring are necessary where chemical rehabilitation is carried out.

5 References

Black, J.H. (1978) The use of the slug test in groundwater investigations. **Water Services**, March 1978.

Brereton, N.R. (1979) Step Drawdown Pumping Tests for the Determination of Aquifer and Borehole Characteristics. **Water Research Centre Technical Report**, TR 103.

Clarke, L.(1977) The Analysis and Planning of Step Drawdown Tests. **QJEG.**, 10, 2, pp. 125-143.

Golleib, O.J. and Blallert, R.E. (1988) Concepts of well cleaning. **Journal American Waterworks Association**, 80, pp. 34-39

Kruseman, G.P and DeRidder, N.A. (1970) Analysis and Evaluation of Pumping Test Data. **International Institute for Land Reclamation and Improvement**, Wageningen, Bulletin No.11.

Skinner, A.C. (1988) Practical Experience of Borehole Performance Evaluation. **JIWEM**, 2, 3 pp. 332-340.

South West Water Services Ltd. (1989) Report of the Refurbishment of Dotton No. 5 Borehole. Unpublished.

Walton, N.R.G. (1982) A Detailed Hydrogeochemical Study of Groundwater from the Triassic Sandstone Aquifer of South West England. **NERC**, Report 81/5 HMSO.

23 A CASE STUDY OF SCREEN FAILURE AND BOREHOLE REHABILITATION IN THE LOWER GREENSAND AQUIFER OF SOUTHERN ENGLAND

D. BANKS
Norges geologiske underesøkelse, Trondheim, Norway

Abstract
The borehole under consideration is in the Lower Greensand at Netley Mill Pumping Station, Surrey. Inspection of very limited data available gave some indication of a possible decline in specific capacity during the 25 years prior to failure. Sudden failure occurred in July 1987 when the bore started pumping sand. CCTV inspection revealed that the bore had partially infilled with sand, and that the screen was encrusted. The infill was removed, and the screen cleaned by wire brushing. Subsequent CCTV and geophysical work showed the likely problem to be the failure of a vertical/horizontal weld junction. The borehole was successfully rehabilitated by the insertion of an interior string of screen with integral gravel pack. Consideration is given to the role of incrustation and corrosion in the borehole. The desirability of colour CCTV logging in such situations, and of the accurate operational monitoring of yield and drawdown in abstraction bores is emphasized.
Keywords: Screen Failure, Rehabilitation, Closed Circuit Television (CCTV), Incrustation, Corrosion, Biofouling, Specific Capacity, Gravel Pack.

1 Introduction

Much time and money may be expended in the construction and initial testing of a groundwater source. There may also be much consultation with hydrogeologists at this stage. The source represents a considerable capital investment on the part of its owner. It is thus surprising that after completion, in most cases, little consideration is given to the fabric and properties of the borehole, and that longer-term communication between hydrogeologists and operators does not exist, at least until a problem occurs. In the British water industry this may partly be due to the "traditional" divide between hydrogeologists and operational staff. The time is ripe for change, however, and already an increased commitment to improved operational monitoring of boreholes is perceived.

The current case study does not "break new ground" but it does provide a concise example of the problems of data availibility for operational sources, and of the methods available to investigate and treat a problem when it arises. The type of failure in this case is considered to be largely "mechanical" in the absence of evidence of any direct involvement of biochemical processes in the problem. Such involvement cannot be ruled out, however, and consideration is given to the possible role of such processes in the borehole in relation to data which suggest a probable decline in yield during the 25 years prior to failure.

2 The site

Netley Mill Pumping Station is located near the village of Shere, Surrey, and consists of seven similar operational water abstraction boreholes and a treatment works, owned and operated by Thames Water Utilities Ltd (Fig. 1). Geologically the site is situated south of the North Downs chalk scarp, on Cretaceous Lower Greensand deposits. The bores take their water from the lower "Hythe Beds" part of the Lower Greensand. Accounts of the geology of the site differ, and according to recent maps (Shephard-Thorn, 1982) the geology is complicated by the Tillingbourne Fault running E-W through the site less than 100 m south of No. 4 borehole. It seems, however, that the aquifer is largely confined or semi-confined, at least to the north of the fault, by the more clayey Sandgate Beds which overlie the Hythe Beds aquifer (south of the fault, some outcropping Hythe Beds are mapped). It also seems that the aquifer has some multilayer character, due to clay layers in the Hythe Beds. One such clayey horizon was recorded at 58 - 61 m depth in No. 4 borehole. Artesian conditions were recorded on completion of this borehole in 1962.

Borehole No. 4 penetrated the entire thickness of the Hythe Beds to the underlying impermeable Atherfield Clay.

The site is adjacent to the River Tillingbourne, and it is possible that some degree of induced recharge via the Sandgate Beds (which are unlikely to be wholly impermeable) may help to support the boreholes' yield. A test in 1981 failed, however, to detect depletion of the river's flow due to increased pumping at the site (Banks & Waters 1988).

Boreholes in the Lower Greensand are renowned for developing problems with incrustation and corrosion. The former is partly due to the frequently high iron content of the water which can precipitate out as iron (III) compounds when mixing with oxygenated water occurs (e.g. in the vicinity of the well where shallow oxygenated groundwater and deeper iron-rich water may mix). Biological activity may "catalyse" or take part in incrustation processes, particularly if increased supplies of nutrient flow are induced by pumping (Cullimore & McCann 1977, Howsam 1988a, Van Beek 1989). Corrosion may be a result of the often electrochemically aggressive nature of confined Greensand groundwater, but it too can be enchanced by microbial activity. Howsam (1988a) notes the potential for anaerobic, corrosive microenvironments to establish themselves beneath encrusting layers. Hutchinson & Ridgway (1977) have made a similar observation when considering iron water distribution pipes.

The Hythe Beds themselves are a somewhat variable deposit consisting of glauconitic sands with varying proportions of finer grade material and clay horizons. The degree of cementation is also variable but in many instances it is sufficient to impart a large component of "fissure flow" character to the aquifer, rather than homogeneous intergranular flow.

The term "Hythe Beds aquifer" is usually taken to include the thinner Bargate Beds sand sequence, which occurs stratigraphically immediately below the Sandgate Beds, and above the Hythe Beds proper.

3 The borehole failure

Borehole No. 4 was drilled in 1962, and details of its construction are summarized in Fig 2. In July 1987, it began pumping sandy material and had to be shut down, as filters were becoming clogged within about 10 minutes of being backwashed. The material was described as being "black and sludge like, only partially comprising normal sand". Howsam (1988b) has suggested that the colour may have been due to iron sulphide or to iron or manganese oxides, and notes that a higher than normal level of manganese in the water was reported at the time of failure.

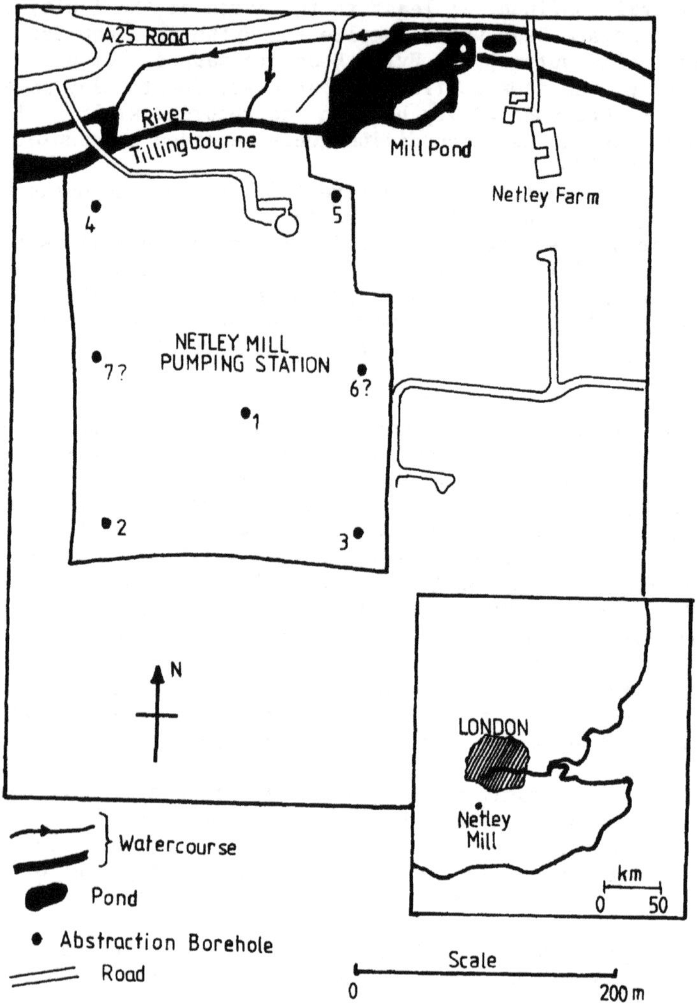

Fig. 1. Plan and location map of Netley Mill site.

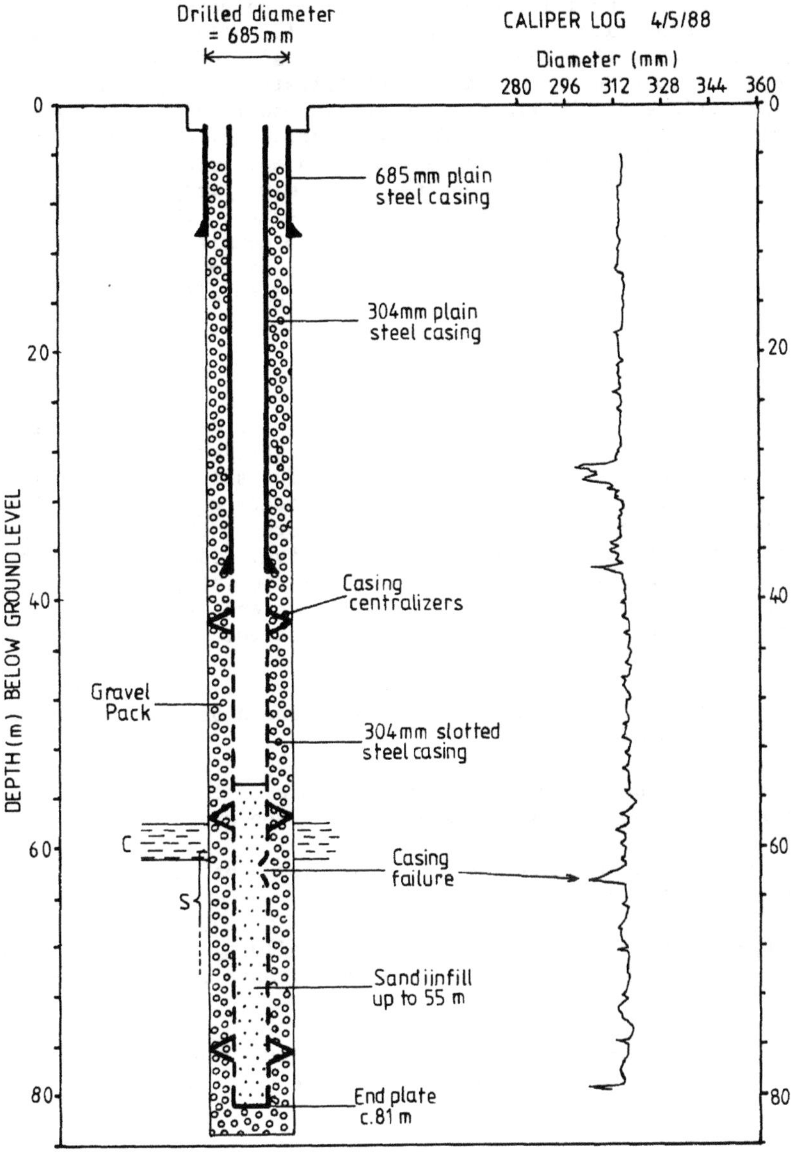

Fig. 2. Schematic diagram of construction of Netley Mill No. 4 borehole
C= clayey layer (possibly semi-confining)
S= collapsing sand strata encountered during drilling
Ground level= ca. 76.0 m o.D.

The pump and rising main were removed from the borehole and examined. The rising main was found to be significantly encrusted with a fine orange deposit (Fig. 3). In places, corrosion of the underlying metal appeared to have taken place below "blisters" in the deposit. On the inside surface of the rising main was a deposit of a fine pale yellow material, which became powdery when dry (Fig. 4). This pale yellow deposit had a cationic content of iron and calcium in the ratio 35 to 65% respectively "as carbonate" , with minor manganese, magnesium and acid-insoluble components. The orange deposit on the outside of the rising main was not analysed, but is likely to consist of various ferric iron compounds (hydroxides, carbonate), maybe calcium carbonate and possibly also biological/organic components (biofilm).

4 Initial investigation

Following the removal of pump and rising main, a closed circuit television (CCTV) survey was carried out by the Water Resources Section of Thames Water using a black-and-white camera. The survey showed that the borehole had filled up with material to 55 m below well top (original as-constructed depth was 81 m). The exact nature of the material could not be discerned from the CCTV survey. The survey did not locate the reason for the ingress of material, though it did indicate a high degree of incrustation on the casing and the screen, and also cast doubt on the integrity of some of the joints between casing sections (Banks & Waters 1988).

It was felt that the excessive quantity of observed incrustation was undesirable, and that its immediate removal should be undertaken for the following reasons:

i) It might be harbouring micro-environments of corrosion, which had already been observed on the rising main - see above.

ii) Removal of incrustation would allow close inspection of the screen for signs of failure.

iii) Apart from the question of borehole failure, the incrustation may have been impairing the borehole performance by clogging of slots.

The removal of the infilling material was thus undertaken, together with the removal of incrustation by wire-brushing methods.

5 The rehabilitation

The infill material was bailed out from the bore and the original depth of 81 m restored. The infill material was found to be mainly Greensand. Howsam (1988b) reports that manganese was only a minor component, though the top few metres of the infill were relatively rich in iron. The borehole screen was cleaned by wire brushing, and a few days later a CCTV survey was undertaken using a colour camera. The survey revealed that wire brushing had been a rather effective means of removing

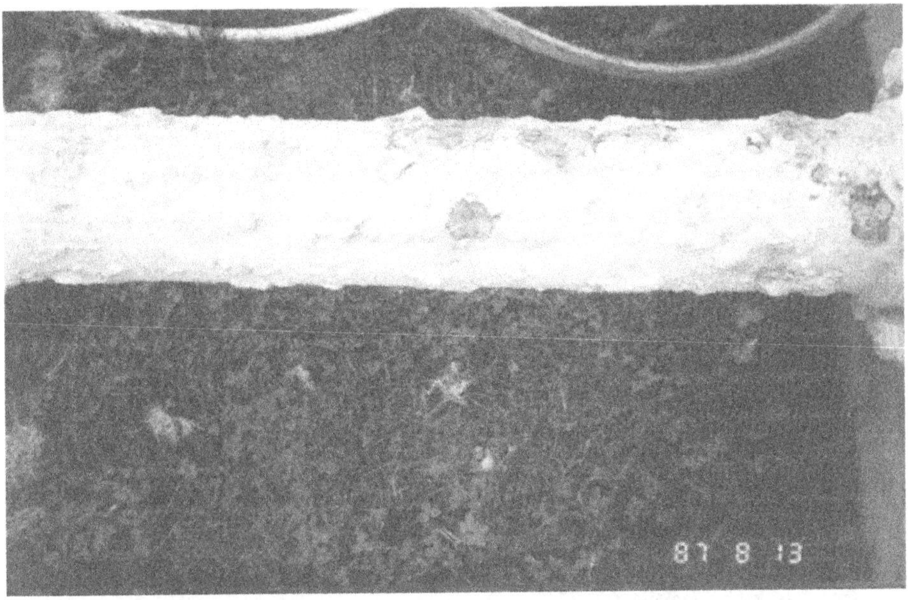

Fig. 3. Incrustation on rising main from Netley Mill No. 4 borehole

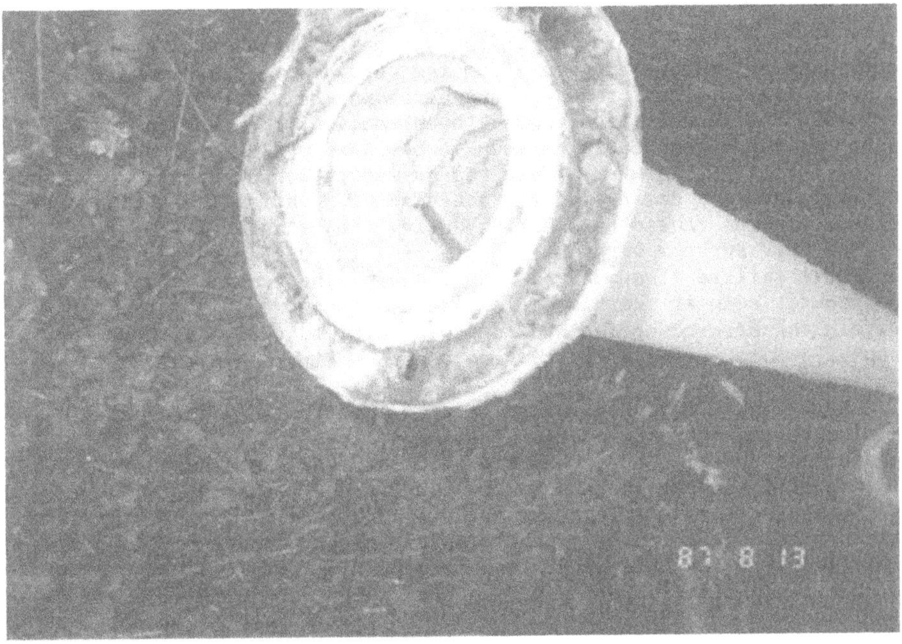

Fig. 4. Encrusting deposits within rising main from Netley Mill No. 4 borehole

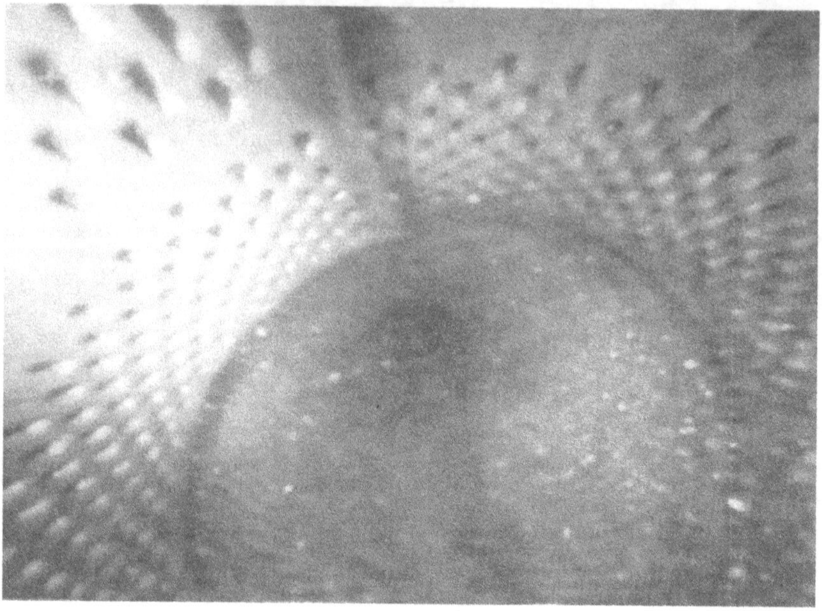

Fig. 5. CCTV picture showing screen failure. Netley Mill No. 4 borehole

incrustation from the inside of the screen (although such a method can obviously not remove clogging from the outside of the screen, the gravel pack, or the formation, and hence may be rather cosmetic in nature). The incrustation having been removed, the vertical bridge slots of the screen could be seen to be largely clean and undamaged. In a few locations, however, there was an indication of a little corrosion of the slots and casing joints, and also of occasional clogging of the slots, perhaps caused by smearing during the brushing exercise.

The CCTV surveys were able to pinpoint the probable cause of the borehole failure at around 62 m down. Here, the junction between a horizontal seam at the base of a screen section, and a vertical seam within the screen section had completely failed (Fig. 5). Other horizontal joints in the vicinity also seemed a little dubious (Banks & Waters 1988).

Geophysical logging was also carried out. The only parameter to give a clue to the failure location was the caliper log which showed a pronounced spike at 62 m (see Fig. 2). This could not have been interpreted without the aid of the CCTV survey, however.

Following these surveys, a string of 200 mm I.D. plastic "Hagusta" slotted and plain lining, with integral (glued-on) gravel pack, was inserted into the hole within the original lining.

The rehabilitated bore was then test pumped.

6 Test pumping

This took the form of six continuous step tests. Steps 1,2,3 and 6 were of two hours' duration; step 4 was 24 hours and step 5 was 22 hours. The results are shown graphically in Fig. 5. The other boreholes on the site were pumping at their "normal" output during testing.

A maximum rate of 1310 m^3/d was achieved with a 2-hr corrected drawdown of 20.79 m. Even at this rate the 2-hr yield-drawdown curve (Fig. 6) is still fairly linear and does not show any marked tendency towards increasing well-loss at high rates.

7 Continuous decline and sudden failure - an assessment of available information

In order to try to understand the reasons for the borehole's failure, and to ascertain whether incrustation/clogging had led to declining borehole yield, a certain amount of desk study was undertaken to assess available information. When considering a public supply source in Britain the sources of information are four-fold.

i) British Geological Survey archives at Wallingford, Oxon, U.K. or their published well inventories. Information often relates to initial drilling and testing.

ii) Drillers. Information also relates to initial drilling and testing.

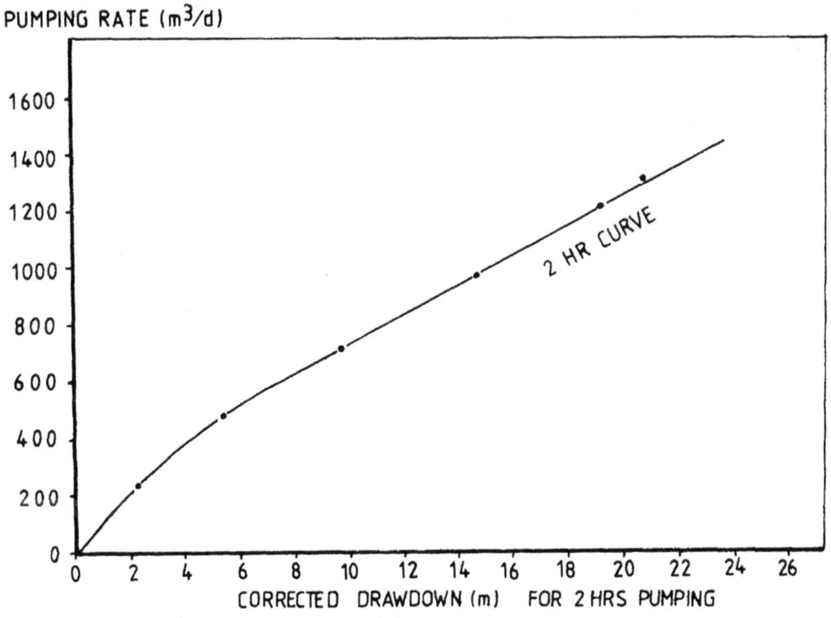

Fig. 6. Yield-drawdown curve after rehabilitation of No. 4 borehole

iii) Regional Water Authority p.l.c./ National Rivers Authority regional hydrogeology/water resources groups. Information usually relates to specific investigatory periods of test-pumping.

iv) Water Authority / Company operational staff. Such information relates to day-to-day operational monitoring of abstraction bores, and may consist of water quality, yield and pumping water level (PWL) data. In the case of yield and PWL, data may be very variable in quality. Problems may arise due to lack of commitment to monitoring, or lack of accuracy of instrumentation.

In the case of Netley Mill No. 4 bore, information was available from all four sources.

7.1 Borehole construction

Information from the drillers suggests that No. 4 borehole experienced a number of problems during construction. Reverse circulation (RC) drilling methods were used to 26 m, when deviation from vertical necessitated backfilling with hardcore to 21 m, and continuing by percussion methods to 29 m. RC methods were then resumed, but below 60m collapsing sand strata (possibly related to artesian inflow below the semi-confining clay horizon between 58 and 61m ?? -Howsam 1988b) caused difficulties. Thus, 608 mm diameter temporary casing was inserted, and the hole completed by percussion. On completion, the temporary lining reached to full depth. The 305 mm plain & slotted casing string was inserted, the gravel pack was emplaced in the annulus between the 305 mm string and the 608 mm temporary casing, and the 608 mm casing was withdrawn. The drillers believe that the gravel pack installation and temporary casing removal were carried out in such a way as to ensure complete gravel pack emplacement. However, with reports of collapsing sand strata and possible artesian inflow at depth, there could have been the possibility of development of large washouts in the formation during drilling, and some doubts as to the integrity of the subsequent gravel pack.

The slotted section of the 305 mm casing string was of stainless bridge-slotted steel. It is acknowledged as possible by the drillers that mild steel was used, however, for the plain sections of the string. (This cost-saving practice, it should be pointed out, is no longer recommended by the drillers; wire-wound stainless steel screen being the preferred option in the Lower Greensand). Such a practice would result in a galvanic cell between the two metals in the string, but in this case would probably lead to enchanced corrosion of the mild, rather than the stainless, steel in the vicinity of the junction between the two metals, as the former is more "active" in the galvanic series.

7.2 Borehole efficiency

Information on yield-drawdown relationships from all sources is summarized in Table 1, and graphically in Fig. 7.

The information from the published **B.G.S. Metric Well Inventory** (B.G.S., 1977), although useful, is incomplete, as it does not specify duration of pumping, or pumping in other borholes on the site. For example, the artesian RWL in April 1962 probably occured when there was

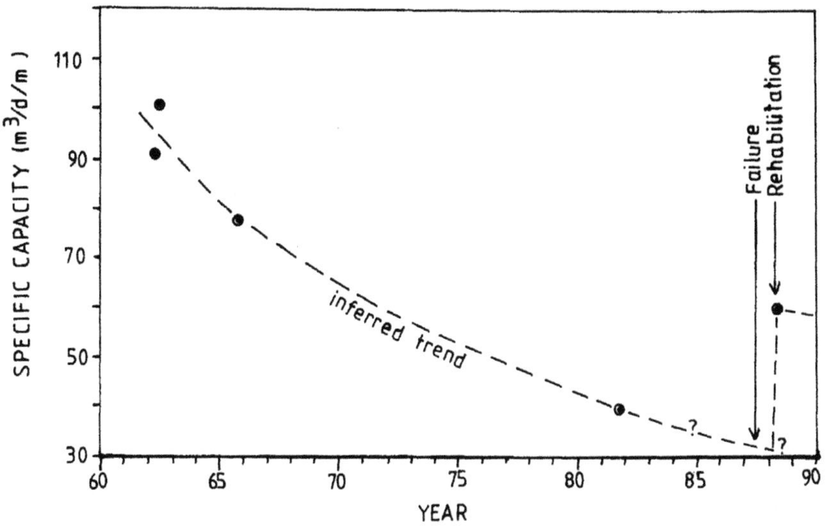

Fig. 7. Trend in specific capacity. Netley Mill No. 4 borehole (Calculated at c.1100-1300 m^3/d)

no other pumping on the site, whereas that of May 1962 may be lower due to the pumping of other bores.

The Water Resources Section of Thames Water possessed records of a combined test in October 1981 where all seven boreholes were pumped in connection with a licence increase. The results show that for a yield of 11-1200 m^3 the PWL in bore No. 4 was 32.9 to 34.8 m below ground level (bgl) with all six other boreholes pumping. After a given period, boreholes 2, 3, 4 and 5 shut down and numbers 1, 6 and 7 continued pumping. In borehole 4 after this partial shut-down the rest water level (RWL) was 4.6 m bgl. At face value this indicated a drawdown of up to 30.2 m for a rate of 11-1200 m^3/d (although this drawdown may be slightly overestimated as boreholes 2, 3 and 5 switched off at the same time as borehole 4).

Operational staff were only able to provide limited information on yield prior to failure. PWL was not monitored. The pre-failure average daily output of the borehole was established as 767-841 m^3/d in 1987. Abstraction figures from a few years earlier are summarized in Table 2 as the daily average for each month, averaged over the year in question. The data is also summarized in Fig. 8. It is not known whether the observed decline in yield is related to the declining efficiency of the borehole, or to a simple decrease in demand from it because no PWL data was available. Furthermore it is known that in both boreholes 4 and 5 pumps were lowered by some 6 m in 1979 to "increase output". Again it is unknown whether this was done to achieve an absolute increase in yield or to try to restore yield after a period of decline. The lowering of the pumps **may** have been related to declining PWL due to reduced efficiency.

Date	Rest W.L (m.O.D)	Yield (m^3/d)	Pumping W.L. (m.O.D)	Drawdown (m)	Specific capacity $m^3/d/m$
4/1962 (a)	c.76:80[x]	1310	62.40	c.14.4	91
5/1962 (a)	75.20	1090	64.38	c.10.8	101
11/1965 (a)	?	1090	61.34	c.14	78
10/1981 (b)	71.4	c.1100-1200	41.2-43.1	28-30*	37-43
1987 (c)	?	c.767-841	?	?	?
Post-Rehabilitation (5/1988)		1210		19.31+	63
		1210		c.20.5++	59

x Artesian (above ground level)
* Possibly slightly overestimated - see text
\+ 2 hr drawdown
++ Estimated 48 hr drawdown

Ground level = c.76.0 m.O.D
m.O.D = m above Ordnance Datum
W.L. = Water Level

Sources: (a) = B.G.S. Metric Well Inventory - Sheet 285
(b) = Thames Water / N.R.A. Pumping Test Achieve
(c) = Information from Operational Staff (average daily output)

Table 1. Historical yield-drawdown relationships in Netley Mill bore no. 4

Table 2: Trend of daily average output for Netley Mill Borehole 4

Year	Average daily mean yield m^3/d*
1983	961
1984	939
1985	884
1986	843
1987	817

* Defined as the average value over the year of mean daily yield for each month, excluding non-operational days/months

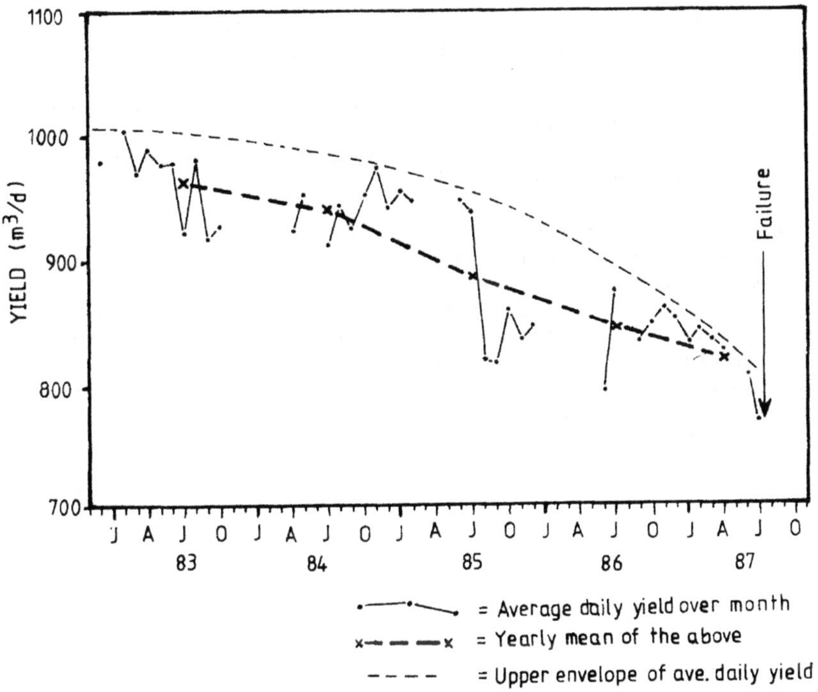

Fig. 8. Trend of daily average output for Netley Mill borehole 4 (N.B. non-operational days/months excluded for calculation of average)

There are also reports of various pump repairs and replacements in this and other boreholes on the site. Might these have been related to incrustation or corrosion problems?

In summary, the available evidence suggests a decline in specific capacity (from 101 $m^3/d/m$ in 1962 to around 40 $m^3/d/m$ in 1981, and possibly lower still by 1987), though it should be noted that some of the data is a little open to question due to lack of supporting information. Specific capacity can be affected by factors other than impaired borehole efficiency e.g. pumping elsewhere on site or seasonal effects. Assuming, however, that the decline was due to decreasing borehole efficiency, it seems that the rehabilitation operation was able to restore part, but not all of the original capacity.

7.3 Water quality

Howsam (1988b) studied the water quality data available from operational staff and notes the following:

i) Content of manganese in water was higher than normal at the time of failure.

ii) Electrical conductivity of abstracted water from No 4 bore has shown a rising trend from around 280 µS/cm in 1980 to 330-340 µS/cm in 1983. Other boreholes on site show similar trends. Such a trend may indicate increasing infiltration of river water. It is also noteworthy that the total hardness, and alkalinity, of abstracted water is highest in boreholes 4 and 5, near the river, and decreases considerably with distance from the river in the other boreholes. This can be taken as further evidence of infiltration of river water (which is itself likely to be hard, being partly fed by springs from the Chalk limestone aquifer).

iii) Routine standard bacteriological plate counts from several of the site's boreholes, including No. 4, showed occasional high values (>300 colony-forming units per ml after 3 days at 22°C) against a background count of 0 - 10. As Howsam (1988a) has pointed out, such excessive values are often assumed to be due to sampling/laboratory error, but they could alternatively represent true values. Such values could be obtained if sampling coincided with a time when a piece of biofilm-incrustation had just sloughed off from a surface within the bore into the water.

iv) In water samples collected during the removal of infill from the borehole, significant numbers of sulphate reducing bacteria were detected. These are often related to subsurface corrosion processes beneath an encrusting layer (Hutchinson & Ridgway 1977, Howsam 1988b).

Typical ranges for various recorded chemical parameters in No 4 bore are shown in Table 3.

Table 3. Range of values recorded for chemical parameters in borehole No.4 during the period 16/4/80 - 7/6/83

Parameter	Units	Range of values
Conductivity	µS/cm	275 - 360
pH		6.4 - 7.0
Total Hardness	mg/l $CaCO_3$	112 - 150
Alkalinity	mg/l $CaCO_3$	74 - 96
Chloride	mg/l	16 - 24
Iron	mg/l	0.08-0.82 (typically c.0.3)
Manganese	mg/l	0.01-0.06 (typically c.0.04)

8 Synthesis

From the available information, one can construct a plausible picture of processes within the borehole.

i) A long 25 year history of declining specific capacity and declining borehole efficiency.

ii) A sudden mechanical failure of borehole construction leading to ingress of formation material.

The two may not be entirely unrelated.

8.1 Long-term decline

Yield and drawdown information indicated that a decline in efficiency has occurred. This may be due to clogging by either biofouling / incrustation or particle redistribution in the formation or gravel pack. Evidence of incrustation and subsurface corrosion is provided by the CCTV survey and direct observation of the rising main. Evidence for microbial activity in both processes is summarized in section 7.3. Microbial activity may have been enhanced by increasing proportions of river infiltrate, and hence of nutrients, during the course of pumping. Clogging appears to have reduced the borehole's specific capacity by 60% between 1962 and 1981 (and probably more by 1987).

8.2 Sudden failure

Evidence for the cause of sudden failure is not as clear. The failure of the screen seams may have been partially caused by any of three factors:

i) Possible defects in construction/welding of seams and joints.

ii) Corrosion of seams, possibly aided by microbial activity in anaerobic environments beneath layers of incrustation. Welds and joints, being irregularities in a metal surface, may be particularly susceptible to corrosion, due to galvanic activity.

iii) Incomplete installation or incorrect grade of gravel pack, leading to ingress of sand and thus to cavities behind the screen, possible instability and collapse. Such collapse could induce screen failure. Ingress of sand and formation of cavities could also have been caused by locally increased entrance velocities at one part of the screen due to heavy incrustation of other parts.

It is interesting to note that Bouwer (1978) envisages a combination of incrustation and corrosion as important factors in screen failure.

8.3 Rehabilitation

This appears to have been achieved successfully by installation of an interior string of casing with integral gravel pack.

An increase in specific capacity of 50% (from 40 to 60 $m^3/d/m$) over the 1981 value occurred, presumably as a result of wire-brushing the interior of the screen. The value is still only 60% of the original specific capacity, however. This is not surprising as there is also likely to be clogging of the exterior of the screen, the gravel pack, and the formation, which could not be removed by wire brushing.

It remains to be seen whether the specific capacity of the borehole will be maintained as the annulus between the new Hagusta screen and the old 305 mm screen fills with sand (as it presumably will). It is also likely that, because no biocidal treatment was employed during rehabilitation, biofouling/incrustation will re-establish itself with time on the old, and possibly the new screen. Due to the emplacement of the new interior string of screen it will now be very difficult to remove incrustation from the original screen or beyond.

8.4 Alternative rehabilitation methods

One must consider the other available options to rehabilitate such a borehole with screen failure, the chief of these being to remove the original screen and gravel pack entirely. The hole would have been supported by surcharge of water and the screen repaired by welding stainless steel plates around failed areas. Such a method may have resulted in a greater increase in specific capacity after renovation, as it allows for cleaning the inside and outside surfaces of the screen, and the installation of a clean gravel pack. Such a method is expensive, however, and carries certain risks. The hole may collapse while the screen is being repaired; the welding on of plates over failed areas induces further irregularities in the metal surface and enhanced opportunity for galvanic corrosion. Finally, although visible holes can be patched up by such a method, there may be plenty of other areas where corrosion has occured to some degree. It may only be a matter of time, after reinstallation of the screen, before corrosion at these other sites leads to another screen failure. Despite these criticisms, it appears that such a method has been used with success in other Lower Greensand holes in South England.

The only entirely satisfactory alternative option would have been to replace the failed screen completely, by a new, preferably wire-wound, screen with a fresh gravel pack. The cost of such an operation, needless to say, would have been considerably higher than the method adopted.

9 Recommendations

In the light of the discoveries of the case study, and of the uncertainties still surrounding the borehole and the site as a whole, the following recommendations are made:

i) That every operational public supply borehole should be regularly monitored for output and pumping water level by an accurate, well-calibrated method. There should be manual dipping access to enable measurements to be checked.

ii) That every such borehole should have a planned programme of step testing performed at regular intervals to quantify any decline in efficiency.

iii) That consideration be given to a planned programme of carrying out CCTV surveys in vulnerable boreholes at regular intervals to check for incrustation, corrosion, and integrity of the screen.

10 Acknowledgements

The above investigation was carried out while the author was employed by Thames Water Authority, and he is most grateful for the information supplied by all staff involved in the Netley Mill incident. The views expressed in this article are purely those of the author, however, and do not necessarily represent those of the Water Authority or its successor bodies, the National Rivers Authority (N.R.A.) or Thames Water Utilities Ltd.

Thanks are due to Alan Waters of N.R.A. Thames Region for performing the black and white CCTV surveys, Telespec Ltd. of Send for performing the colour survey, and Mr. E. Ainscough of George Stow & Co. Ltd. for information concerning construction of the borehole.

A debt of thanks is particularly owed to Peter Howsam of Silsoe College, who acted as consultant to Thames Water in this study. He has kindly permitted the author to use much of the data he collected during his work, and also some of his interpretations thereof!

11 References

Banks, D. and Waters, A.J. (1988) The failure, rehabilitation and test pumping of Netley Mill Pumping Station borehole No. 4, 1987/1988. **Thames Water (Water Resources)** Internal Report I.R.157 (unpubl.).

B.G.S. (1977) **Record of wells in the area around Aldershot: inventory of one-inch geological sheet 285, new series.** Metric Well Inventory, British Geological Survey, 1977.

Bouwer, H. (1978) **Groundwater Hydrology**. International Student Edition, Mc Graw-Hill Kogakusha, Tokyo (see p 193).

Cullimore, D.R. and McCann, A.E. (1977) The identification, cultivation and control of iron bacteria in groundwater, in **Aquatic Microbiology** (eds. F. A. Skinner & J. M. Shewan), Academic Press, New York, pp 219 - 261.

Howsam, P. (1988a) Biofouling in wells and aquifers. **Journal of the Institution of Water and Environmental Management**, Vol 2, No. 2, 209 - 215.

Howsam, P. (1988b) The possible causes for the failure of a joint in the lining screen of No. 4 borehole, Netley Mill. Report to Thames Water, S & W Division (unpubl.).

Hutchinson, M. and Ridgway, J.W. (1977) Microbiological aspects of drinking water supplies, in **Aquatic Microbiology** (eds. F. A. Skinner and J. M. Shewan), Academic Press, New York, pp 179 - 218.

Shephard-Thorn, E.R. (1982) **Geological Survey of Great Britain, Sheet TQO4NE.** Resurveyed at 1:10000 scale (north of northing 46) in 1982. British Geological Survey 1982.

Van Beek, C.G.E.M. (1989) Rehabilitation of clogged discharge wells in the Netherlands. **Quarterly Journal of Engineering Geology**, 22, 75 - 80.

24 THE ALLUVION RIVER COMPLEX OF THE LOIRE, FRANCE

G. GOUESBET
Centre Régionale des Pays de l'Ouest, Compagnie Générale des Eaux, Angers, France

ABSTRACT

In addition to a theoretical presentation of clogging as a phenomenon, this paper gives the results of observations on the operating of 2 catchment wells with horizontal collector drains on the bank of the river Loire, anti-clogging tests, and currently used cleaning techniques.

I INTRODUCTION

The production of potable water from catchment wells with radially disposed horizontal collector drains is a current practice all along the Loire. Compagnie Générale des Eaux operates several catchment fields situated between the towns of Saumur and Nantes, with flow rates ranging between 3,000 and 50,000 m³ daily. Certain of these structures are affected by heavy clogging, and sometimes very quickly. Since 1978, we have been engaged in a theoretical study of clogging, and experiments in the field that have enabled us to operate the catchments to suit our needs.

2 GENERAL REMARKS

We briefly call to mind the different types of aquifers and catchment systems. We then proceed wth a summary description of the operating of alluvial water-bearing layers on the Loire and, more particularly, the running of the Montjean and Coutures sites, forming the subject matter of this article.

2.1 Types of aquifer

Aquifers are classified as:

- confined aquifers
- free-flowing aquifers
- alluvial aquifers

2.1.1 Confined aquifers

These are aquifers that have no natural point of discharge: the groundwater is held captive (fig.1)

2.1.2 Free-flowing aquifers

These are aquifers where the water may or may not surface, depending on their depths. (fig. 2).

2.1.3 Alluvial aquifers

These are groundwater tables of which the distinctive feature is that they are quite closely connected hydraulically with a watercourse. (fig.3).

This paper is concerned with the latter type of groundwater resource.

2.2 Types of catchment system

strainer wells
weep hole wells
radially disposed horizontal collector-drain wells

2.2.1 Strainer wells

These wells consist of a string of blind tubes and strainer tubes (fig.4).

2.2.2 Weep hole wells

These consist of a conventional casing string except that they have peripheral perforations, each hole being fitted with a pipe sloping towards the bottom and on the outer face. (fig.5).

2.2.3 Wells with radial type horizontal collector-drains.

Here again casing is of the conventional type but is fitted at one or more levels with drains. These collector-drains are pushed horizontally into the ground from the well shaft. (fig. 6).

It is precisely on this kind of well that the clogging study was performed.

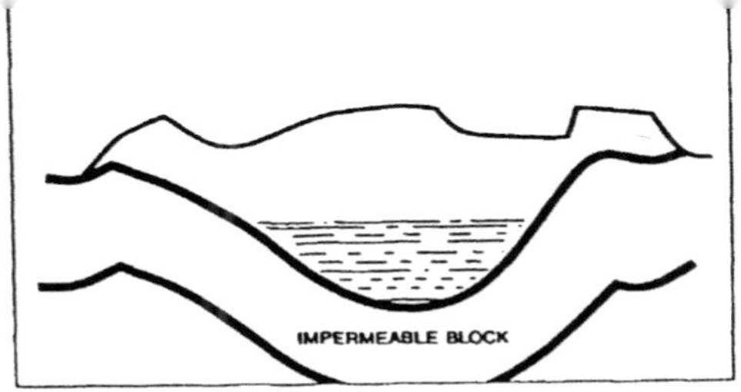

Fig. 1: Confined aquifer

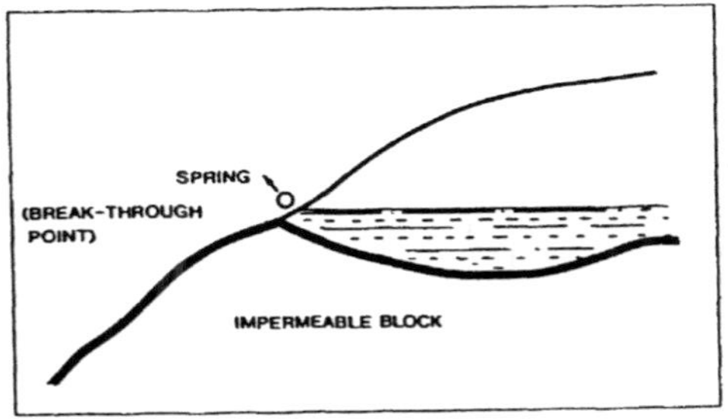

Fig.2: Free flowing aquifer

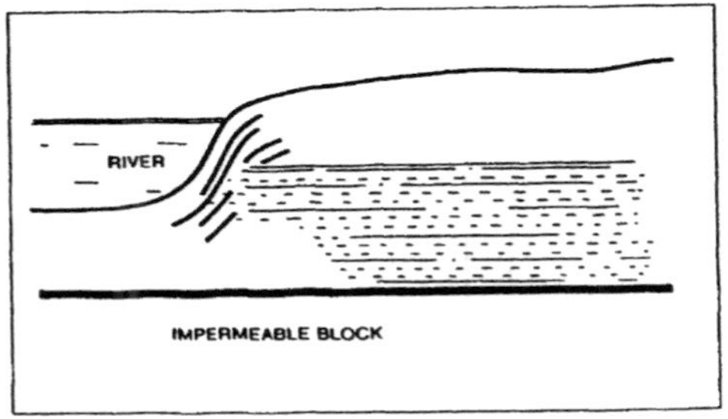

Fig.3: Alluvial aquifer

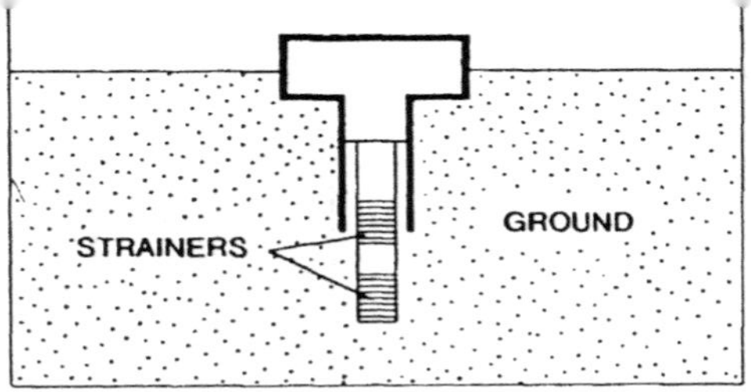

Fig.4: Strainer wells

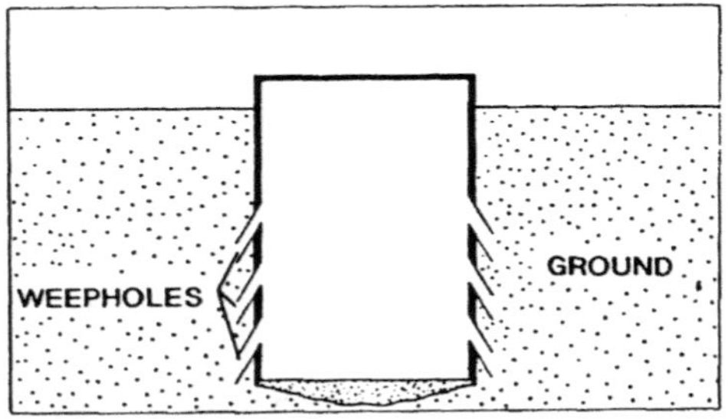

Fig.5: Weephole wells

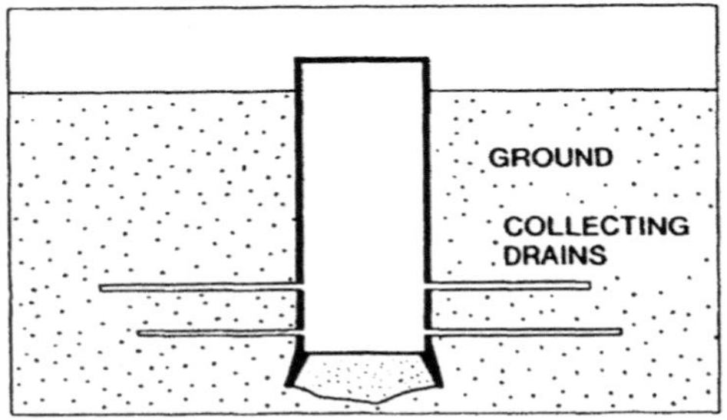

Fig.6: Well with horizontal radiating levels) collecting drains(2 levels)

2.3 Operating the alluvial aquifer of the Loire river

2.3.1 Catchments on the alluvial water table between Saumur and Nantes

2.3.1.1 Location on the Loire between Saumur and Nantes, including 5 main catchment sites on the alluvial layer. (fig 7).

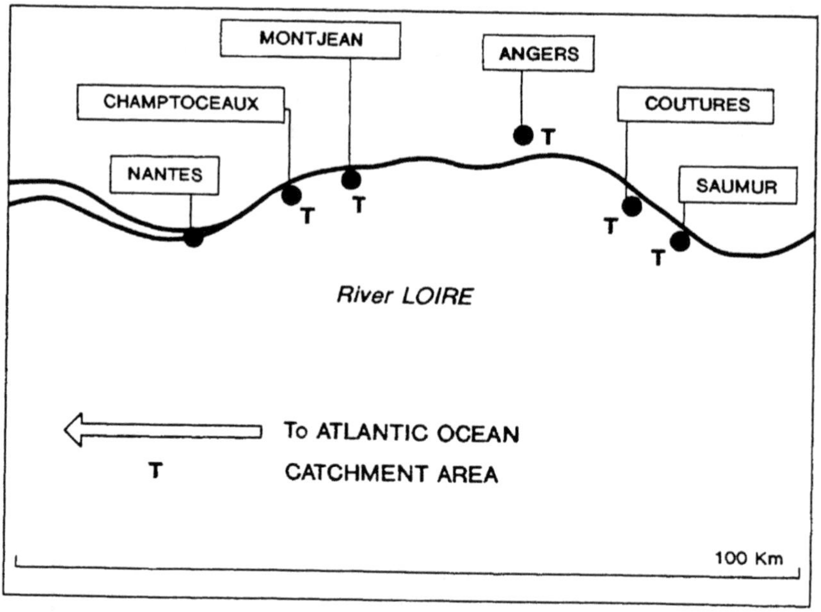

Fig.7: The 5 main catchment sites on the Loire alluvion layer between Saumur and Nantes

2.3.1.2 Production

The daily output of these sites are (or were in the case of Angers) as shown in table I

TABLE I

CATCHMENT	PRODUCTION (m^3/day)
Champtoceaux	3,000
Montjean	50,000
Angers	40,000 (formerly)
Coutures	9,000
Saumur	15,000

3 of these sites are managed by Compagnie Générale des Eaux i.e., Champtoceaux, Montjean and Coutures.

2.3.1.3 Clogging

Each of these catchments suffers operating problems due to clogging. The most severely affected are Coutures and Angers but also Montjean and Saumur to a lesser extent. Champtoceaux is only slightly concerned.

The sites dealt with below are Montjean and Couture.

2.3.2 Montjean wells

2.3.2.1 Location of wells

4 out of the 9 wells sunk are still in production and 2 have clogging problems: wells Nr 2 and Nr 8. They are all located within a rectangle measuring 160m x 300 m (fig.8).

2.3.2.2 Description of wells

Tables II and III give a summary description of the Montjean wells:

TABLE II

GENERAL DATA CONCERNING MONTJEAN WELLS

well Nr	yr.built/ altered	type	distance from LOIRE	present status	nominal output (m^3/h)
1	56	weep hole		dismantled	
2	56/78/86	weep hole/ drains	110m	working	600
3	56	weep hole		dismantled	
4	56	"	160m)	closed	
5	56	"	110m)	due to	
6	56	"	60m)	clogging	
7	65/77	drains	60m	working	500 (800
8	72/78/86	"	60m	"	800
9	87	"	70m	"	800

TABLE III

CHARACTERISTICS OF MONTJEAN WELLS Nr 2 and Nr 8

well	diameter of well (m)	depth	total drain length (m)	Number of drains (m)	drain characteristics
2	2	16	67 (78)	3 (78)	3 steel chased ribs.ID:200mm
			152 (86)	6 (86)	3 st.steel oblong peforations ID: 200mm
			67 (72)	3	4 steel oblong perforations ID: 230 mm
8	3	15	67 (78)	3	1 st. steel oblong holes ID: 300 mm
			152 (86)	5	

() year constructed or altered

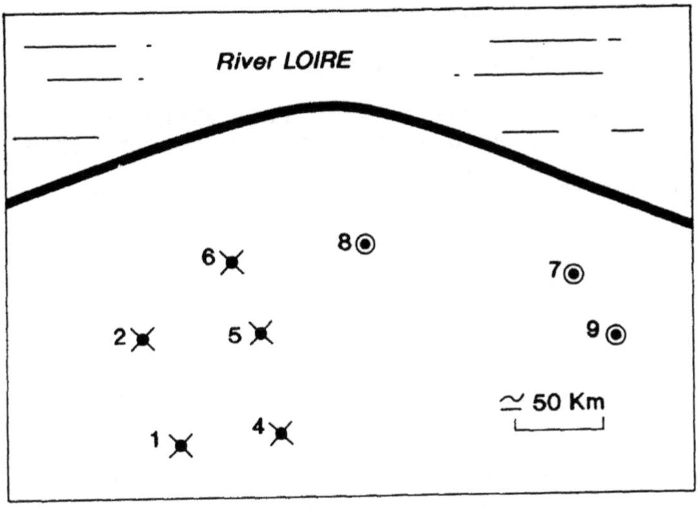

Fig.8: Layout of Montjean wells

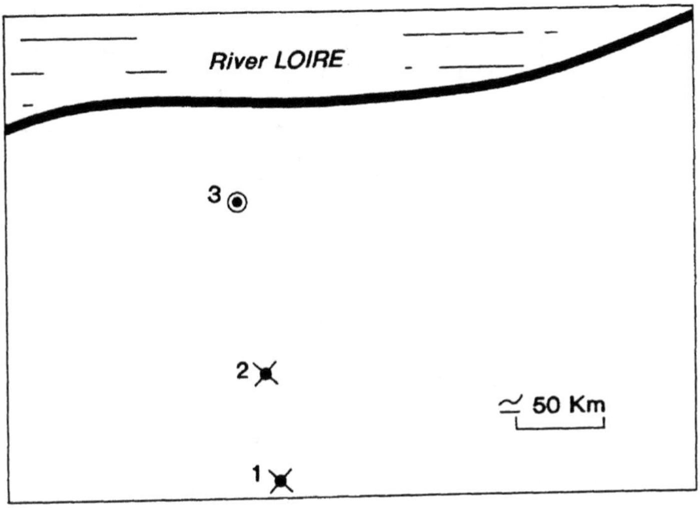

Fig.9: Layout of Coutures wells

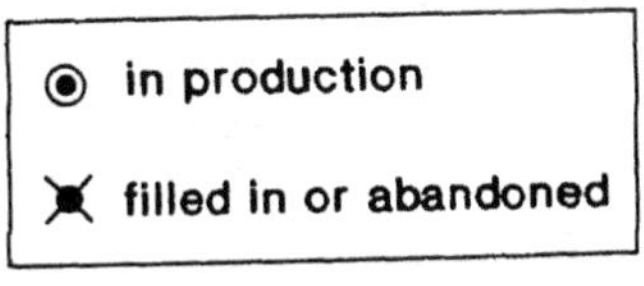

2.3.3 Coutures catchments

2.3.3.1 Location of wells

Only one of these 3 wells is still working. However, all three are still operable. The wells were sunk in alignment, increasingly close to the Loire (fig.9). all three give clogging problems, or have done so in the past.

2.3.3.2 Description of the wells

Tables IV and V give a summary description of the wells in Couture:

TABLE IV. GENERAL DATA ON COUTURES WELLS

well Nr.	yr.built/ altered	type	distance from LOIRE	present status	nominal output (m^3/h)
1	69	weep hole	270 m	closed in 70	100m^3/h
2	71	drains	200 m	closed in 88	200
3	75	drains	20 m	working	350

TABLE V. CHARACTERISTICS OF COUTURES WELLS

Well	diameter of well (m)	depth	total drain length (m)	Number of drains (m)	drain characteristics
1	7	8			
2	3	12	90	5	5 steel w/holes ID 200mm
3	3	12	60 (75) 127 (78)	3 6	3 steel, chased ribs ID: 200 mm 3 PVC, w/holes ID: 200 mm

2.3.3.3 Cross section of ground layers round the Coutures wells

The ground round Coutures consists of a clayey layer to a depth of 0 to 2m (fig.10). It is more in the form of scattered sills rather than in a compact layer. This clay, rich in organic substances, is known in France as "Jalle"

The Montjean terrain also includes a few highly organic clay sills.

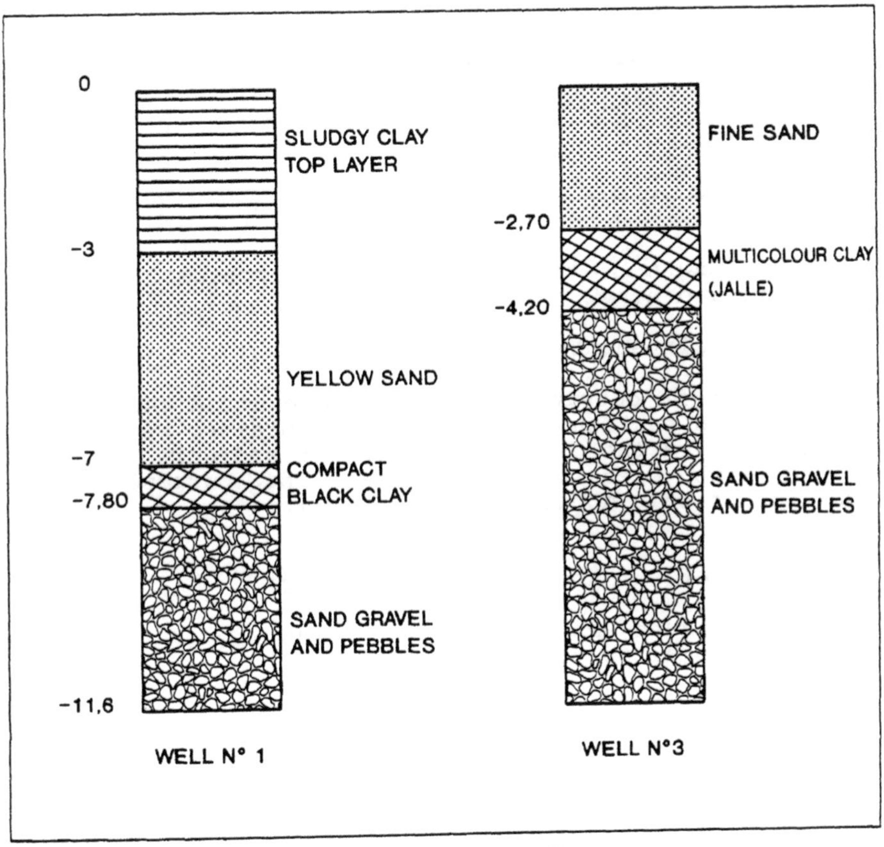

Fig. 10· Geological cross–section of wells n°1 and n°3 in Coutures

3 OPERATING PROBLEMS

3.1 Evolution of specific flow rates

The difficulty experienced on the Montjean and Coutures wells is the speed at which the flow rates decrease in time. As can be seen from the flow rate graphs, and flow rate speed reduction graphs, the specific flow rate on drawdown height can be divided by 2 and even by 4 within a year. (fig 11 to 16).

3.2 Observation on drain management

3.2.1 Falling flow rates

All these types of drains are affected by a reduction in flow rate:

- steel collector-drains, whether perforated or with pushed ribs;

- PVC drains

- chased ribs

We note, however, that the output decreases faster on ribbed steel drains.

3.2.2 Maintenance of drains

The decrease in outputs induced us to examine the state of the wells and particularly that of the drains. We were thus able to observe the clogging phenomenon and were obliged to proceed with cleaning operations. The purpose of the remainder of this document is to describe:

. the state of our knowledge of clogging;

. the techniques applied in the fight against clogging, whether remedial or preventive.

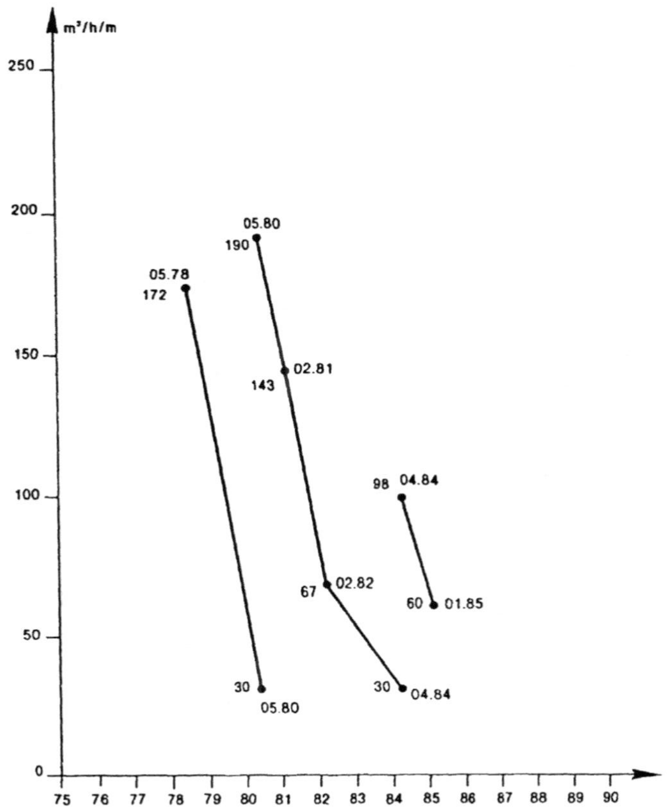

Fig. 11: Specific flowrate reductions in well n°2 _ Montjean

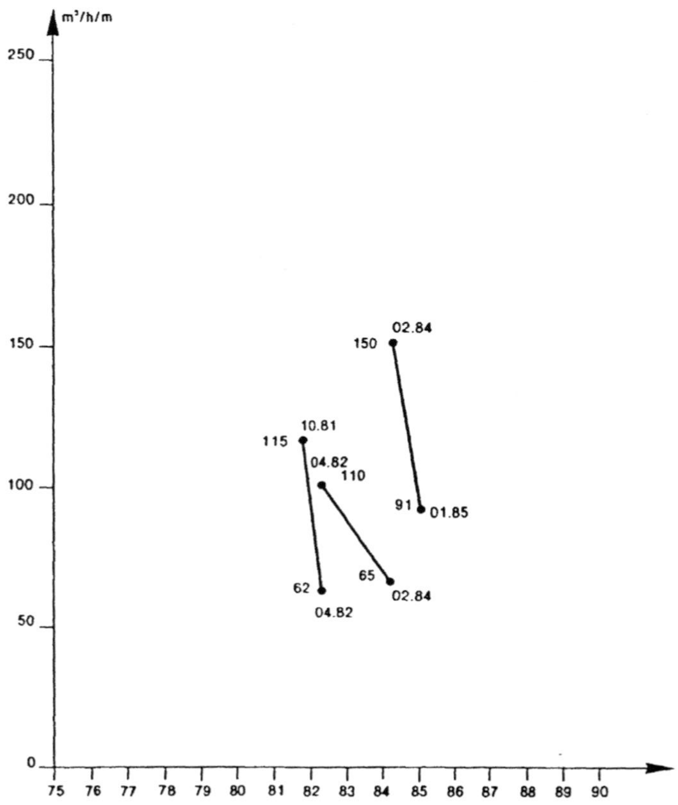

Fig. 12: Specific flowrate reductions in well n°8 _ Montjean

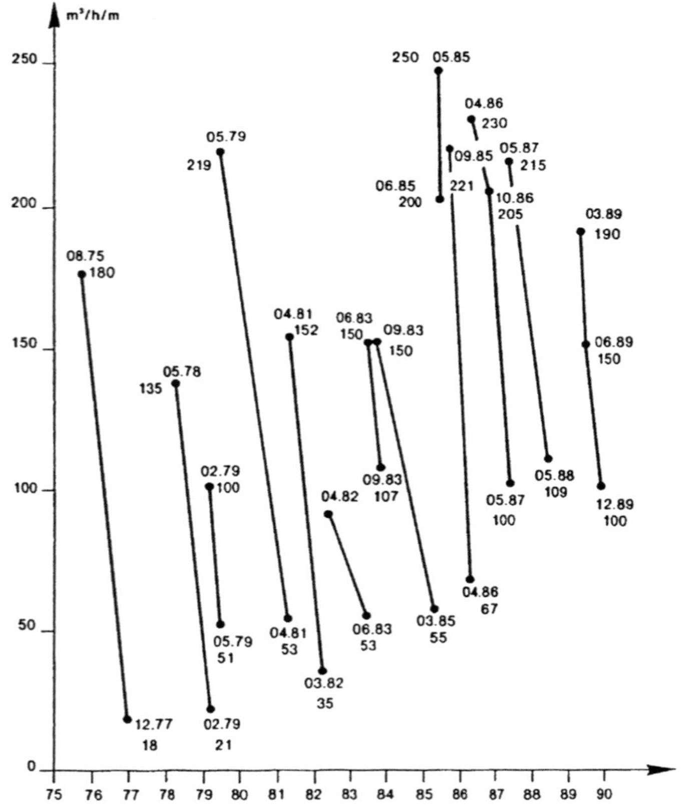

Fig. 13: Specific flowrate reductions in well n°3 _ Coutures

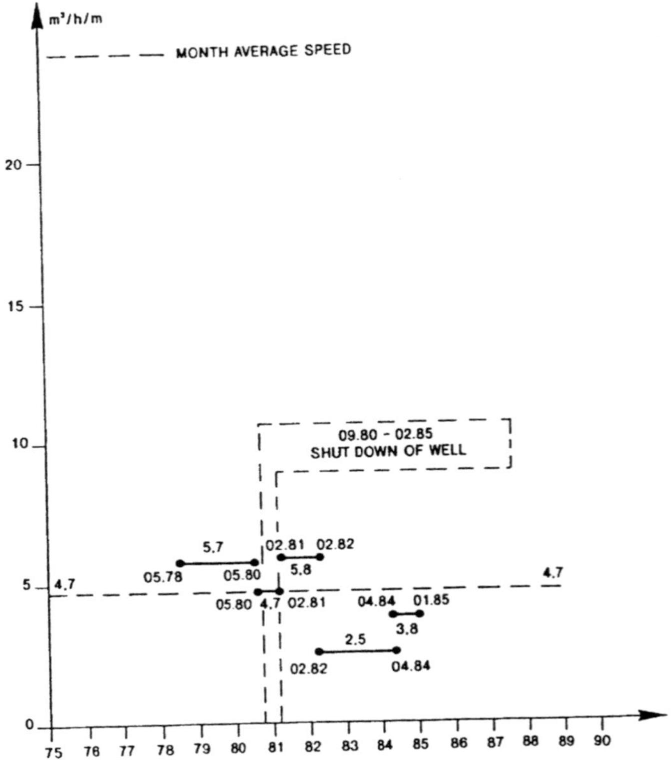

Fig. 14: Flow rate reduction speed in well n°2 _ Montjean

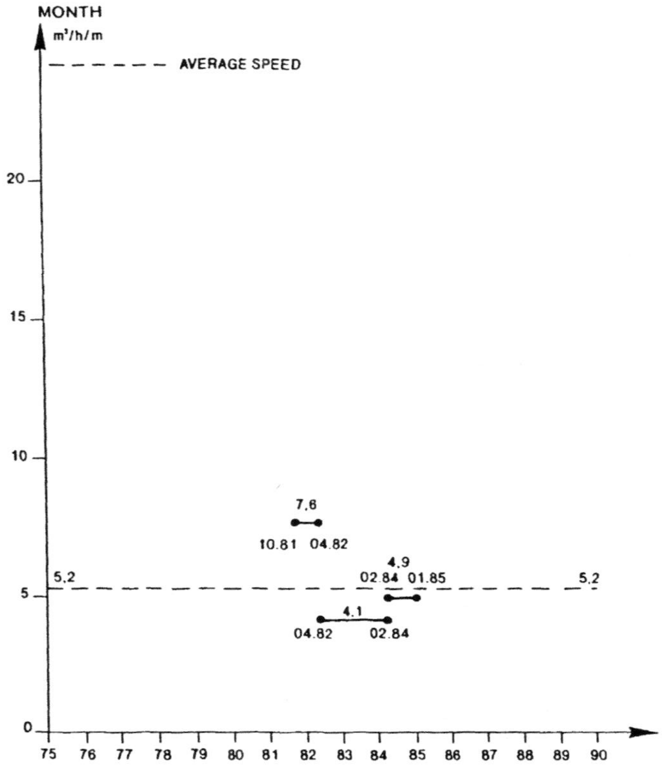

Fig. 15: Flow rate reduction speed in well n°8 _ Montjean

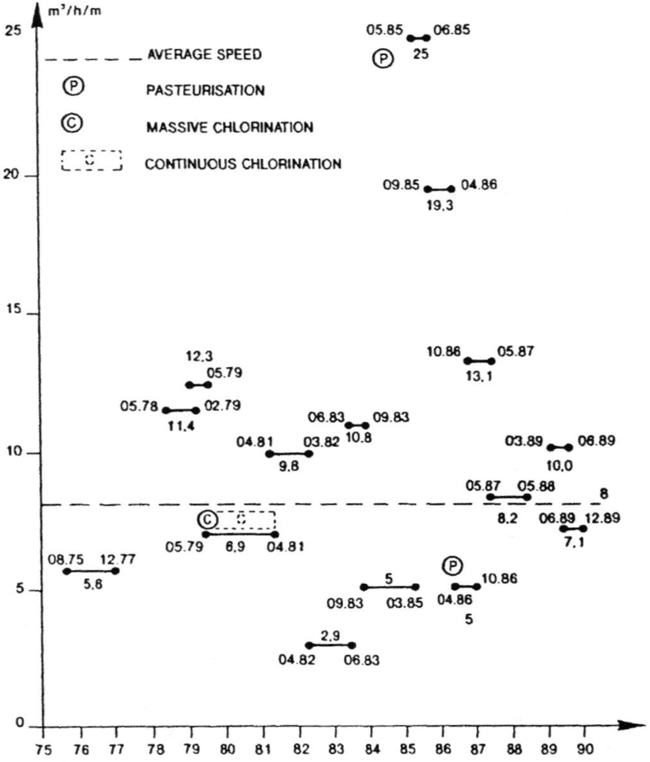

Fig. 16: Flow rate reduction speed in well n°3 _ Coutures

TABLE VI RECAPITULATES THE DATA SHOWN IN FIGURES 11 to 16

WELLS	Date of highest production	Date of lowest production	Interval between dates (months)	High flow rate (m3/h/m)	Low flow rate (m³/h/m)	Discrepancy between flow rates (m3/h/m)	Flow rate reduction speed (p/month)	Average reduction speed (p/month)
MONTJEAN n° 2	05.78	05.80	25	172	30	142	5,7	4,7
	05.80	02.81	10	190	143	47	4,7	
	02.81	02.82	13	143	67	76	5,8	
	02.82	04.84	15	67	30	37	2,5	
	04.84	01.85	10	98	60	38	3,8	
MONTJEAN n° 8	10.81	04.82	7	115	62	53	7,6	5,2
	04.82	02.84	11	110	65	45	4,1	
	02.84	01.85	12	150	91	59	4,9	
COUTURES n° 3	08.75	12.77	29	180	18	162	5,6	8
	05.78	02.79	10	135	21	114	11,4	
	02.79	05.79	4	100	51	49	12,3	
	05.79	04.81	24	219	53	166	6,9	
	04.81	03.82	12	152	35	117	9,8	
	04.82	06.83	15	90	53	43	2,9	
	06.83	09.83	4	150	107	43	10,8	
	09.83	03.85	19	150	55	95	5,0	
	05.85	06.85	2	250	200	50	25,0	
	09.85	04.86	8	221	67	154	19,3	
	04.86	10.86	5	230	205	25	5,0	
	10.86	05.87	8	205	100	105	13,1	
	05.87	05.88	13	215	109	106	8,2	
	03.89	06.89	4	190	150	40	10,0	
	06.89	12.89	7	150	100	50	7,1	

This precipitation is caused by the mineralization of the water and a shift in its physico-chemical characteristics (pH, oxidation reduction potential). The latter depend on parameters such as pressure, temperature, oxygen content.

4.1.1.3 Biological clogging

Two factors are involved:

- the incrustation of biomass
- the build-up of sediments.

a) Incrustation of biomass

As they grow bacteria incrust themselves in the ground and on collector-drains, creating a clogging effect.

b) Build-up of sediments

Bacteria, such as those that live on iron or manganese, generate hydroxides of these metals. The hydroxides settle, thus bringing a source of clogging.

4.1.2 Clogging prevention

The problem is to create conditions in which the above mentioned causes will not appear.

4.1.2.1 Mechanical clogging

Depending on its characteristics, a formation is more or less subject to erosion. The idea of prevention must therefore be uppermost when choosing a location for the catchment. Prevention also means defining the characteristics of the material (length of collector-drains, empty drain ratio), and drain operating conditions (draw-off pump suction pressure, flow rate in drains), from which the mechanical characteristics of the stream of water will be derived (pressure of water in the formations, water velocity at drain inlets). Bearing in mind, therefore, the formations and the catchment material, we can define a maximum drawdown height and a specific flow rate (flow rate divided by the drawdown height), that will be characteristic of the well and its condition. Respecting these two parameters avoids making the erosion effect, hence the clogging potential more acute, when clogging is due to mechanical factors.

4.1.2.2 Physico-chemical clogging

a) Corrosion of collector-drains

An obvious preventive action is to use drains that are more or less corrosion-proof, such as stainless steel or plastic collector-drains.

a1) Chemical corrosion

In fact, the conditions that are responsible for chemical corrosion are of two origins:

a1.1) origins related to the hydrogeological conditions of the aquifer that account, for instance, for the degree of aeration or mineralization of the water. It is very difficult to act on these conditions first for lack of knowledge concerning the interaction between the aquifer and the surrounding formations (aquifer and exterior of water-bearing layer) and secondly because of the weight and cost of the works involved (drilling, followed by monitoring the water characteristics, etc.).

a1.2) Origins related to the presence of bacteria in the water. This explains how the oxygen, nitrogen and H_2S that influence the physico-chemical characteristics of the water (pH, ORP), are themselves influenced by the bacterial transformation of the organic and mineral matter contained in the water (denitrification of the nitrates into nitrites, transformation of sulphates into H_2S).
Hence the actions taken to counteract biological clogging can also have a preventive effect on chemical corrosion. These actions will be described in the paragraph concerning biological clogging below.

a2) Electrochemical clogging

The method here is to respect the usual rules on the prevention of this type of corrosion (control and maintenance of drain surfaces, taking particular care of welds.

b) Precipitates of various salts, hydroxides and oxides.

What was said about chemical corrosion also holds good here.

c) Biological clogging

It has been observed that biological clogging has 2 aspects to it. One is connected with the presence of bacteria as such, the other with their power to generate precipitates. Hence the destruction of bacteria can be considered as a preventive action as well as a remedial measure. We therefore describe the destruction of bacteria in the part on remedial action against clogging.

Bacterial ecology shows us that the conditions of their survival and proliferation depend mainly on the geological and hydrogeological characteristics of the aquifer. We saw earlier that it is difficult to influence these characteristics, but they also depend on catchment operating conditions. On the one hand, the bacteria can derive their nourishment from the water, on the other, they swarm when water velocity is above certain thresholds. Hence, we can hope to destroy the bacteria completely or control their numbers respectively by stopping the catchment for a while and controlling the velocity of the water.

4.1.3 Remedial action against clogging

There are three types of action:

- thermal cleaning
- chemical cleaning
- mechanical cleaning

4.1.3.1 Thermal cleaning

The purpose of this method is to destroy the bacteria. It is a pinpoint operation whereby the temperature of their environment is increased far beyond the bracket in which they thrive, roughly between 25 and 28°C. The safest is to reach 100°C by injecting hot water or steam for a given period of time.

4.1.3.4 Chemical cleaning

Two types of chemical cleaning method exist:

- one aims at destroying the bacteria
- the other is to dissolve the sediments.

a) Chemical destruction of bacteria
This is done by injecting chlorine either in the catchment well in the direction of its periphery, or from the periphery towards the centre of the well.

Chlorine is used in different forms: hydrochloric acid, hypochlorite.

b) Chemical dissolution of the sediments

The method in this case is to dissolve the precipitated particles formed mainly by iron and manganese oxides and hydroxides. This is achieved by injecting a chemical such as ascorbic acid.

4.1.3.3 Mechanical cleaning

This cleaning principle is based on loosening the clogging products (by brushing, suction or the application of high pressure) and removing the residues. The technique used will be described as part of the presentation of the cleaning techniques used at the Montjean and Coutures sites.

4.2 Clogging observed on the Montjean and Coutures sites.

4.2.1 Analysis of sediments in collector-drains

The analysis of sediments was performed during cleaning operations on samples of 2 different origins:

- residues contained in effluents (liquid samples)

- residues obtained after brushing drains (solid samples).

4.2.1.1 Effluents

The analysis of these waters is complete, and so defines the organoleptic, physical, chemical, ionic, calcium-carbon, corrosion potential and microbiological characteristics, as can be seen from the attached analysis report (see Cleaning water from Montjean well Nr 2 (June 85) in appendix I).

The comparison of results obtained before and after cleaning, shows that, while it is in progress, cleaning causes an increase in the iron content, and especially iron bacteria. Consequently, we are now sure that iron bacteria are among the factors involved in the clogging process.

4.2.1.2 Brushing residues

Analytical results show that there is a difference between the sediments on steel and those on PVC drains. (see analysis of cleaning residues Coutures well Nr 3 (April 85) in Appendix 2).

a) Stainless steel drains

Deposits consists mainly of ferrous hydroxides $(Fe(OH)_2$. In addition to a strong iron concentration, we also observed a high concentration of manganese, a small iron bacteria count and very little organic matter.

b) PVC Collector-drains

Compared with the sediments in steel drains, PVC collectors have only a small iron and manganese content, whereas the iron bacteria count and the organic content are significantly higher.

c) Remark:

i) we note for each of these drains a good deal of aluminium but no significant amount of calcium.

4.2.1.3 Grit in drains

In some drains, such as those of well Nr 2 in Coutures, the quantity of grit removed was considerable: it represents a layer several centimeters thick all along the drains.

4.2.2 Other remarks

a) Cross section of formations

When there is no covering of clay, clogging is practically nil. On the opposite, when the water sheet is confined by a clayey layer, the catchment structures are invariably clogged.

b) Water analyses

b1) water pumped up is always rich in iron and manganese, except on well Coutures well Nr 3 that is very srongly affected by the nearby Loire waters. In some areas, the concentration decreases in old structures in the course of their operating life.

b2) H2S is found in the samples collected on the edge of the clogged structures.

b3) Operating

Clogging can occur very shortly after commissioning. And it can continue to grow in an idle period. (This was the case when well Nr 2 at Montjean was shut down from September 80 to February 81).

4.2.3 Conclusions drawn from observations

The observations made between 78 and 90 on the Montjean and Coutures sites are not really very plentiful except studies connected with declogging tests that, on the contrary, were very numerous between 78 and 85. These observations can be summed up as follows:

- observation of the geological cross-section of the formations;

- analyses of pumped water;

- analyses of residues obtained after cleaning;

- observation on the efficiency of preventive and cleaning operations (which will be reverted to hereunder).

- observations of operating conditions

However, these observations lead to 2 important conclusions: one on the compatibility between the "reduced area model" and the case of Montjean and Coutures sites, the other on the complexity of the case.

4.2.3.1 Reduced area model

This model was described by Mr Bize of BURGEAP.

It takes into account the role of the aquifer formation, the mechnical, physicochemical and biological behaviour of the water and the influence of the tapping operation.

It turns out that the observations collected confirm the necessity of this overall calculation. Thus the presence of aluminium leads us to suspect that there the interaction between the content and water exerts an influence, while the difference between steel and PVC collector-drains shows that each of the 3 causes of clogging produces its effect to different degrees under the influence of operating conditions.

4.2.2.2 A combination of causes

We would like to emphasize the point approached in the previous paragraph i.e., clogging, according to our study, is the fruit of a combination of mechanical, physico-chemical and biological factors This combination, linked to the difficulty of knowing the geological and hydrogeological phenomena involved, accentuate the complex character of clogging.

5 DECLOGGING

The different techniques used are described in relation to the sites of Montjean and Coutures, ending with the one currently operating. The relative efficiency of these techniques will be compared.

5.1 The techniques used

5.1.1 Preventive

5.1.1.1 Physico-chemical clogging

The latest collector-drains installed were made of stainless steel, then PVC.

5.1.1.2 Biological clogging

We have tried decontamination with chlorine and pasteurization. Except in the case of continuous injection lasting several months, each test treatment including mechanical cleaning, either beforehand or afterwards.

a) Decontamination with chlorine

a1) Mass injection

This is done in 2 stages:

. a central injection saturating the well with chlorine (163 ppm), then establishing contact with the aquifer, one drain at a time.

. a peripheral injection until stabilization of the chlorine content in the well (18 ppm).

a2) Weak continuous injection (June 79 to May 81)

This was a peripheral injection regulated to obtain a chlorine concentration of 0.2 to 0.3 ppm.

b) Pasteurization (85-86)

This treatment was performed with hot water, on a drain by drain injection basis. The water temperature was between 95 and 100°C for the steel drains and from 55 to 60°C for PVC drains. The temperatures reached in the collector-drains was respectively:

85°C for water between 95 and 100°C (steel pipes)

50°C for water at 60°C

c) Shutdown of water supply

The water was shut off as if for an operational reason, such as works on the well for the installation of new pumps, from September 80 to February 81 on Well Nr 2 at Montjean. This made it possible to study the effects on clogging if the flow of water is cut off.

d) Effectiveness of preventive treatment

d1) Immediate effect

No injection treatment has an immediate effect. The gain of specific flow is obtained almost exclusively during mechanical treatment, as is shown by the pasteurization test of March-April 85 on Well Nr 3 at Coutures (see Table VII).

TABLE VII: PASTEURIZATION TEST ON WELL Nr.3 at COUTURES

	Before pasteur-isation	After pasteurization	After mechanical cleaning
specific flow rate of the 3 steel drains ($m^3/h/m$)	12	14	172
Specific flow rate of the 3 PVC drains	50	54	252

d2 Preventive efficiency

Whatever the preventive treatment used, the speed at which the specific flow rate decreases in the following months is comparable to that of any other period, including the one before the treatment (fig 14 and 16) - preventive attempts to suppress bacteria are the more disappointing since they are believed to play an important part in the build-up of clogging.

5.1.1.3 Mechanical clogging

There is no such technique, as such. We respect the thresholds for the specific flow rate and the drawdown height.

5.1.2 Cleaning

Cleaning always includes a preliminary mechanical treatment. We have tried adding to this treatment a chemical cleaning stage, the aim of which was to dissolve the iron and manganese oxides and hydroxides.

5.1.2.1 Chemical treatments

These follow cleaning with a brush and/or pressurized flushing, in a central injection of:

- either polyphosphates (78-79),
- or a product essentially made up of ascorbic acid such as HERLI RAPID TWB (80-83).

The tests conducted on well Nr 3 at Coutures from 78 to 83 began by yielding interesting results and then turned out to be disappointing. Considerng the chemical part of the treatment only, the gain in flow rate is only about 10%. Because of operating complications (safety during handling of the products, downtime on the facility) and the cost of implementation, we focused our research on perfecting the mechanical cleaning process.

5.1.2.2 Mechanical cleaning

We implemented 3 mechanical cleaning techniques:

a) tube brushes with or without flushing.
This consists of inserting a metal brush in the collector-drain and twisting it from side to side round the centre line. It is a very effective technique enabling 75 to 100% of the specific original flow rate to be restored. But it can be dangerous because of the hydrostatic pressure. Moreover, it requires installing very cumbersome equipment and as a manual operation it is extremely hard work.

b) Filling the drains with water

The principle of this technique is to create a backwash effect in the drains, obtained in 2 different ways:

. through the pressure obtained by filling the well with water. The maximum pressure reached is limited by the height of the well, i.e;, in the region of 600g to 1 kg (fig 17).

. or with the help of a pump immersed in the well and directly connected to the drain (fig 18).

The principal purpose of using a pump is to increase the water flow rate (up to 100 m^3/h).

This treatment yields only a 40% gain on the specific flow rate, a very mild rate of efficiency due to the fact that the impact of the jet of water is dispersed over a large area that increases as the cleaning proceeds along the drain, thus slowing down the peripheral water velocity on the collector-drain. We therefore tried to improve the effectiveness of the jet by increasing the water flow rate and reducing the jet impact area.

5.2 The present technique

We have developed a technique already used for cleaning sewers. It consists in using a nozzle on the end of a hose.

5.2.1 Description of equipment (fig 19)

It consists of the following components:

- a high pressure pump (150b)

- a high pressure hose (diameter 20 mm).

- an air-chamber fixed at the entrance to the drain and fitted with a drain valve to discharge the cleaning products.

- a water injection nozzle.

5.2.2 Operating data

This is a traditional type of process:

when the pump is discharging, the nozzle sinks towards the bottom of the drain. The geometry of the nozzle enables it to direct jets to a limited portion of the drain.

5.2.3 Effectiveness and limits of the currently used technique.

The gain on specific flow rate achieved is 100%. On PVC drains, this technique is altogether satisfactory. However, on steel drains that often become damaged,it accentuates the effects of wear and tear. The velocity of the jets must therefore be limited, which automatically implies a loss of efficiency.

5.3 Frequency of intervention

On the Montjean and Coutures sites, the year is divided into 2 periods characterized by the demand in water:

- a period of low demand: October to April

- a period of high demand: May to September.

At the present time, the loss of flow rate is acceptable all through the low demand period. Hence a

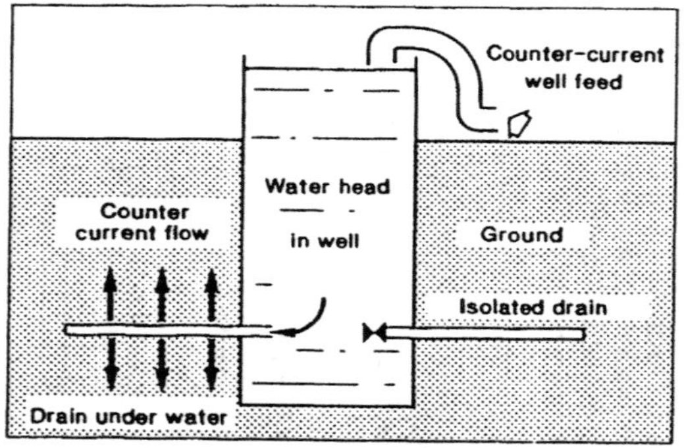

Fig. 17: Natural priming of a drain

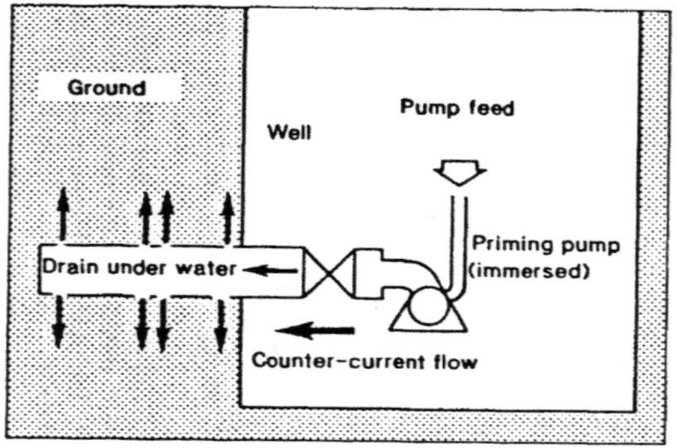

Fig. 18: Priming a drain by pumping

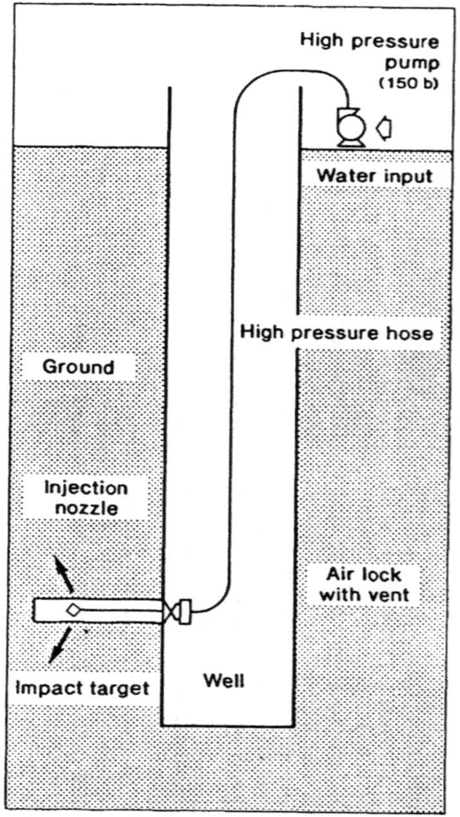

Fig. 19: Cleaning with high pressure water jet

yearly clean-out in April to May is sufficient to meet requirements all year round.

However, it is obvous that increasing needs will mean that these rates of cleaning must be more frequent.

6 CONCLUSION

From 78 to 86 on the Coutures and Montjean sites, we conducted academic studies and practical tests on clogging, its prevention and the scouring of collector-drains.

From a practical point of view, they showed up the sophisticated nature of the clogging effects occurring on these sites.

The foreseeable increase in the needs for water, the will to make the best of a very high quality resource and the research for lower production costs, prompts us to continue our efforts to achieve better understanding of these highly complex phenomena.

BIBLIOGRAPHIE

BIZE (J.)- Exploitation du complexe alluvions-rivière en bordure de Loire. Etude du colmatage des puits et recherche de remèdes.- Rapport BURGEAP de 1ère phase, R.573-E.1383, Juin 1984, 51 p., Compagnie Générale des Eaux, C.R. Ouest

BOUDOU (J.P.).- Ecologie des bactéries aérobies des eaux souterraines. Fixation du fer et du manganèse. Colmatage des forages d'eau, Octobre 1984, 85 p., Compagnie Générale des Eaux, Anjou-Recherche

MANSUY (N.).-Analyses chimiques et microbiologiques des dépôts et de l'eau à Couture et Saint Jean, Août 1985, 29 p., Compagnie Générale des Eaux, Anjou-Recherche

MANSUY (N.).- Pasteurisation. Traitement du puits n°3 à Saint Rémy La Varenne en bordure de Loire, Juin 1985, 132 p., Compagnie Générale des Eaux, Anjou-Recherche

APPENDIX 1

ANALYSIS OF CLEANING RESIDUES FROM WELL Nr 2 at MONTJEAN (June 85)

COMPAGNIE GENERALE DES EAUX

Report Nr
date: 16/07/85

CENTRAL LABORATORY

Mr MANSUY

on bahalf of: Company A.R.
Site Job No 635 504
Department

Sample: Nature MONTJEAN 1 sand specimen

Date and conditions of sampling: received on 1/07/85

Percentage of dry solids	89%
Fraction of volatile D.S.	0.7 %
Fraction of inorganic D.S.	99.3 %

as mg/g of D.S.

Carbonates	as	CO3	12
Silicates	as	SiO2	950
Iron	as	Fe	14.4
Manganese	as	Mn	0.15
Calcium	as	Ca	2.4
Aluminium	as	Al	3.9

MICROBIOLOGICAL ANALYSES MONTJEAN Nr 2
19 JUIN 1985

	Before	During	After
microscopic analysis on membrane filter 0.45 ppm	No filamentous bacteria	100% galliomella	no filamentous germs
Total aerobic germs 1/4 TSA 5 days at 20°C	57,000/ml	43,000/ml	27,000/ml
Heterotrophic iron bacteria Winogradsky's 7.4 5 days: 20°C	250/ml	34,000/ml	5/ml
Pseudomonas AGAR isolated Pseudomonas 5 days: 20°C	0/ml	9,000/ml	0/ml
Autotrophic iron bacteria Wolfe's medium 10 days 20°C	negative	positive	negative

APPENDIX I

CLEANING WATER FROM N°2 WELL AT MONTJEAN (JUNE 85)

		BEFORE	AFTER	BEFORE
ORGANOLEPTIC	Aspect	clean	clean	=
	Colour			
	Colour (mg/l-Pt-Co)	< 5	< 5	=
	Residual chlorine (mg/l Cl2)			
	Total iron (μg/l Fe)	280	340	190
PHYSICAL	on the spot/in lab.	/20	=	=
	pH on the spot/in lab.	/7,4	=	=
	Resistivity at 20° C (microhm/cm/cm2)	3640	=	3650
	Turbidity NTU	2,05	5,50	1,30
	Total SS (mg/l)			
CHEMICAL	Immediate dissolved O2			
	Oxidation potential			
	(KMl in alkaline environment (mg/l O2)	1,0	1,3	1,0
	Total TA (French degrees)	12°80	12°64	12°6
	TA (French degrees)	0	0	0
	TAC (French degrees)	10°6	=	10°4
SET POINTS	OH - mg/l			
	CO3=			
	HCO3-	129,32	=	127,49
	HSi O3-	14,63	=	13,86
	Cl-	16,90	17,10	16,60
	SO4--	24,00	=	23,00
	NO2--	<0,02	=	=
	NO3-	<1,00	=	=
	PO4---	<0,10	=	=
	F-	0,10	=	=
	S--			
	Total-	184,95/3,291	185,15/3,297	181,05/
	H+			
	Ca++	41,50	40,60	=
	Mg++	5,89	5,95	=
	Na+	12,50	12,20	11,60
	K+	2,60	2,80	2,70
	NH4+	0,02	0,05	<0,02
	Fe++	0,280	0,340	0,190
	Mn++	0,778	0,748	0,550
	Al+++	<0,060	=	=
	Total+	63,57/3,209	62,69/3,164	61,59/3,120
CALCO-CARBONIC BALANCE	plus (labo)	+7,85	=	+7,87
	saturation coefficient "	-0,42	-0,48	-0,47
	Free carbonate alkalinity	+7,30	+8,20	+7,70
	Aggressive carbonate alkalinity	+4,20	+5,10	+4,70
CONDUCTIVITY	Rysmann coefficient	+8,28	+8,35	+8,34
	Comments	averagely aggresive	=	=

APPENDIX 2

ANALYSIS OF CLEANING RESIDUES FROM WELL Nr 3 at COUTURES (April 85)

COMPAGNIE GENERALE DES EAUX

Report Nr 85/1801
date: 24/06/85

CENTRAL LABORATORY

Mr MANSUY

Sample: Nature — 2 water specimens containing sediments

Date and conditions of sampling: Raw water - steel and PVC collector-drains
Coutures 49 - 17/04/85
Well No 3

	PVC	STEEL
Suspended solids (mg/l) (FC) Analysis of sediments was performed on suspended solids (after centri-fugation)	700	295
Percentage of volatile dry solids	7.1	6.6
" " inorganic " "	92.9	93.4
Absence of carbonates in mg/G dry solids		
Calcium as Ca	16	22.9
Aluminium as Al	20.7	16.7
Manganese as Mn	1.8	3.2
Fer as Fe	150	200
Silica (residue after acid aggression) SiO_2	560	470

COUTURES WELL Nr 3

MICROBIOLOGICAL ANALYSES OF WATER AFTER CLEANING OF COLLECTOR-DRAINS

	Steel drains	PVC drains
Microbiological analysis on filter membrane P.45 ppm	gallionella 98% (Ferruginia) Crenothrix 2%	Gallionella 98% (Ferru-ginia) Crenothrix 2%
Total anaerobic germs 1/4 TSA, 5 days: 20°C	250,000/ml	183,000/ml
Heterotrophic iron bacteria Winogradsky's 7.4 5 days: 20°C	270/ml	59,000/ml
Pseudonomas Agar isolated Pseudonomas 5 days: 20°C	100/ml	39,000/ml
Autotrophic iron bacteria Wolfe's medium 10 days: 20°C	Nil	Nil

25 REHABILITATION OF DETERIORATED TUBEWELLS IN PUNJAB SCARPS

K.G. AHMAD
Water and Power Development Authority, Lahore, Pakistan

Abstract
A large number of drainage tubewells have been installed in the Indus Plain of Pakistan since 1960; many of these have subsequently deteriorated badly. The history of tubewell development is described and the serious problem of deterioration is reviewed and discussed in terms of causes and remedial measures adopted. Various experimental rehabilitation programmes are described in detail, particularly that of seven badly deteriorated wells which were treated to a combination of chemical and physical processes which resulted in a six fold sustained improvement in specific capacity.
Keywords: Indus Plain, Tubewells, Rehabilitation

1 Introduction

To combat the menace of waterlogging and salinity in the Indus Plain, Pakistan about 13,000 tubewells have been installed by the Water and Power Development Authority (WAPDA) since 1960, when the Salinity Control and Reclamation Programmes (SCARP) were started. Initially, these tubewells were used to depress the water table and to provide additional water for crops in fresh groundwater (FGW) areas. Seeing these benefits, farmers installed small capacity tubewells rapidly in FGW areas. However, within 10 - 12 years deterioration or breakdown of tubewells had become a serious problem, causing waterlogging and salinity to recur.

Efforts to evaluate the scale of the problem, its causes and appropriate rehabilitation techniques make an interesting case history which has relevance elsewhere. A vast quantity of information is available and this paper describes the findings and conclusions.

2 Brief History of Scarp Tubewells in Punjab

2.1 Development Period 1959 - 72

In this first development phase about 7000 large capacity tubewells were constructed and commissioned. Mild steel (MS) screens were used mostly as the tubewells were installed in FGW zones. Reductions in well discharge were observed after a few years and the causes of this were investigated with the help of foreign experts. A major cause was attributed to unsuitable well screens and all efforts were directed towards finding a suitable material. Thus fibreglass came to be used extensively in successive SCARPs. However, a uniform design for tubewell construction continued to be applied.

Even with the use of fibreglass screens, well deterioration did not cease and the experience gained during this development period crystallised as follows:

a) The need for inert screen material for saline groundwater zones was established.
b) Construction procedures were standardised.
c) Regular monitoring of project wells was introduced.
d) Construction was methodically supervised and data recording improved.
e) No attention was given to the need for variability in individual well designs.
f) Rehabilitation of tubewells was not done scientifically and poorly recorded.

2.2 Development Period 1974 - 1984

This phase of development was characterized by a slower rate of installation owing to financial stringency and operational problems of tubewells which led to a re-examination of national policies. Foreign consultants were divested of design and construction supervision and the responsibility entrusted to local engineers. Just over 1700 public tubewells were installed in this period. Numerous attempts were made to identify the causes of tubewell deterioration.

In 1975 Mr T.P. Ahren was engaged through HARZA International to review the problems and suggest remedial measures.

Mr Ahren's report outlined the following causes for the rapid deterioration of tubewells in Pakistan:

a) A standard tubewell design had been adopted in all SCARP Projects thus not taking into account the widely varying conditions of the aquifer in different areas.
b) There were deficiencies in tubewell construction practices, such as placing of gravel pack,

inadequate tubewell development and unsuitable screen material.

c) There had been a virtual complete lack of maintenance during the operation of tubewells.

d) Construction of tubewell projects proceeded so rapidly that it was not possible to fully introduce improvements based on the observed performance of installed tubewells.

Mr Ahren recommended improvements in tubewell design, construction, maintenance, rehabilitation and in monitoring research. He estimated that if a tubewell had lost less than 25% of its specific capacity, there was a 90% chance that it could be fully restored, but this chance reduced to 10% if more than 50% of the specific capacity had been lost.

These recommendations were not vigorously pursued initially, perhaps because a fairly large number of tubewells had deteriorated beyond the rehabilitation stage. Instead, two replacement programmes of 800 and 650 tubewells were completed in 1978 and 1982 respectively.

However, a first attempt at partial adoption of his recommendations was made in the design of replacement wells belonging to SCARP-I, III, IV and the Minchinabad Pilot Project. It was recognised that the length of screen should be adequate to keep the entrance velocity within limits, and the gravel pack was matched with soil formation data resulting in a reduction of slot size from 0.063 and 0.05 to 0.04 and 0.03 inch. Unfortunately, no headway was made towards identifying the causes of tubewell deterioration.

2.3 Development Period 1985 -90

During this third phase improvements to the design of tubewells and the efficiency of construction methods have been made. Improved design parameters were used for saline drainage tubewells in the SCARP-VI project; these were related to well field design, setting of pumps and length of pump housing, screen length and diameter, and the screen entrance velocity. Longer screen lengths were used to exclude finer strata and larger diameters to lower the entrance velocity. The well field designs were based on a triangular grid system. Automatic control devices were fitted to 50% of the tubewells on an experimental basis to avoid operator error. For the Haddali Project the gravel pack and slot size were matched to the various formation sizes. In the Shorkot Kamalia Project, tubewell discharges were further reduced to 2 cusec because of aquifer conditions and deeper saline water. Well rehabilitation on more scientific lines is now planned in the Scarp Transition Project for Scarp-I.

3 Rehabilitation Works

3.1 General

Rehabilitation studies commenced in 1964 when SCARP-I tubewells started showing alarming declines in specific capacity. WAPDA constituted a team from the then Groundwater Reclamation Division, Water and Soils Investigation Division, Tipton & Kalmbach and Harza Engineering Consultants to examine the problem and suggest remedial measures.

Nine tubewells of the Zafarwal Scheme and 54 tubewells of the Shadman Scheme were examined. The MS screens were heavily corroded and encrusted. The affected wells were periodically treated with acidizing and prima cord blasting before abandoning them. Improvements were neither satisfactory nor long lasting and the programme was discontinued.

A more scientific approach was adopted for developing metallic screens in the Mona Reclamation and Experimental Project by WAPDA. Seven wells were redeveloped in three foot sections after identifying the blocked portions by means of velocity surveys. The average specific capacity which had been 95 gpm/ft originally but had deteriorated to 42 gpm/ft improved on development to 97 gpm/ft (Table 1); unfortunately the tubewells deteriorated again after one year.

Table 1. Rehabilitation of 7 tubewells in Mona Reclamation Experimental Project

Tubewell Number	Specific Capacity (gpm/ft) Original	Before Treatment	After Treatment
MN-42	72	48	83
MN-52	96	51	101
MN-53	90	48	67
MN-69	106	47	104
MN-136	92	60	118
MN-138	102	14	98
MN-140	108	28	109
Average	95.1	42.3	97.1

Fibreglass screens were used extensively in SCARP-II, III and IV to overcome corrosion. However, this did not stop well deterioration. The use of explosive devices for decrustation became risky and rehabilitation could only be done by chemical or mechanical means.

The Punjab Irrigation Department (PID) also carried out rehabilitation in Alipur and Kot Adu units (Scarp III) but a proper scientific approach was not used and the results were discouraging (Table 2).

Table 2. Tubewell Rehabilitation in Alipur Unit

Tubewell Number	Specific Capacity (gpm/ft) Original	Before Treatment	After Treatment
AP-403	78	38	57
AP-404	83	42	49
AP-408	90	61	59
AP-409	96	17	32
AP-420	86	30	32
AP-423	67	22	36
AP-425	78	22	26
AP-431	83	21	54
AP-443	78	45	47
AP-444	107	28	35
AP-448	79	36	46
AP-449	98	31	35
AP-452	98	39	62
AP-453	121	38	38
AP-458	113	25	26
Average	90.3	33.0	42.3

WAPDA formed a committee to examine rehabilitation methods; its conclusions are reproduced below:

a) Although conclusive and corroborative evidence is not available, the identification of iron bacteria, the presence of iron deposits in pumps and the success achieved in treatments using bleaching powder indicate that the major cause of well problems may be incrustation and/or slimes, produced by bacterial activity.
b) Carbonate incrustation is also a problem, although not as widespread. This conclusion is supported by the positive action of acid which was used in two tubewells. However acid treatments do not offer a general solution to well problems and may prove successful only where the conditions for acid reaction are right. The identification of such conditions requires investigation and research.
c) Clogging of the gravel pack by silts and clays is not considered to be a significant problem.
d) Rehabilitation treatments need to be developed through research. Those employed, although usually giving positive results, did not achieve the values of discharge and specific capacity which were recorded at the acceptance tests.
e) The use of a dispersing agent alone is not likely to produce positive results unless effective surging methods are also employed.
f) Chlorine treatments have shown the most

encouraging results and these need to be developed.

g) Surging methods using a surge plunger and compressed air need to be evaluated.

h) Chemical dosages in the treatments need to be fixed on some rational basis.

In SCARP IV, an abandoned tubewell No. MDK-504 and its new replacement were selected for investigation under the directions of Grey, Beath and D'Oliver. The study showed that the 100 foot fibreglass screen was clear and that gravel had mixed with the formation and had vanished at various depths, having been replaced by the formation next to the screen. It was hypothesized that the rotation of micaceous sand to a vertical orientation had restricted the movement of water towards the well but that further research was needed.

20 tubewells were recently rehabilitated by the Servicing Division of SCARP-I using sulphamic acid, chlorine and polyphosphates at an approximate cost of Rs. 37000 per tubewell. Table 3 gives the results.

Table 3. Chemically rehabilitated tubewells

Screen Material	Number of Tubewells	Specific Capacity (gpm/ft) Before Treatment	After Treatment	After One Year
Brass	7	82.1	72.5	60.1
Mild Steel	7	34.2	40.2	35.0
Fibre-glass	5	49.0	67.9	55.5
PVC	1	64.0	-	75.1

The data in Table 3 indicate that improvements were not only low but also short-lived.

3.2 Rehabilitation by SCARPs Monitoring Organisation, WAPDA

About 54 tubewells were rehabilitated by the SCARPs Monitoring Organisation of WAPDA on a systematic basis during 1983-84 and 1984-85. Investigative procedures included the use of TV cameras, and velocity and geophysical logging. Preliminary tests indicated that the cause of deterioration was mainly clogging by fines. Therefore, only mechanical rehabilitative means were employed. In SCARPs I and IV, the wells were rehabilitated by air surging using double packers. In SCARP II only mechanical surging with a surge block was used. Although the pack and the formation were choked by fines, the intakes of the pumps in many tubewells were blocked by ferric oxide deposits produced by iron bacteria. In SCARP III blasting followed by rawhiding from the lower portion upwards was used. The trials disclosed choking of the screen and its close vicinity by fine sand and silt. In SCARP V, rehabilitation was carried out with a surge block followed by rawhiding. After rehabilitation, a 48 hour test was run on each well using the original motor/pump and carefully recording hydraulic data and sand contents. Special care was taken to replenish the gravel pack and wells were regularly monitored.

The SCARP Monitoring Organisation has carried out valuable work on rehabilitation and its findings are significant. Those factors responsible for rapid deterioration of tubewells are summarized below:

a) A universal filter pack design was adopted for each programme without considering the variation in grading of the natural formation. The pack was generally too coarse to check the movement of fines into the wells; initially the wells discharged sand which then choked the pack.
b) Filter pack thickness has relevance to rehabilitation efficiency. A thin pack allows removal of all undesirable fine material during development and rehabilitation. Most SCARP wells had an annular thickness of 10 inches.
c) Coarse pack results in a wider screen slot size, causing an enhanced pressure drop which triggers chemical incrustation encouraged by the rapid change in porosity and material composition across the formation/gravel interface.
d) Old wells with short screens are less efficient than their replacements with longer screens.
e) Fibreglass screens have only 8 - 10% open area, which is low for development/rehabilitation using the water jetting technique; sufficient energy is not available for removing the fines in the vicinity of the formation. This problem has been overcome by better slotting methods.
f) Many tubewells have performed poorly because screens were placed incorrectly, for example in fine sand close to the pump where pumping and surging actions are high.
g) The free fall technique adopted for placement of gravel pack is faulty as it causes heavier gravel to settle earlier than lighter gravel. The resulting segregated pack allows the formation to move into the well. Furthermore, caving of the formation generally occurs where the water table is shallow and adequate differential head is not maintained during reverse circulation duration.
h) A conglomeratic deposit develops around the housing pipe in the depth range between the static and pumping water levels so that pack

replenishment during well operation is difficult.

i) During construction, the drilling fluid is enriched in clayey material due to repeated circulation between the hole and the pit which contains loose clay. This clayey fluid penetrates into the wall of the hole and the aquifer under hydrostatic pressure. If the hole is left idle after drilling, colloidal clay settles and is difficult to dislodge with mild development techniques. Dispersion occurs during pumping over a period of time. The clayey particles tend to reunite and deposit in the gravel pack, so reducing the permeability near the screen. Many tubewells pumped out clay in fairly large quantities during sectional development of the screens and their performance was much improved subsequently.

j) Development by "rawhiding" alone was used in the SCARP wells. This has proved to be an ineffective method, serving only to clean the well rather than develop it vigorously.

k) The tubewells installed under the supervision of qualified inspectors/sub-engineers have a longer life span and perform better than those supervised by unqualified staff.

l) Low lying pump houses are vulnerable to muddy rain water or irrigation breach water entering the well, which ultimately obstructs replenishment of the pack.

m) Iron bacteria promote precipitation of iron, which plugs the screen, formation pores and the pump intake. It is not yet clear whether these bacteria are originally present in groundwater and simply multiply as the amount of iron increases during well construction or whether they are introduced into the aquifer from the subsoil or back-siphoning from an affected to an unaffected well.

3.3 Rehabilitation of seven tubewells by WAPDA Construction Works Organisation (Central)

Before starting rehabilitation, the complete construction records, past performance data, and present hydraulic and mechanical data for seven tubewells were studied.

The present data include well discharge, pumping and static water levels and the condition of the pumping equipment. Discharges from three of the tubewells were measured using an orifice plate, whereas the smaller discharges from the remaining four tubewells were measured using 55 gallon drums. Water samples were collected two hours after pumping had stopped and sent to M and E Laboratory, WAPDA. Static water levels were also measured. The turbine pumps were then pulled out and samples of the incrustation on the bowl assembly collected in a bottle and immediately air-tightened for sending to PCSIR laboratories.

Rehabilitation then proceeded as follows:

a) The internal depth of each tubewell was measured and the base of the tubewell then cleaned with a bailer followed by compressed air, via an airline.
b) A chemical solution of 112 lbs of sodium hexametaphosphate and 10 lbs of hydrochloride in water to make 110 gallons was poured in the gravel parts and around the casing of each tubewell and left for 12 hours. About 300 cu.ft. of well graded gravel was kept at the site to pour into the annular space when required.
c) The tubewells have upper and lower screened sections of 10 and 8 inch diameter respectively. Systematic redevelopment using a 10 inch diameter surge block was carried out in the upper section first, moving downwards with an hour of surging spent on each individual length of screen. Initially, depth soundings were taken at hourly intervals but this reduced once less than 2 feet of sediment were settling out per hour. The surge block was then removed and the base cleaned using compressed air. This procedure was then repeated for the lower section, using an 8 inch surge block.
d) The smaller surge block was lowered to the bail plug and a chemical solution double the strength of that used in b) poured into the well. The surge block was run for 15 minutes per hour for a period of 12 hours taking hourly depth soundings, and the procedure repeated using the larger diameter surge block in the upper screened section. Thereafter, cleaning was carried out using compressed air and the whole sequence repeated until the rate of sediment deposition became negligible. A depth sounding was retaken.
e) The depths of which screen required mild blasting with prima cord were determined from construction records. The blasting operation was carried out in perforated seamless pipe of 6 inch diameter with a 0.375 inch diameter wall thickness to act as a muffler. The 10 feet long pipe was perforated for 8.5 feet of its length, the perforations being 1 inch diameter at 2 inch centres. A detonator was fixed with the prima

cord and installed in the perforated pipe tightly, facing the perforated portion. The pipe was lowered along with the cord and cable at such a depth that the lower end of the muffler was 8 feet above the top of the bail plug. The cable was connected to battery leads and starting from the second lowest individual length of screen two blasting operations were carried out per length. After blasting, depth soundings were taken.

f) Surging using compressed air was then carried out at intervals of 5 feet and the airline moved up when the water was found to be comparatively clear. Discharge and sand contents were measured at each step.

g) The 8 inch surge block was again lowered to the bottom and the double strength chemical solution applied. After this a double pocket eductor pipe and airline were lowered to the bail plug and surging and pumping operations were repeated at intervals of 5 feet in the screen. Discharge and sand contents were recorded at each step and the operation continued until the water became relatively sand free.

h) The final stage of rehabilitation involved rawhiding using a turbine pump, which was continued until the water became sand-free. Discharge, sand content and specific capacity were then determined after 12 hours of pumping, and the static water level was measured at 2 hours after cessation of pumping.

Table 4 shows the results of this detailed rehabilitation work; before treatment the average specific capacity was only about 10% of the original, but after treatment this had improved to 60%.

Table 4. Tubewells rehabilitated by WAPDA Construction Works Organisation (Central)

Tubewell Number	Specific Capacity (gpm/ft) Original	Before Treatment	After Treatment
MGT-247	102	13	63
MGT-252	113	16	79
MGT-255	148	10	90
MGT-256	112	4	69
MGT-257	116	15	85
MGT-287	105	8	45
MGT-281	111	9	52
Average	115	11	69

4 Discussion

The results of the foregoing rehabilitation projects can be summarised and compared with other programmes (Table 5).

Table 5. Effectiveness of rehabilitation by various agencies

Project	Number of tube-wells	Average Specific Capacity (gpm/ft) Original	Before Treatment	After Treatment	Percent Improvements *	
Mona Reclamation Project	7	95	42	97	230	102
Scarp III (Alipur Unit)	15	90	33	42	127	46
SCARP I	20	-	56	59	105	-
WAPDA Construction Works Organisation	7	115	11	69	627	60
Irrigation & Power Dept.	20	112	48	58	120	51
Water Resources Scientific Information Centre, USA	900	-	-	-	145	43

* Specific capacity after retreatment as a percentage of that before treatment (first figure) and as a percentage of the original (second figure).

The original specific capacity is rarely reattained once the well has significantly deteriorated and performances of 40 to 60% of the original are likely. There appears to be a linkage between the scale of improvement and the degree of investigation and experimentation used in the rehabilitation work. Furthermore the Central Works Organisation's tests indicate that significant improvement can be achieved even when wells have deteriorated to an extremely poor condition. The degree of deterioration and scale of improvement are largely independent of screen material.

5 Conclusion

Careful documentation of well construction and detailed knowledge of general rehabilitation techniques is vital to selecting the appropriate methods for rehabilitation of any particular well and completing this work effectively. Furthermore, various conclusions are presented above in which the role of high quality design, construction and supervision is clearly emphasised if large replacement programmes are to be avoided.

26 MONITORING, MAINTENANCE AND REHABILITATION OF WATER SUPPLY WELLS IN INDIA – STATE OF ART

P.N. PHADTARE and G. GHOSH
National Drinking Water Mission, New Delhi, India

Abstract

The high seasonal water level fluctuations in this tropical region with a limited wet period, makes it necessary to collect long term data on water level and ground water quality variation, to identify problems in declines in yields of wells. The Majority of well failures are due to high declines in water levels. Non-availability of slotted pipes of suitable size results in sand pumping. Indiscriminate construction of wells in the absence of ground water legislation, has not only resulted in well interference but has caused salinity ingress in certain vulnerable areas.

The Presence of iron in ground water is related to the high corrosion of the well and pump assemblies but the problem has now been solved to a great extent the by use of non-corrodable material in well construction.

Maintenance of water supply wells is receiving greater attention from water supply engineers in view of the very high growth rate of these structures during the International Water Decade. Several systems of maintenance are being attempted. The paper discusses their merits and demerits.

Very little organised work on rehabilitation of tubewells/borewells has been carried out in India. However, as these have started replacing the older shallow open wells, rehabilitation activity has increased. The paper describes the experimental work carried out in areas with problems of salinity and high iron content in ground water.

1 Introduction

Monitoring of the water supply wells to identify the reasons for their failure, needs thorough knowledge of the hydrogeological conditions, in India. This is because of the high seasonal water level fluctuations on account of limited wet (rainy) season and prolonged dry season in this tropical region. The problem has become more severe with frequent droughts.

In the drought prone areas where the severe drought is experienced once in almost every decade, a problem of progressive dewatering of the aquifers has created a very complex situation. Even though the well hydrographs indicate that the ground water recharge due to normal rainfall is just sufficient to compensate the discharge due to ground water abstraction plus the normal sub-surface outflow, the normal rainfall is not able to compensate the declines in water levels which take place during severe droughts. The zone of seasonal water level fluctuation thus continues to decline with every severe drought.

This decline in water levels is the main cause of well failure (wells go either dry or yields reduce considerably), especially in the normal hard rock areas where the phreatic aquifer comprising weathered zone and a few metres of fracture porosity zone immediately underneath the same, is the only water bearing zone. In the multi-layered aquifer systems in the hard rock areas (the basaltic flows with interbedded sedimentary formations or interbedded zone of paleoweathering and the sedimentary rocks with alternate sandstone beds), although the deeper aquifers continue to supply water even after partial or complete dewatering of the phreatic aquifer, the yields decrease considerably. This is also the case with the wells in alluvial areas. The monitoring of wells in these areas therefore needs long term observations for a decade of seasonal variations in yields of wells.

The common reasons for the well failure are given in the table No. 1.

2 Maintenance of Water Supply Wells

In India, where 81% of the rural water supply is provided by borewells/tubewells, from ground water resources, maintenance of wells has become an important factor in water supply schemes. Further, with the majority of the wells being fitted with hand pumps, finding out an efficient methodology for the maintenance of hand pumps has received great importance. The following are some of the procedures adopted in hand pump maintenance and each is having its own advantage depending upon the socio-economic background of the users. Varying combinations of the following maintenance systems were followed.

Table 1. **SHORT LIFE OF TUBEWELLS/BOREWELLS**

Problem	Reasons for failure	Remedial Measures
I. Non-availability of water bearing zones	Unproper Site Selection.	1) In the case of alluvial area, the site has to be abandoned and a better site selected. 2) In case of hard rock areas, blasting or hydrofracturing may help; otherwise the well has to be abandoued.
II. Pumpage of fine sand.	i) Inadequate well development ii) Lack of proper gravel pack iii) Unsuitable screen iv) Unproper screen placement v) Over pumping vi) Damage to well assembly due to corrosion	1) Pump development has to be carried out with discharge slightly higher than the proposed discharge. 2) Electrical logging is essential for proper screen placement. 3) In the north eastern region, the fine Tertiary sands need screens of 0.8 or 0.4 mm slots which are not generally available, Coir rope wrapping does help but is not a permanent measure. Only remedy is to carry out development for a longer period of time to provide a thicker gravel pack, if proper screens are not available. 4) Pump of proper specifications should be selected on the basis of long duration aquifer performance test carried out preferably during late winter or early summer.

		5) Clogging due to bacterial growth or encrutation, can be removed by treatment with bleaching powder/sulfamic acid/or other standard chemicals.
		6) Corrosion and damage of well assembly means mechanical failure of well and there is no other alternative but to redrill the well.
		7) In the coastal saline tracts and areas with high iron, clogging, corrosion and encrustation are common problems. Use of pipes of non-corrosive material (high density PVC pipes) help in preventing these to a great extent but in the initial well construction, care has to be taken to prevent leakage of saline water.
III) Reduction in well yield	i) Clogging of screens ii) Lowering of water levels during drought or due to overexploitation of ground water	1) The clogging due to bacterial growth (in area with high iron and sulphide in ground water) is a major problem in theeastern and north - eastern regions whereas clogging due to encrustation is a common plenomenon in the semi-arid western and northwestern regions; proper and frequent chemical treatment is the only remedy.
		2) Lowering of yields due to declining water levels has beocme a common phenomenon in the semi-arid and drought prone areas; control on ground water exploitation (legislation) and artifical recharge of ground water are the only preventive measures.
IV) Deterioration in quality of water	i) Salinity Ingress	1) This is mainly due to ground water over exploitation and there are no remedial measures except artificial recharge of ground water.
	ii) Leakage of water from adjoining saline water aquifers	1) This is normally taken care of at the time of construction of wells by providing slotted pipes against fresh water aquifer zone and by providing either Portland cement seal or grannular Bentonite (3 to 9 mm grain size) clay seal at the top and bottom of the fresh water aquifer.

iii) Contamination from surface water sources.	1) This takes place normally through the annular space around the casing pipes and is worst for shallow tubewells. In the case of boreholes in hard rocks, this may also result from surface water infiltrating from adjoining areas through permeable (primary or secondary fracture porosity) over-burden. There are no simple remedial measures for the regional contamination but the contamination through the annular space around the borewell can be prevented by providing a thick cement platform around the bore well/tube well.
	2) In the case of deep confined aquifers, a cement seal just below the overlying phreatic aquifer, which is vulnerable to contamination, is all that is needed.

i) Government Sector.
ii) Training village blacksmiths or cycle mechanics in maintenance work.
iii) Training unemployeed village youths/tribal people/women in maintenance work.

Except when it is carried out by the Government agency, the maintenance is of a routine type such as lubrication, nut-bolt tightening and replacement of leather washers. These tasks involve work above the ground level. Any major repairs and work with the rising main assembly and pump cylinder are carried out by the Government organisation.

The use of the India Mark II hand pump, has itself reduced the failure rate of handpumps from 70% to 12%.

The maintenance entirely at community level or without involvement of a Government agency or maintenance entirely by Government agency without involvement of the user public has not been found feasible. This is true not only in the case of hand pump wells but also applicable to any rural water supply scheme. The mixed nature of the rural community with diverse financial capabilities does not make it possible to allow complete village level maintenance. However, experience has shown that no water supply scheme would be 100% successful unless there is full community participation. Awareness camps and health education programmes are a must but more than these, part financing by the community, either in kind or cash is found to be more useful in ensuring sustainability of the source.

2.1 The different systems followed in India are as follows:

Three-tier-system: The first level comprises a trained person from the village who is provided with the necessary repair kit and some honorarium. Then for every 100 hand pumps or say a group of 20 villages, a regular government fitter or mechanic is appointed who visits the hand pump installations periodically or as and when called out. A team at district level with mobile van renders help when there is a major problem.

Two-tier-system: Is almost on the same lines but instead of a district level mobile van, block level mobile vans are provided. This method was found to be more effective as the mobile van at block level (a district normally comprises 9 to 12 blocks) helped in speedy completion of repair works. On an average, there are about 500 hand pumps in a block.

The Denmark assisted DANIDA PROJECT in coastal Orissa is having village blacksmiths as self employeed mechanics who not only attend to minor repairs of pumps but report about well performance and the quality of water on the basis of enquiries from the user public. This method was found to be very effective and informative.

3 Monitoring Of Variations In Quality Of Water

The main system in quality monitoring is the regional well inventory carried out at the State level or Central Ground Water Departments. They have established regular observations wells (one observations well for every 100 to 250 sq.Km area) where water level data is recorded every quarter of the year and a water sample is collected in the pre-monsoon season (when the water levels are at their lowest) for detailed chemical analysis. Although this gives the regional effect of ground water exploitation, such as the regional declining trends and variation in quality of ground water, problems of certain individual wells, which are often critical, are not brought out clearly. The Following methodology may help in identifying such problems.

Immediately after the construction of the well, a detailed basic data report indicating yield test results and complete chemical analysis of water sample should be prepared. Thereafter, a rapid survey of the wells with the help of block level mobile units as in the three and two-tier systems of maintenance, should be carried out to collect information about the performance of the well. A simple Electrical Conductivity (EC) measurement of the water sample with the help of a portable EC meter once every 3 months is more than sufficient to find out any significant variation in water quality.

4 Rehabilitation Of Wells

Not much organised work on rehabilitation of wells has been carried out in India. Normally, a failed tubewell is replaced by a new tubewell wherever there is problem of sand pumping, drastic reduction in yield or quality problem. However, as the tubewells and borewells have started replacing the age old system of construction of shallow open wells, the rehabilitation work is getting more attention.

Experiments on a small scale have been started recently, as pilot projects. These are described below:

4.1 Danida Rural Water Supply Project, Coastal Orissa

Hundreds of hand pump wells were constructed in this coastal tract occupied by 100 to 300 metres of thick alluvial deposits which are mainly estuarine deposits. The formations have inherent brackishness but the high ground water recharge from the western hilly and forested areas have a flushing effect on some of the aquifers, thereby producing some fresh water zones. The skill in well construction lies in identifying the fresh water aquifer and tapping it with proper cement sealing to prevent salinity inqress from overlying or underlying aquifers. Another problem with the aquifers, in this area, is the high content of iron in ground water. The problems arising out of these two factors, that is high salinity and iron are as follows:

a) Development of iron bacteria and sulphide reducing bacteria which has resulted in clogging of well screens and subsequent reduction in yield of well.

b) Verticle leakage from adjoining saline aquifers: Decrease in yield of fresh water aquifer due to clogging, with incoming constant leakage from saline aquifers, resulted in quality deterioration.

c) Corrosion of well assembly due to saline and iron rich water and action of iron bacteria.

The remedial measures implemented and described below, showed very satisfactory results.

1) As the discharge from handpumps was very insignificant, (ie. around 0.20 litres per second), it was decided to avoid gravel packing around the well assembly. The absence of gravel packing and the provision of a cement seal above and below the fresh water bearing zone prevented leakage of saline water, as the main leakage was through the gravel pack.

2) PVC riser pipes and brass cylinders for the hand pump assembly and PVC well assembly decreased incidences of screen clogging due to bacterial growth. Morever, there were no problems of corrosion.

3) Treatment with 5% hypochlorite solution at the final stage of well construction helped in reducing the growth of iron and sulphate reducing bacteria.

4) Frequent well development with slightly higher discharge also retarded clogging action due to growth of bacteria.

4.2 Western Saline Tracts: In the western arid and semi-arid parts of India, where the high rate of evaporation, limited annual ground water replenishment, inherent salinity of the marine sedimentary formations and salinity ingress from inland saline marshy lands of the famous desert of "Little Rann of Kachchh" has resulted in the high salinity of ground water resources. Although interbedded or lenticular fresh water aquifers are available for limited exploitation, salinity ingress is often a problem. Encrustations of carbonates and sulphates often cause the clogging of wells. Experiments carried out jointly by Gujarat Engineering Research Institute and Central Ground Water Board have established the effectiveness of sulfamic acid (commerical product - Aquamor) in cleaning the wells. Treatment for 10 hours with solutions of 20% concentration, followed by mechanical surging and compressor development showed improvement of 12.8% in well yields. The sulfamic acid did not have any adverse effects on the metal of the tubewell casing pipes.

4.3 Clogging Due To Silt Deposition And Aquiffer Formation Readjustment This was observed in the water supply wells of Sami-Harij regional domestic water supply schemes, Mehsana District, Gujarat State. The wells were constructed in such a way that they served as dual purpose wells - pumping cum recharge wells. The over exploited deeper aquifers between depths of 40 to 225 metres were connected to the highly saturated and under-utilised phreatic flood plain aquifer of the Saraswati river. This was effected by tapping the phreatic and confined aquifers in a single tubewell of 225 m depth, (connector well) thus allowing free fall of water from phreatic aquifer (depth to water level, 4 m below ground level) to the confined aquifer zone (depth to piezometric level, 30 m below ground level), at the rate of 280 cubic metres per day. After 100 days of continuous recharge, the recharge rate started reducing due to clogging and by the end of 250 days recharge, the well became clogged to the extent of 60%.

Treatment with bleaching powder (10% concentration) for 12 hours, followed by 24 hours treatment with dilute hydrochloric acid (5% concentration) and mechanical surging, improved the well efficiency, bringing it back to 85% of the initial efficiency. The hydrochloric acid however reacted with the steel casings, producing iron chloride. As such, in subsequent cleanings, sulfamic acid, which does not effect the metallic parts, was used. The T.V. bore-hole camera did not indicate any bacterial growth on the inside of the slotted pipes. There was no possibility of clogging due to air bubbles as the cascading water has to pass through 80 metres of water column in the well before entering the slots, thus allowing the vertical escape of air bubbles. It was therefore, presumed that the silt (8 parts per million turbidity) from the phreatic aquifer and the aquifer clogging due to the rearrangement of aquifer grains due to the back pressure of recharged water were responsible for the clogging. Hydrochloric acid was used for the removal of silt which was mainly calcareous whereas bleaching powder was used for disinfection, as some bacterial encroachment from the shallow aquifer was suspected. Redevelopment with pumping, removed aquifer clogging due to readjustment of grains as well as due to the precipitates resulting from acid action. It was therefore decided to use the well as dual purpose well so that daily pumping could do the job of well-cleaning. The well thereafter maintained the same efficiency even after 3 years' use without any acid treatment.

4.4 Decrease In Yields Of Wells Due To Lowering Of Water Levels: This has become a wide spread phenomenon in the entire hard rock terrain due to increased exploitation. Normally the boreholes in hard rocks are of 60 to 90 m depth. However, in most of the wells, the high yielding zones are present down to 40 m depth. The lower zones have poor yields due to poor intensity of the fractures. The only remedy in case of such wells was found to be development of the fractures in the lower zones. The normal practice was to extend the fractures by blasting or by acid treatment (in the case of calcareous rocks) and these treatments helped in improving yields of wells in the case of over 70% of the wells. The failure in the case of the 30% remaining was mainly due to the absence of any high yielding major fracture in the immediate vicinity. Since 1989, the hydrofracturing units supplied by TGB Water and Energy, Sweden are also being used. These units operating at 340 litres per minute flow and under 2175 p.s.i. (150 bar) pressure, had showen very encouraging results in the development of fractures. However, the cost of treatment, by hydrofracturing which works out to nearly Rs. 15000/- per well is much higher than Rs. 2000/- per well by blasting and acid treatment.

References

Carl Bro International Denmark (1989): Project Review Report, DANIDA PROJECT, GOVT. of Orissa, Bhuvaneshvar, India.

Ghosh G.(1988) Management of Drinking Water in Drought. 14th WEDC Conference, Water and Urban Services in Asia and Pacific, Kaulalumpur.

Phadtare P.N.(1988): Artificial Recharge studies in Mehsana and Coastal areas of Gujarat, India. Seminar Volume, International Seminar on Artificial Recharge, Ahmedabad, India.

Mudgerikar R & Phadtare P.N.(1989): Rehabilitation of failed wells by hydrofracturing, Seminar Volume, Statewise Seminars - Maharashtra, Gujarat, Rajasthan, Madhya Pradesh, India.

27 VILLAGE BASED HANDPUMP MAINTENANCE – EXPERIENCES FROM SRI LANKA

J. HETTIARACHI
National Water Supply & Drainage Board, Sri Lanka
S.W. ERIKSEN
Kampsax Kruger/Danida, Copenhagen, Denmark

ABSTRACT
This paper summarizes experiences from five years of maintaining & monitoring 1600 Deep tube wells installed in central Sri Lanka during 1985-1988, and the efforts made in establishing a village based Maintenance Organization.
Keywords: Handpump Maintenance, Organization, Monitoring, Classification Of Failures, Failure Rates, Sri Lanka.

1 INTRODUCTION

Since the beginning of the Water decade, considerable efforts have been vested in installing Handpumps in developing countries in places where the demand for potable water is high and the sources are scarce or at times nonexistent.

In this regard, one major concern is to ensure that the handpumps - once installed - are maintained.

Experience indicates that maintenance can be ensured only if the users are directly involved, and that this involvement is likely to increase in proportion to the ratio of total repairs that can be carried out by the users.

In deep tube wells, a number of "failures" are caused by factors that are not related to pump performance, such as bore hole collapse, and situations requiring flushing.

Such major repairs require assistance from agencies outside the village, or even the nearest Administrative unit to which the village belongs.

For minor repairs, assuming that the handpump caretaker appointed by the consumer society is capable of carrying out the majority of such repairs, she/he would still need outside assistance to purchase spares and services which are not available at village level. Therefore, the capacity to cater for this need must be established at the nearest administrative level, and maintained there until private entrepreneurs are able to fill this need at a reasonable cost.

When this happens, handpumps at village level may be obsolete. Making the consumers understand that they themselves are responsible for maintaining their drinking water facilities and enabling them to find the resources to do so will accelerate this process.

2 VILLAGE BASED HANDPUMP MAINTENANCE

In Sri Lanka, the capability to carry out major maintenance operations such as redrilling and flushing, rests almost exclusively with the Donor Assisted projects and the Regional Support Centres (RSC) of the National Water Supply and Drainage Board (NWSDB).

When requesting a major repair, e.g. flushing, the users (Consumer Societies & Caretakers) can approach the RSC only through the local governmental office (Pradeshiya Sabha). Therefore, since the presence of Donors is not permanent, the only feasible maintenance organization (for the time being) seems to be the "Three Tier" type. Village (1), Pradeshiya Sabha (2) and Regional level (3), as shown on Fig. 1.

At present the capability of carrying out maintenance at levels (1) and (2) is limited, except for routine maintenance. Therefore, efforts are continuously made to enable level (2) to assist the users (caretakers) to carry out all repairs, except major ones (Table 2).

This entails a considerable amount of training and human resources development, the nature and extent of which is determined by type and frequency of repairs, and the performance of the O & M Organization at all levels.

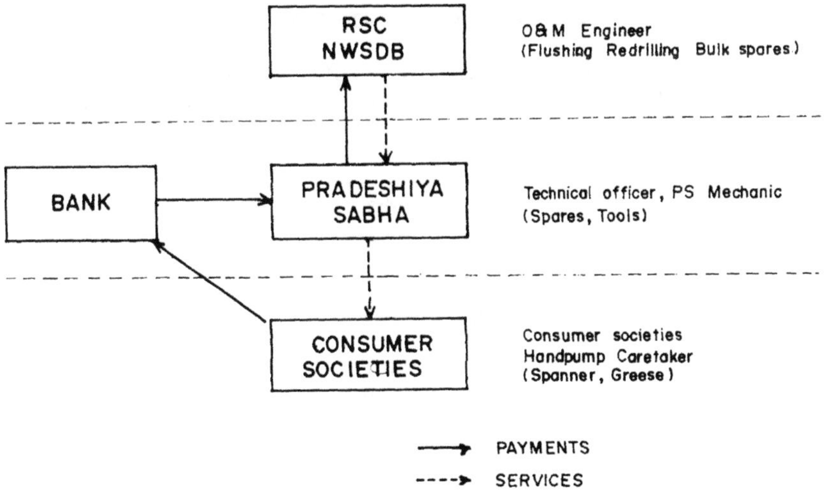

Fig. 1. Maintenance organization.

The strategies employed in this regard are:

- Strengthening the regional offices of the National Water Supply and Drainage Board, the assistance of which will be required in the event of major repairs such as bore hole flushing.
- Strengthening of the local government structure. In this regard, requirements such as workshops, tools and equipment, motorcycles, and technical training have been provided.
- Training and motivating users, to make them understand that maintenance of the handpump is their responsibility. Therefore, formation of consumer societies has been encouraged, and villagers, preferably women living near the installations, are trained as volunteer caretakers.
- The consumer societies pay for services (supply of spares and assistance of PS Mechanic) rendered by level (2), and level (2) pays for services (Spares, Flushing & Redrilling) provided by RSC level (1).
- Close monitoring of maintenance activities at all levels and modification of present procedures on basis of experiences thus gained.

These strategies are the results of experiences gained during the Implementation Phase and from the constant dialogue with all parties involved, users, government and donor agencies. It must be admitted however, that they have not yet been fully implemented, in fact there is still a long way to go before the system becomes "self-sustaining".

3 MONITORING OF MAINTENANCE ACTIVITIES

The handpump failures are divided into 12 main categories, each of which is subdivided according to status as shown below.

Table 1. Failure Codes.

Main Category Codes		Status Codes
HP - Handpump	CY - Cylinder	1 - Worn
HS - Handle Shaft	CW - Cup Washer	2 - Inoperative
BO - Bolts	FV - Foot Valve	3 - Disconnected
CH - Chain	AP - Apron	4 - Broken
CR - Connecting Rod	BH - Borehole	5 - Disappeared
RM - Riser Main	WQ - Water Quality	6 - Fallen into the well
		0 - Other

This data is stored in the Databank, using the codes shown in Table 1.

Failure of each component is determined by several parameters, such as the number of users, depth of well, Water consumption, strength parameters of component, diligence of caretakers quality of installation and quality of water.

For each failure event, there is a linear relationship between the time lapse T until failure, and the inverse INV (N) of the normal distribution:

$$T = S * [INV (N)] + Tav \qquad (1)$$

Using (1) the standard deviation, S, and the average service life, Tav, can be calculated using linear regression. The result of this exercise is summarized in Table 2 below.

Table 2. Main category failure rates.

Code	Item	Average service life (years) Tav	Standard deviation (years) S	Corr.	Rep category	NOTE
CR	Connecting Rod	24.1	7.4	0.90	m	AP, WQ Categories insignificant.
RM	Risermain	5.1	1.9	0.93	m	m : Minor Repair can be carried out at village level by the caretaker, occasionally assisted by the Pradeshiya Sabaha Mechanic level II.
CW	Cup washer	5.2	1.7	0.93	m	
CY	Cylinder	6.0	2.1	0.88	m	
BH	Borehole	6.5	2.3	0.88	M	M : Must be carried out with assistance from NWSDB RSC level
BO	Bolts	9.6	3.1	0.94	m	
CH	Chain	8.5	2.8	0.88	m	* : Excluding "Childhood diseases (e.g. CR3)
FV	Foot Valve	11.8	3.5	0.87	m	
HB	Handle Bearing	7.3	2.2	0.97	m	

Based on 2531 failure events during 5 years of monitoring 1631 installations.

Using the above data, it is possible to calculate the future failure events of any component, as shown on Fig. 4 below.

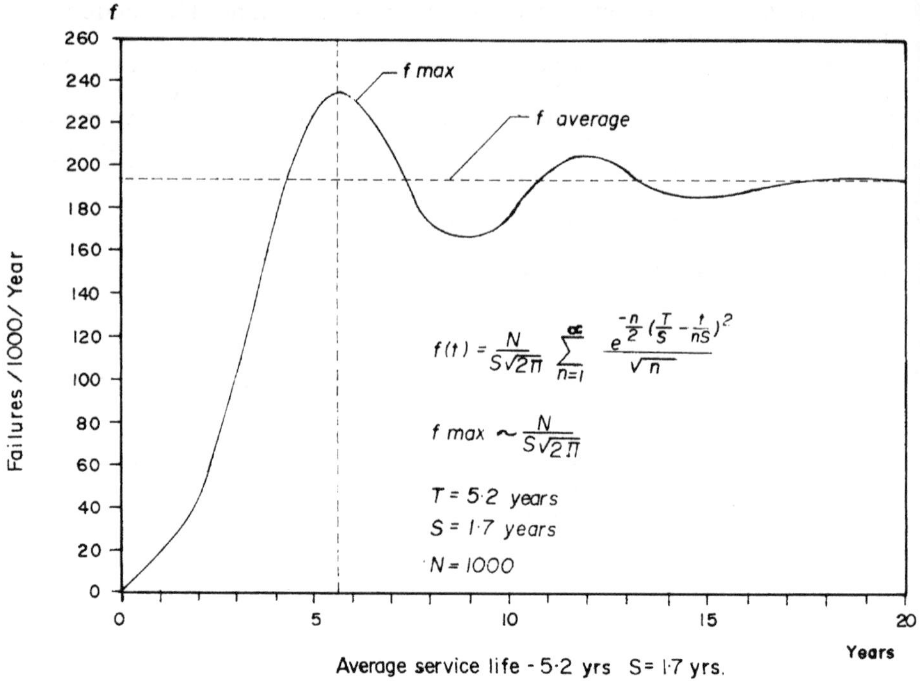

Fig. 2. Prediction of cup washer failures (CW1)

It is seen that the peak failure rate is reached after a time lapse approximately equal to the Average service life, (MTBF) and that after a period f ≈ 3 x MTBF, the failure rate is practically constant (1000/5.2=192). It is further seen, that the "average failure rate" during the first two years is ≈ 20 ie only 10% of the real steady state failure rate.

It is further seen (Table 2) that when the individual failure rates have all stabilized, the total failure rate (TFR) pr pump pr year is:

$$\mathbf{TFR \approx \Sigma\ 1/Tav \approx 1.2/pump/year.} \quad (2)$$

As mentioned earlier, the above estimates are based on the assumption that the failures occur according to a Normal process.

This Hypothesis has been Chi Square tested on 70 installations having a cylinder setting at 28m, and the outcome is shown on Fig. 3. below.

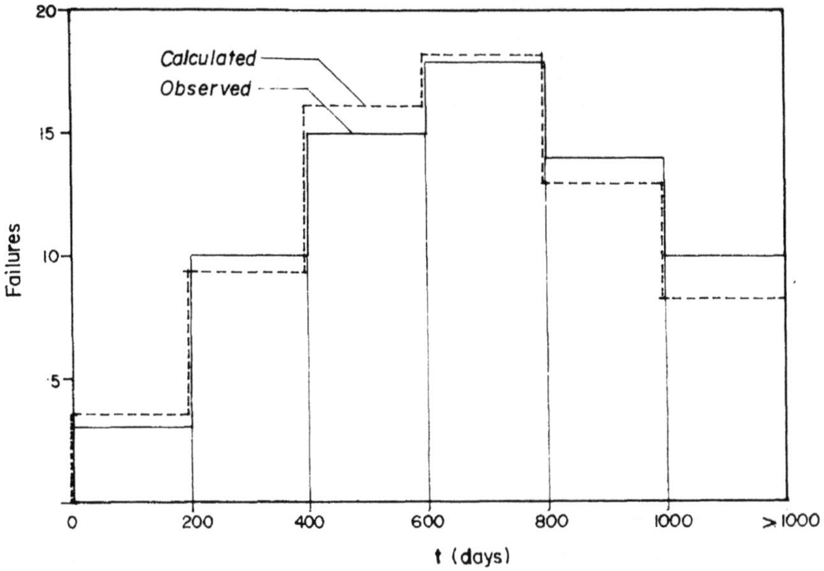

Fig. 3. Calculated and observed RM4 failure frequencies of installations at 28 m.

4 MONITORING THE PERFORMANCE OF THE MAINTENANCE ORGANIZATION

From the users point of view, the maintenance is satisfactory, if the time during which the pump is out of order is kept at a minimum.

In case of major failures (M) level 1 has to be notified through level 2. further, in case of a flushing operation, the repair has to be fitted into the overall flushing programme. For this reason, a "down time" Td of about one month would be the minimum.

For minor failures (m) the minimum down time approximately 7 "days, ie 3 days to notify level 2 (The PS TO) and 4 days for mobilization.

In order to monitor Td and other aspects of HP performance - monthly surveys are carried out on basis of a 20% random sample of the installations. The relationship between the observed failure rate, α_O and the real failure rate α_R is:

$$\alpha_O = \alpha_R * Td/Ts \qquad (3)$$

Where Td is the "down time" (days), Ts the duration of the survey (days) and α_R is the rate of repair equal to the real failure rate, α_{rf}

If $\alpha_R > \alpha_{rf}$, then after a certain period, Td will approach zero, and no failures would be observed. If $\alpha_R < \alpha_{rf}$, Td will increase from survey to survey.

Table 3 below summarizes surveys carried out during Jan., Feb., and March 1990.

Table 3. Failure rates observed during first quarter of 1990.

Month	January	February	March
Ts (days) α_o (%)	28 0.01	25 0.01	30 2.9
Td* (days)	-	-	6

It is seen that the performance was satisfactory during the quarter, albeit with a downward trend.

In the above, it is tacitly assumed that preventive maintenance is not carried out, which of course it is. The effect of preventive maintenance is to reduce α_{rf}, and since a standard procedure is followed, the ratio of this reduction is constant, so that Td is an upper bound, and will vary as the real Td. Therefore (5) is appropriate when measuring trends in maintenance performance.

5 CONCLUSION

The maintenance of Deep tube wells at village level entails operations that cannot be carried out without services from agencies outside the village.

In order to ensure that such services are delivered, it is necessary that the consumers organize themselves into consumer societies, capable to pay for such services and -if need be - able to enforce the delivery. Similarly, it is important that the persons involved in O & M at higher levels in the organization, be made conscious of their rôle in serving the community to which they themselves belong.

The creation of such ideal conditions is a long process, and constant feed back regarding performance is needed at all levels.

Therefore, close monitoring of all aspects of operation and maintenance is an important activity.

In the long run, the activities of the consumer societies may diversify into covering other aspects of improvement of living conditions in the villages, thereby accelerating overall development.

28 REHABILITATION OF PROBLEM WELLS IN ORISSA DRINKING WATER SUPPLY PROJECT

R. KUMAR DAW
Orissa Drinking Water Supply Project, Bhubaneshwar, Orissa, India

Abstract

The Orissa Drinking Water Supply Project, assisted by Danida and working in 20 blocks of the coastal saline tract of Orissa, had constructed about 4000 tube wells fitted with hand pumps in 8 blocks of the project area by December 1989. A significant fraction of the 1650 wells completed in Phase I of the project, during 1985-87, showed varying degrees of rejection by user communities in early stages of pump life. The cause for rejection was deterioration of water quality, which was confirmed to some extent by laboratory analysis of water samples from problem wells. Subsequently, a special study of Phase I tube wells indicated some of the problem areas which may have caused water quality deterioration with time.

Since mid-1989 about 50 problem wells in one block of the project area have been chemically treated, redeveloped, and reinstalled with pumps some of which have noncorrosive components.

This paper attempts to document the experience from this well rehabilitation effort, and tries to draw inferences from the data which may have bearing on future well rehabilitation programmes in the project.

1 Background

The Orissa Drinking Water Supply Project is a bilaterally aided project for 20 blocks in the coastal saline tract of Orissa. It is being implemented by the Public Health Engineering Department-PHED, of the State Government of Orissa with financial and technical assistance from Danish International Development Agency-Danida.

During Phase I of the project, which commenced in August 1985, approximately 1400 new tube wells were constructed and 250 existing tube wells were rejuvenated in 3 blocks of the project area. All these tube wells were fitted with hand pumps. Generally India Mark II hand pumps were installed in deep well configuration on new wells and suction pumps were installed on existing tube wells.

As construction of tube wells and installation of hand

pumps progressed, completed installations were handed over to the village-based hand pump maintenance system established by the project to ensure sustained operation of the pumps through regular maintenance.

2 Water Quality Problems

Problems of high iron content and salinity were anticipated in groundwater of the project area and these two factors were used to set up the basic water quality criteria of an acceptable well. Wells delivering water with Total Iron content up to 3 mg/l and Chloride content up to 1000 mg/l were considered as acceptable wells for installation of hand pumps. Geophysical techniques were used to delineate fresh water zones from saline zones and laboratory analysis of water samples was done for each well for measuring iron and chloride content. Almost all newly drilled wells were fully cased with PVC pipe, with a 4 m length of PVC well screen placed in the aquifer to be tapped.

Within one year after handover of pumps to the maintenance system, consistent reports began to be received by the maintenance system that pumps were falling in disuse due to worsening of the quality of water from an increasing number of wells. Internal evaluatory studies and laboratory testing of water samples confirmed thesereports to some extent, as did external studies by consultants. By October 1989, users' reports and subsequent laboratory confirmation indicated that between 20% to 30% of the 1650 wells of Phase I of the project showed varying degrees of water quality problems in the categories of taste, color, and odor. By the criteria of iron and chloride content mentioned earlier, the incidence of such problem was, ofcourse, much lower.

The occurrence of problem wells showing water quality deterioration, had far reaching implications since this phenomenon had not been anticipated, could not be easily understood or explained, and brought the credibility of the maintenance system to question. If and when hydrogeological, hydrochemical, microbiological, etc., solutions were found for this problem , they would be applied to new wells. But the maintenance of 1650 wells of Phase 1 and an additional 2200 wells of Phase II A (completed by late 1989), with their share of problem wells, still needed solutions.

The purpose of this paper is to document the experiences gained by the project in its attempt to rehabilitate problem wells and attempt to find the degree of validity and relevance of these rehabilitation attempts.

3 Nature of Rehabilitation Efforts

Delang block was one of the 3 blocks where wells were constructed in Phase 1 of the project. By late 1989, 507 hand pumps were under the maintenance system in Delang block, of which 438 pumps were installed by the project on newly drilled wells, 50 pumps were installed on existing wells and 19 wells with serviceable pumps were already in the block. Inspection of pump components from wells with reports of water quality deterioration, showed that riser pipes, pump rods and pump cylinders of such wells generally had deposits, scaling and corrosion. The deposits and scale material were light colored clayey, brown brittle layers like rust, hard black layers, powdery chalk white deposits and brown and black slime from algae. Sometimes this material shook loose or peeled off from the rods or pipes of the pump and came up with water drawn from pumps.

When such a problem pump was opened and reinstalled after repairs, cleaning or replacement of below-ground components, users generally reported an improvement in water quality, and the total iron content in water samples from such wells dropped. This was one of the main reasons why total iron content became a yardstick for quantifying the general acceptability of water from a well and for monitoring the behavior of water quality in a problem well.

While water quality improved in a well with cleaning or replacement of pipes and rods, it was also observed that water quality of such problem wells gradually deteriorated again with time, recording rejection from users a corresponding increase in total iron content.

The above observation led to the need to find non-corrosive below ground pump components. Favorable experience with PVC riser pipes in a similar project in central Sri Lanka, resulted in the choice of a similar design. Four installations were converted in early 1988 in Delang block on identified problem wells with PVC pipes, bright steel electro-galvanised rods, and the standard India Mark II pump cylinder.

During late 1988 and early 1989, a special study of problem wells in the Orissa project indicated that inadequate development, sedimentation, leakage in the casing assembly, and a very complex hydrogeological situation were contributing factors to the deterioration of water quality in once-successful wells. The study also pointed out that groundwater quality in the project area was time-dynamic.

From mid-1989 to early 1990, 50 problem wells in Delang block were treated with a variety of well rehabilitation measures. These measures were:

3.1 Chemical Treatment

Initially a few wells were treated with a solution of bleaching powder in an attempt to sterilize the wells and retard bacterial activity associated with high iron content. Later a more refined procedure of chemical treatment was formulated using Sodium Tripolyphosphate, Sodium Hypochlorite, and Calcium Carbonate. This combination of chemicals was expected to breakup residual drilling mud and sterilize the wells, killing existing bacteria.

3.2 Well Development

After the chemical treatment, wells were developed by using compressed air to lift water out of the well at a high rate. This air lift method was expected to remove accumulated sediments, leftover drilling mud and fines from the formation around the well screen.

3.3 Pump Reinstallation

Reinstallation of pumps on developed wells was done using a combination of PVC and new GI pipes, and stainless steel or new bright steel electro-galvanised rods. Standard India Mark II cast iron jacketed cylinders were used on these wells.

4 Methodology of Data Analysis

For the purpose of data analysis, the original well construction records for each well provided the basic information of well depth, date of construction, depth, position of strainer, etc. Results of water sample analysis following reports of water quality problems were incorporated in each well's history. In some cases, when the exact date of water sampling was not available, the last day of the month was used against the particular sample's results. After identification of a well as a problem well, results of water quality analysis, detailed records of chemical treatment, and well development were maintained for each well. Though detailed records of maintenance of each pump was available, most of this data was disregarded for this analysis. The maintenance data,however, provides very significant information on the condition of below-ground components at the time of maintenance interventions and reinstallations.

After completion of development and reinstallation, water samples were drawn regularly from wells and analysed for the total iron and chloride content. Total iron content was used as an indicator for the acceptability of water quality of a well. Consequently, the behavior of iron content in a well before and after interventions of chemical treatment, development and reinstallation, was to determine the manner in which the quality of water in that well changed.

For the purpose of data analysis, it was necessary to establish a " Date of Intervention" before which and after which the behaviour of iron content could be compared. However, interventions such as chemical treatment, development and reinstallation did not always occur in quick succession, or in a set sequence. Therefore, to establish a date of intervention, data for each well was examined and where possible, the date of the last intervention, usually that of a reinstallation, was taken as the Date of Intervention for data comparison. In some cases, it became necessary to establish two separate Dates of Intervention for the same pump, since the first attempts of rehabilitation with bleaching powder treatment and reinstallation was followed much later (sometimes about a year later). before chemical treatment and well development was attempted.

Detailed well history sheets, called "Data Summary Sheets-Well Rehabilitation", have been compiled for each of the 50 wells giving details from the basic well records, and from the date-wise record of water quality observations, static water level measurements, chemical treatment, development, and pump reinstallation.

4.1 Total Iron Content Data

Corresponding to each well's record, a graph of time in "Agedays" vs. Iron content in mg/l has been plotted. In most cases the record of Agedays has started from 1st. October 1985 since some initial observations of iron and chloride contents were available from that time. These two documents form the basis for all further analysis of data. Examples of Agedays vs. Iron Content graph are given in Annexure 1.

The Iron Content graphs were further elaborated for each well by plotting similar graphs for each well for the period after the date of intervention. By this method a clearer picture of the behavior of iron content after intervention was possible. Some examples of these graphs have also been shown in Annexure 1.

To understand if there are any geographical patterns to the behavior of iron content, graphs of wells falling within one Gram Panchayat (a group of villages) were plotted together. One such graph is presented in Annexure 2.

4.2 Qualitative Categorisation of Iron Content Changes

A tabulation of the highest and lowest values of iron content in each well before and after the date of intervention was then made. In considering the high values of iron content, some times uncharacteristically high values were observed. These have been eliminated from the analysis. This analysis was then used to arrive at a qualitative categorisation of the maximum level of iron in each well before and after intervention. For the purpose of making this qualitative assessment, the following guide-lines were applied to the maximum value of iron content before or after the date of intervention:

Table 1: Qualitative Categorisation of Total Iron Content

Maximum Value of Total Iron Content	Qualitative Categorisation
$\geq$10 mg/l	Very High
$\leq$ 5 mg/l to $<$10 mg/l	High
$\leq$ 2 mg/l to $<$ 5 mg/l	Medium
$<$ 2 mg/l	Low

This qualitative categorisation has been applied to the iron content data. The above categorisation has been made to facilitate classification of the vast number of readings of iron content of water samples drawn from problem wells.

Having determined the qualitative category of iron content of a well before and after intervention, the categorisation of change in behaviour of iron content before and after intervention was done in the following manner:

Table 2: Categorisation on Change in Iron Content before and after Date of Intervention

Qualitative Categorisation of Iron Content		Categorisation of resultant change in Iron Content
Before Intervention	After Intervention	
Very High	Very High	No Improvement
" "	High/Medium	Partial Improvement
" "	Low	Improved
High	Very High	Deteriorated
"	High	No Improvement
"	Medium	Partial Improvement
"	Low	Improved
Medium	Very High/High	Deteriorated
"	Medium	No Improvement
"	Low	Improved
Low	Very High/ High/Medium	Deteriorated
"	Low	No Improvement

The tabulation of resultant changes in iron content / iron content behaviour is illustrated in Annexure 3.

4.3 Well Development Data

An extract from the records of well development indicating the results of development efforts, where multiple attempts of well development was necessary, the duration of development on each occasion, and wherever possible, the nature of unsatisfactory development, was compiled from each well's development records. Since chemical treatment was a precondition to development, only treatment with bleaching powder, which was not followed by development, was indicated (as in Annexure 4) with the symbol "B". The symbol "Y" denotes "yes" to signify those cases of chemical treatment where development was undertaken on the same day.

The qualitative evaluation of the behavior of iron content before and after intervention and the results of development were then combined together in Annexure 3 to give the final result of the rehabilitation efforts for each well. By this method, it became possible to make qualitative assessmentsof the degree to which iron level changed before and after intervention and whether development yielded satisfactory results.

5 Results

Data as illustrated in the Annexures have led to the emergence of three significant correlations. These are:

5.1 Iron Content Monitoring & Well Development Results

The results of behaviour of iron content and of well development are summarised in Table 3 below.

Table 3 : Summary of Results of Iron Content Monitoring & Development

Results from Iron Content Monitoring	Results from Development: Single Attempts of Development					Multiple Attempts		TOTAL
	Satisfactory Development	Results Not Known	Bleaching Powder Treatment Only	Unsatisfactory Development	Not Developed	Ending Satisfactory	Ending Unsatisfactory	
Improved	8	2	-	4	-	5	1	**20**
Partly Improved	3	1	2	3	-	2	2	**13**
Not Improved	3	1	-	4	-	1	-	**9**
Deteriorated	-	-	1	-	2	1	2	**6**
Multiple Reinstallation	1	-	-	-	1	-	-	**2**
TOTALS :	**15**	**4**	**3**	**11**	**3**	**9**	**5**	**50**

From Table 3 the following conclusions can be drawn:

20 wells (40%) showed an improvement in iron content behaviour.

13 wells (26%) improved partly, 9 wells (18%) did not improve and 6 wells (12%) became worse by the yardstick of iron content.
Development of 15 wells (30%) was satisfactory. An additional 9 wells (18%) recorded satisfactory development after more than one attempt.
Bleaching Powder treatment was done in 3 wells (6%) without development. An additional 3 wells (6%) were neither treated nor developed.
Unsatisfactory development was recorded in 9 wells (18%) Also, 5 wells (10%) recorded unsatisfactory development even after multiple attempts.
Only 8 wells (16%) yielded the most desireable results of both appreciable improvement in iron content behaviour and satisfactory development.
An additional 6 wells (12%) improved partly in iron content behaviour and did not have any development result recorded. To this group could also be added 8 wells (16%) which need multiple attempts of development before satisfactory results were obtained, and showed appreciable or partial improvement in iron content behavior and 1 well (2%) which was developed satisfactorily and had multiple dates of intervention.

5.2 Depth of Wells & Development Results

In Table 4 shows the relationship between well depth and development results, From which the following inferences can be drawn·

31 wells (62%) were less than 45 m deep. The next largest group were 15 wells (30%) which over 100 m deep.
In the group of wells less than 45 m deep, 18 out 26 wells (69%) were satisfactorily developed. In the above-100 m depth group 5 out of 15 wells (33%) could be successfully developed.
This result was expected since the air lift method of well development has severe limitations in deep wells with high water tables.

Table 4 : Correlation between Well Depth & Development Result

Depth Ranges of Wells	Results from Development: Single Attempts of Development: Satisfactory Development	Results Not Known	Bleaching Powder Treatment Only	Unsatisfactory Development	Not Developed	Multiple Attempts: Ending Satisfactory	Ending Unsatisfactory	TOTAL
Less than 30 m	-	1	1	-	-	-	-	2
30 m - 45 m	13	1	-	6	3	5	1	29
45 m - 60 m	-	-	-	-	-	-	1	1
60 m - 100 m	-	-	-	-	-	2	-	2
More than 100 m	3	2	2	3	-	2	3	15
Not Known	-	-	-	-	-	-	1	1
TOTALS :	**16**	**4**	**3**	**9**	**3**	**9**	**6**	**50**

5.3 Iron Content Behavior & Non-corrosive pump components
The relationship between the above two factors is given in Table 5 below.

Table 5 : Correlation between Iron Content Behaviour & Below-Ground Pump Components

Results from Iron Content Monitoring	Below-Ground Pump Components				TOTALS
	PVC Pipe & SS Rod	PVC Pipe & BS Rod	GI Pipe & BS Rod	Multiple Reinstll. ending PVC-SS	
Improved	12	6	2	1	21
Partly Imoroved	7	2	4	-	13
Not Improved	4	1	4	2	11
Deteriorated	1	2	1	1	5
Multiple Reinstallation	-	-	-	-	-
TOTALS :	24	11	11	4	50

Table 5 indicates that :

With the combination of PVC riser pipes and Stainless Steel rods, 12 out of 24 installations showed appreciable improvement in iron content behavior, and 7 wells show partial improvement, i.e., 79% wells with PVC-SS combination showed some improvement.

As compared to this 8 out of 11 wells (73%) using PVC pipes and Galvanised bright steel rods showed some improvement.

In the case of GI pipes and BS rods,6 out of 11 wells (55%) showed some improvement.

All 3 combinations of pipes and rods showed one or two cases of deterioration of iron behavior.

6 Conclusions

From Table 3 it is evident that the method of well development chosen for the rehabilitation work, i.e., the air lift method has resulted in about 30% cases (15 out of 50 wells) of clear satisfaction on the first attempt. An additional 18 % satisfactory results were recorded upon attempting development by air lift more than once.

A clear improvement of iron content behavior, and satisfactory development was recorded in only 16% wells, and could be extended to 26% with multiple attempts of development.

Table 4 indicates that proportion of satisfactory development was significantly higher for wells less than 45 m depth (with 69% satisfactory results) than wells more than 100 m depth (with 33% satisfactory results). This is an inherent limitation of the air lift method of well development.

PVC pipes with Stainless Steel rods seemed to have produced the most favorable results in improving the fluctuations of iron content and generally keeping it low. Though PVC pipes and Galvanised bright steel rods also indicated favorable results, these reinstallations were not very old as compared to PVC-SS reinstallations. It is known that BS rods will eventually begin to corrode, but the extent to which it will affect water quality in the long run can not be inferred from the present data.Reinstallation with GI pipe and BS rods after development presents the least desireable reinstallation option.

From the above conclusions, it can be inferred that the air lift method of well development can have very limited success in rehabilitation of tube wells in the Orissa Drinking Water Supply project. Also reinstallation of pumps on rehabilitated wells should be with non-corrodible pump components. In this regard, the cast iron cylinder of the India Mark II pump needs an non-corrodible alternative since PVC pipes and SS rods already show remarkable results in controlling water quality deterioration.

Lastly, the validity of setting total iron content and chloride content as the only criteria for judging water quality in an area such as coastal Orissa needs serious reconsideration.

Annexure 1

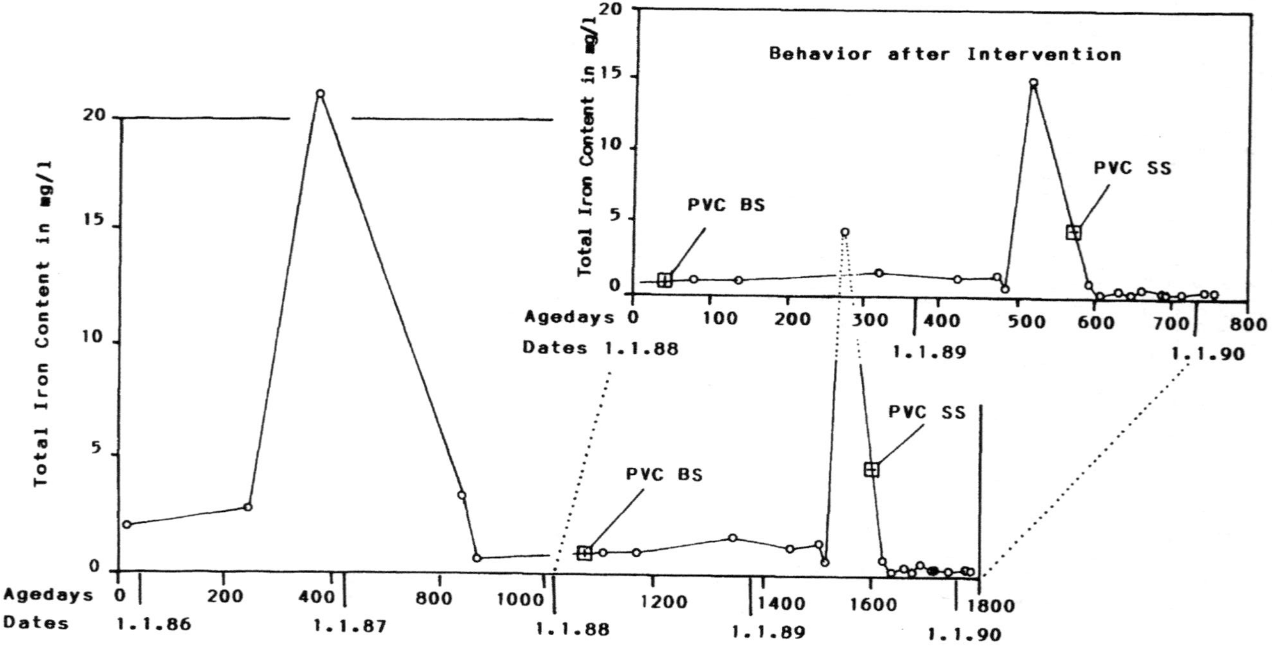

Behavior of Total Iron Content during the entire pump life and after intervention. Site 13122404502, Sujanpur

Annexure 1

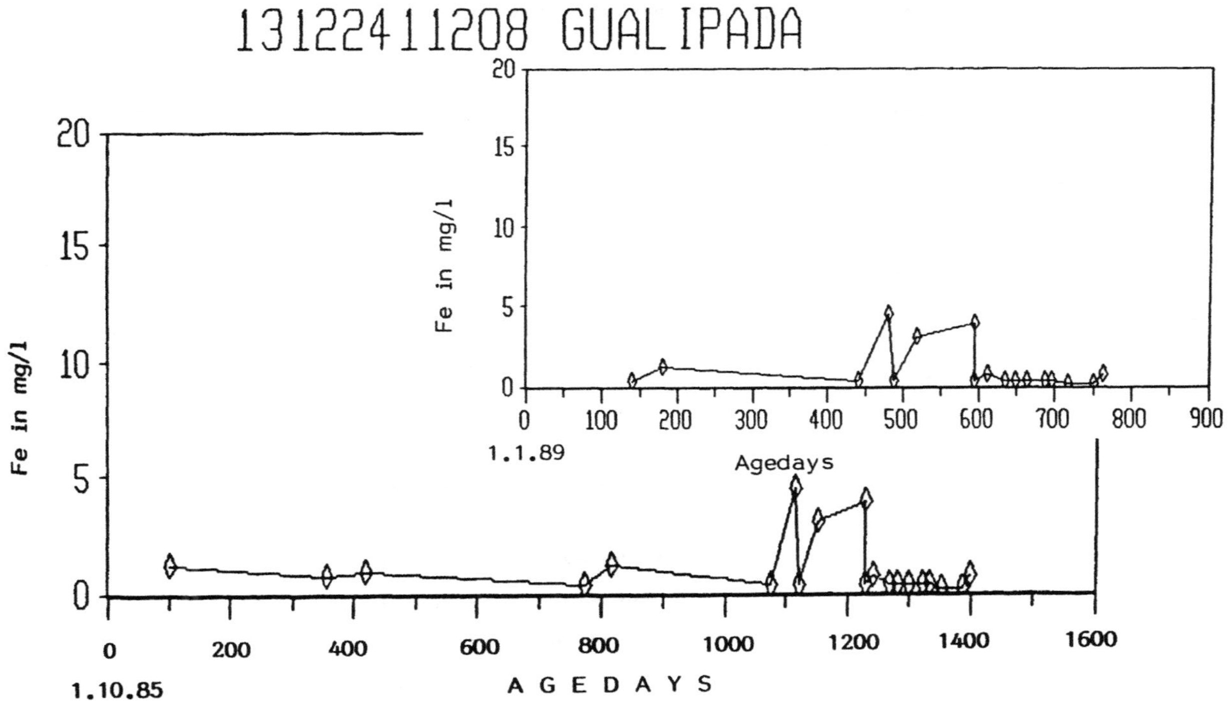

13122411208 GUALIPADA
Fe in mg/1
Agedays
1.1.89
AGEDAYS
1.10.85

Annexure 2

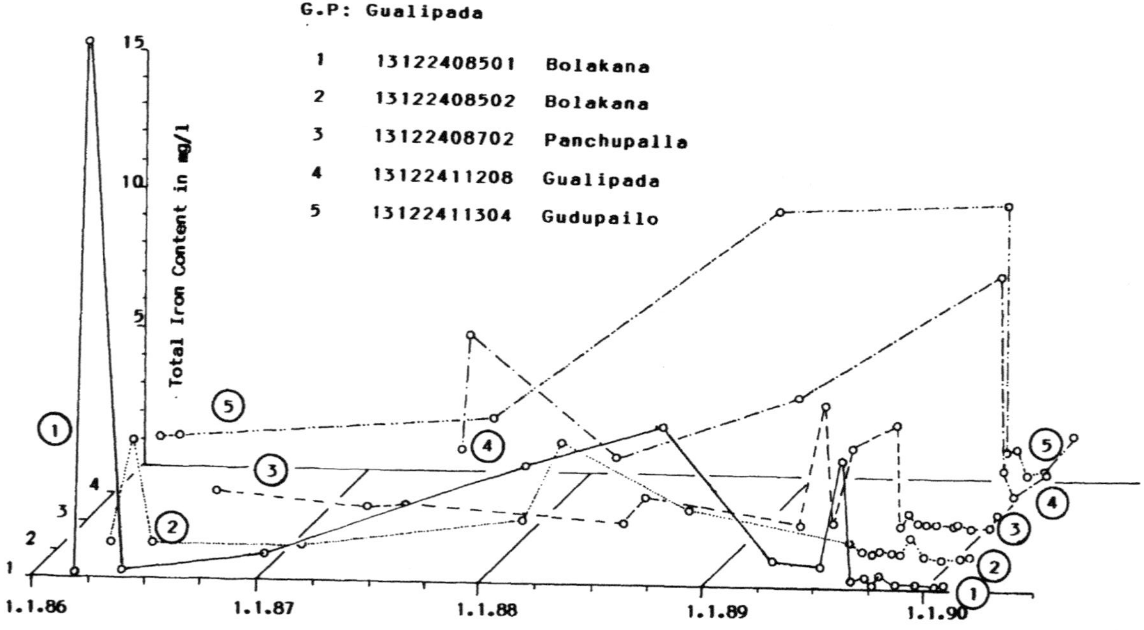

- Behavior of Iron Content in Problem Wells in Gualipada Gram Panchayat

Annexure 3

Regn. No	Village	Date Drilling	Depth (m)	Fe Level Before Intervention	Fe Level After Intervention	Resultant Change in Fe Level	Development Result
G.P: GODIPU	T MATIAPADA						
13122400204	G-M'PADA	23/4/86	38.85	VERY HIGH	MEDIUM	PARTLY IMPROVED	BLEACHING POWDER
							UNSATISFACTORY
							SATISIFACTORY
GP: SAURIA							
13122403603	SAURIA	23/11/86	66.00	HIGH	LOW	IMPROVED	BLEACHING POWDER
							UNSATISFACTORY
							UNSATISFACTORY
							UNSATISFACTORY
G.P: SUJAN	PUR						
13122404101	JAGDALPUR	8/5/86	35.50	HIGH	MEDIUM	PARTLY IMPROVED	UNSATISFACTORY
13122404102	JAGDALPUR	1/10/86	34.00	HIGH	HIGH	NOT IMPROVED	SATISFACTORY
13122404103	JAGDALPUR	18/8/86	38.00	HIGH	LOW	IMPROVED	SATISFACTORY
13122404502	SUJANPUR	18/10/85	34.00	VERY HIGH	VERY HIGH	NOT IMPROVED	NOT DEVELOPED
						IMPROVED	
13122404503	SUJANPUR	28/10/85	40.00	MEDIUM	MEDIUM	NOT IMPROVED	SATISFACTORY
						IMPROVED	
13122404601	RATANPUR	22/10/85	36.00	VERY HIGH	LOW	IMPROVED	SATISFACTORY
13122404602	RATANPUR	22/10/85	36.00	HIGH	LOW	IMPROVED	UNSATISFACTORY
13122404603	RATANPUR	27/10/85	36.00	VERY HIGH	MEDIUM	PARTLY IMPROVED	UNSATISFACTORY
							UNSATISFACTORY
13122406101	PATANPUR	20/10/85	36.00	MEDIUM	LOW	IMPROVED	UNSATISFACTORY
							SATISFACTORY
13122406103	PATANPUR	28/3/88	34.10	VERY HIGH	VERY HIGH	NOT IMPROVED	UNSATISFACTORY

Regn. No	Village	Date Drilling	Depth (m)	Fe Level Before Intervention	Fe Level After Intervention	Resultant Change in Fe Level	Development Result
GP : BERBOI							
13122408601	DAMAPUR	23/12/85	31.88	VERY HIGH	LOW	IMPROVED	UNSATISFACTORY
13122408901	BERBOI	8/11/85	36.00	MEDIUM	HIGH	DETERIORATED	NOT DEVELOPED
13122409101	OD'T'BOI	8/11/85	35.00	MEDIUM	LOW	IMPROVED	UNSATISFACTORY
13122409102	OD'T'BOI	30/11/85	34.00	HIGH	LOW	IMPROVED	UNSATISFACTORY
							SATISFACTORY
13122409301	RENCHA	18/11/85	40.00	MEDIUM	MEDIUM	NOT IMPROVED	SATISFACTORY
13122409302	RENCHA	25/11/85	36.00	HIGH	LOW	IMPROVED	UNSATISFACTORY
							SATISFACTORY
13122409303	RENCHA	27/11/85	40.00	MEDIUM	MEDIUM	NOT IMPROVED	UNSATISFACTORY
13122409304	RENCHA	26/11/85	40.00	MEDIUM	LOW	IMPROVED	SATISFACTORY
13122409305	RENCHA	25/12/85	36.10	VERY HIGH	LOW	IMPROVED	SATISFACTORY
G.P: GUALIP	ADA						
13122408501	BOLAKANA	8/3/86	32.50	HIGH	LOW	IMPROVED	SATISFACTORY
13122408502	BOLKANA	29/3/86	32.50	MEDIUM	LOW	IMPROVED	SATISFACTORY
13122408702	PAN'PALLA	31/1/86	40.75	HIGH	LOW	IMPROVED	SATISFACTORY
13122411208	GUALIPADA	16/7/88	33.95	LOW	MEDIUM	DETERIORATED	NOT DEVELOPED
13122411304	GUDUPAILO	17/7/87	136.5	HIGH	LOW	IMPROVED	UNSATISFACTORY

Annexure 4

Regn. No.	Village	Date	Chem. Trtm.	Development Result	Duration	Remarks
GP: GODIPUT	MATIAPADA					
13122400204	GODIPUT MATIAPADA	10/6/89	B		0.00	
		5/10/89		UNSATISFACTORY	4.50	
		6/10/89		SATISFACTORY	14.00	
GP: SAURIA						
13122403603	SAURIA	10/6/89	B		0.00	
		29/9/89		UNSATISFACTORY	7.00	
		30/9/89		UNSATISFACTORY	8.50	
		1/10/89		UNSATISFACTORY	6.00	LOW YIELD
GP: SUJANP	UR					
13122404101	JAGDALPUR	9/12/89		UNSATISFACTORY	7.50	TURBIDITY, SAND
13122404102	JAGDALPUR	5/12/89		SATISFACTORY	6.00	
13122404103	JAGDALPUR	8/12/89		SATISFACTORY	5.00	
13122404601	RATANPUR	23/7/89		SATISFACTORY	8.00	
13122404602	RATANPUR	22/7/89		UNSATISFACTORY	6.00	FINE SAND
13122404603	RATANPUR	24/7/89		UNSATISFACTORY	7.00	
		25/7/89		UNSATISFACTORY	9.50	
13122406101	PATANPUR	19/7/89		UNSATISFACTORY	6.25	
		1/8/89		SATISFACTORY	6.50	
13122406103	PATANPUR	31/7/89		UNSATISFACTORY	5.00	SL. TURB. LOW YIELD
13122406303	DELANG KOTHABAD	2/8/89		UNSATISFACTORY	7.00	TURBIDITY
		14/8/89		SATISFACTORY	2.30	
13122406305	DELANG KOTHABAD	16/8/89			8.25	

Regn. No.	Village	Date	Chem. Trtm.	Development Result	Duration	Remarks
13122409303	RENCHA	11/12/89		UNSATISFACTORY	4.50	TURBIDITY
13122409304	RENCHA	15/7/89		SATISFACTORY	4.30	
13122409305	RENCHA	10/11/89		SATISFACTORY	4.50	
GP: GUALIPA	DA					
13122408501	BOLAKANA	12/7/89		SATISFACTORY	4.25	
13122408502	BOLAKANA	13/7/89		SATISFACTORY	5.00	
13122408702	PANCHUPALLA	11/11/89		SATISFACTORY	6.00	LOW YIELD ?
13122411304	GUDUPAILO	19/12/89		UNSATISFACTORY	11.00	
GP: DHANKE	RA					
13122506304	BIRAMUKUNDAPUR	19/6/89	B		0.00	
		23/6/89		SATISFACTORY	4.00	
13122506305	BIRAMUKUNDAPUR	19/6/89	B		0.00	
		23/6/89		SATISFACTORY	4.50	
		25/9/89		UNSATISFACTORY	7.50	
		26/9/89		SATISFACTORY	7.50	
13122506307	BIRAMUKUNDAPUR	19/11/89		SATISFACTORY	7.00	
13122506309	BIRAMUKUNDAPUR	19/6/89	B		3.00	
		17/9/89	Y	UNSATISFACTORY	1.00	
		18/9/89		UNSATISFACTORY	3.25	
		19/9/89		UNSATISFACTORY	4.00	FINE SAND, FLAKES
GP: ABHAYA	MUKHI RAMACHANDRAPUR					
13122506411	INDIPURDEULI	4/1/90		UNSATISFACTORY	8.00	LOW YIELD
13122506413	INDIPURDEULI	21/11/89		SATISFACTORY	6.00	

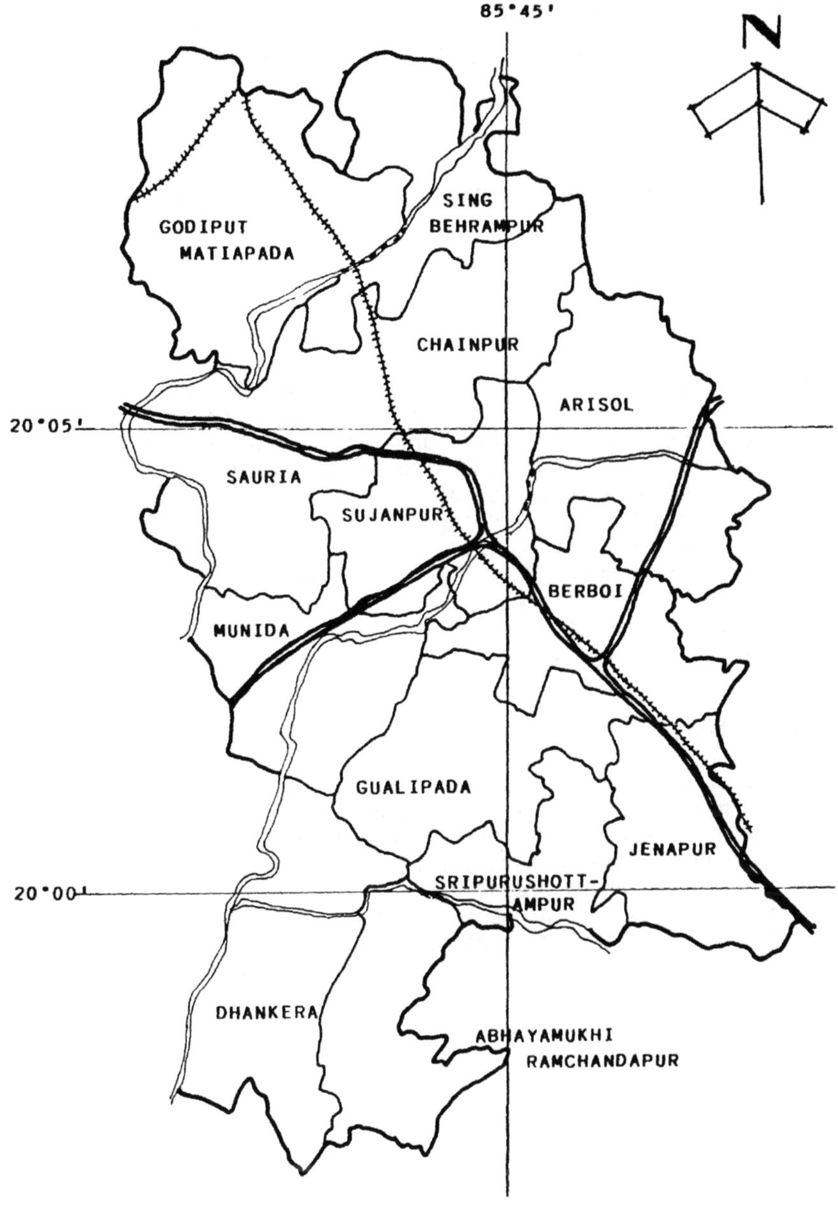

BLOCK MAP OF DELANG SHOWING GRAM PANCHAYATS

29 MONITORING, MAINTENANCE AND REHABILITATION STRATEGIES FOR BIOFOULING CONTROL IN WATER WELLS IN RIO NEGRO, ARGENTINA

R.E. ALCADE
Departamento Provincial de Aguas, Rio Negro, Argentina
M.A. GARIBOGLIO
Universidad Nacional de La Plata, Buenos Aires, Argentina

Abstract
Biofouling in water wells is a widespread problem in Rio Negro province. During last decade it has increased causing the impairment of wells, water quality deterioration and increase in maintenance costs. Diagnoses and rehabilitation treatment have been carried out when the process had reached a noticeable form, mostly when it was in an advanced stage. Information about the actual extent of the problem is lacking, but there is awareness of the importance of controlling its advance. For this end a monitoring, maintenance and rehabilitation program has been established this year. This paper describes the scope of such program which will encompass all the wells used in the water supplies of the province.
Keywords: Biofouling, Monitoring, Maintenance, Rehabilitation, Rio Negro province, Water Wells.

1 Introduction

Impairment of water well supplies of Rio Negro province caused by biofouling has resulted in loss of flow, water quality deterioration and increase in maintenance costs (Alcalde and Castronovo de Knott, 1986). Available evidence suggests that this is an increasing problem. It has taken place where ground water is the only source of drinking water and is widely and increasingly used. It is of primary concern to determine how it is spreading, as well as how to avoid broader disemination in the future. These goals involve the implementation of the following:

Diagnosing biological clogging.
Monitoring biofouling for prediction purposes or early detection of biofilm set up.
Maintenance and rehabilitation practices.
Training of human resources for rigth handling in drilling and operation of water wells.

For the accomplishment of these needs, strategies for monitoring, maintenance and rehabilitation have been designed and are expressed in a similar way to those used in monitoring contaminants in an aquatic environment (Désilets and Kwiatkowski, 1989).

A conceptual scheme of the potential sources of infection and the possible ways for monitoring the biofilm set up in aquifer-well systems, based on current understanding (Cullimore, 1986; Wojcik and Wojcik, 1986) has been developed. This scheme provides the basis to develop a training program and to establish diagnoses and monitoring practices.

2 Monitoring, maintenance and rehabilitation strategies

Fig. 1 shows a scheme of the steps involved in the strategies for biofouling control, which are described in the following sections.

2.1 Inventory of wells

This step comprises an inventory of wells including those being used in water supplies and those out of service due to the biofouling process. Locations are coded and entered into a record in which are included original well data: depth, lithology, design and constructive features, hydraulic characteristics and water quality.

2.2 Performance test of operating wells

Pumping tests of wells are carried out in order to detect any variation with respect to original data.

2.3 Bacteriological tests

Bacteriological studies are carried out to look for iron sheathed and stalked precipitating bacteria, sulphate reducing bacteria and counts of heterotrophic aerobic bacteria. Tests are conducted on pumped water samples or Fe-deposits (SRB and iron precipitating bacteria). Environmental parameters related to biological fouling, such as Eh, pH and tº, are measured at the same time (Smith, 1984).

2.4 Preventive maintenance

Wells where environmental conditions are conductive to clogging and where bacteria are present, are subjected to preventive maintenance and treatment. These steps consist of disinfection of well and pumping equipment.

2.5 Rehabilitation treatment

Wells that have suffered loss of yield are subjected to a rehabilitation treatment. This is carried out using hydrochloric acid and shock chlorination.

2.6 Monitoring

All wells are included in a monitoring program which includes bacteriological tests (iron bacteria, SRB, heterotrophic aerobic bacteria) and environmental parameter measurements (Eh, pH and tº) made monthly and well performance tests carried out annually.

The information generated is stored in the well's record. This information includes: performance tests, bacteriological and physico-chemical analysis, preventive treatments, preventive maintenance, maintenance costs and rehabilitation treatments.

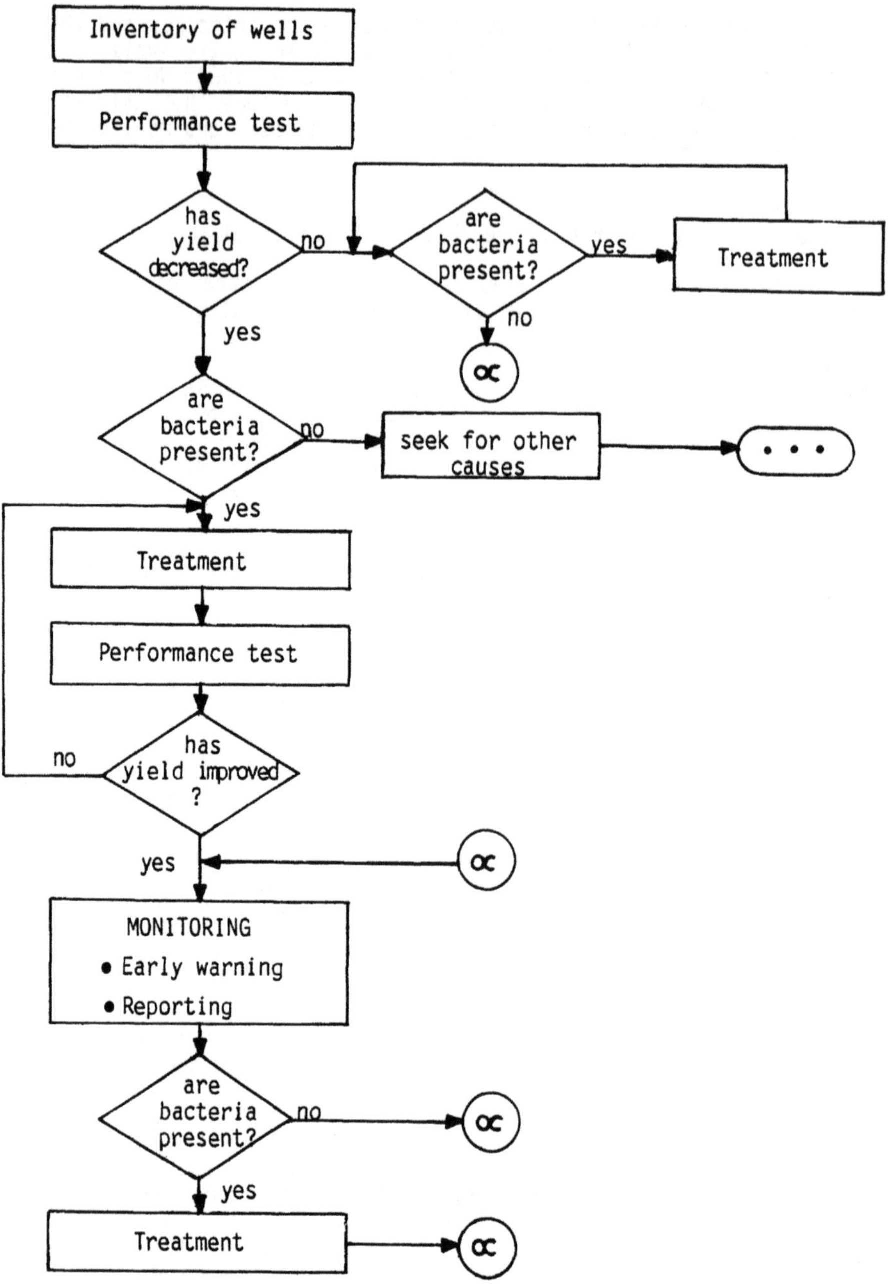

Fig. 1. Flowchart showing the steps of the strategy for control of biofouling in water wells.

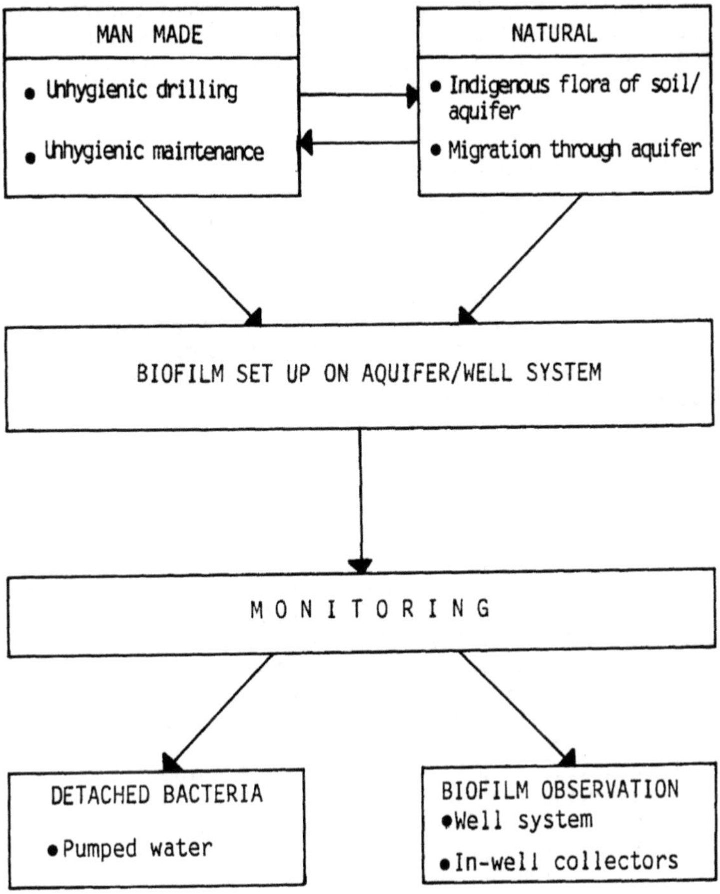

Fig. 2. Conceptual scheme on potential sources of infection and ways for monitoring biofilm set up in an aquifer/well system (based on Cullimore, 1986; Wojcik and Wojcik, 1986)

3 Conceptual scheme on potential sources of infection and monitoring methods

The conceptual scheme intends to link together the causes of biofilm set up in aquifer/well system and the ways for monitoring it to prevent it increasing. This scheme is based on current understanding on the subject. It identifies the main aspects of the problem with the purpose of establishing a training programme and the design of monitoring strategies. This scheme is shown in Fig. 2.

4 Human resources training programme

The success of the monitoring, maintenance and rehabilitation programme in a great measure depends on the degree of understanding of the subject among the people dealing with ground water abstraction. To this end as a complement of the programme mentioned above, Departamento Provincial de Aguas (DPA) and Consejo Federal de Inversiones (CFI) cooperative program includes seminars addressed to managers, well drillers and wells operators and the edition of a handbook in Spanish, since literature in this language is scarce.

5 Available tools for monitoring and treatment

The ways to investigate biofilm occurrence are schematized in Fig. 2. Methods that allow a direct study of biofilm are thought to be more reliable than those giving an indirect indication throught a search for bacteria in pumped water samples. The latter method detects bacteria detached from the biofilm, and this is expected to be a very variable process according to particular conditions of each well.

Nevertheless, direct inspection of full well system is often not possible due to construction features of wells or because the technology needed is not available. For this reason the monitoring strategy has been designed on the basis of bacteriological tests performed on pumped water samples. For enhancing the probability of the earliest detection of biofilm establishment, inspection of pumping equipment is often considered.

For the reasons mentioned above a procedure for bacteriological tests cannot be rigorous and a flexible approach is adopted.

With respect to the rehabilitation treatment used, hydrochloric acid and shock chlorination have been choosen as appropriate.

6 Concluding remarks

The strategies described for monitoring, maintenance and rehabilitation are the first steps towards the control of biofouling in water wells in Rio Negro province. They have been designed to face the problem in its present stage, to watch its evolution and to prevent its spreading. It is expected that experience will require improvements to be made to these strategies for the acievement of the established goals.

7 References

Alcalde, R.E. and Castronovo de Knott, E. (1986) Occurrence of iron bacteria in wells in Rio Negro (Argentina), in Proc. **International Symposium on Biofouled Aquifers: Prevention and Restoration** (ed. D.R. Cullimore), American Water Resources Association, Bethesda, Maryland, U.S.A., pp 127-136.

Cullimore, D.R. (1986) Physico-chemical factors in influencing the biofouling of groundwater, in Proc. **International Symposium on Biofouled Aquifers: Prevention and Restoration**, (ed. D.R. Cullimore) American Water Resources Association, Bethesda, Maryland, U.S.A., pp 26-36.

Désilets, L. and Kwiatkowski, R. (1989) Strategy for monitoring the exposure and effects of contaminants in the aquatic environment, **Scientific Series Nº 172**, Water Quality Branch, Inland Water Directorate, Environment Canada, Ottawa. 25 pp.

Smith, S. (1984) Detecting iron and sulfur bacteria in wells, **Water Well J.**,March, 58-63.

Wojcik, W. and Wojcik, M. (1986) Monitoring biofouling, in Proc. **International Symposium on Biofouled Aquifers: Prevention and Restoration**, (ed. D.R. Cullimore), American Water Resources Association, Bethesda, Maryland, U.S.A., pp 109-119.

30 REHABILITATION OF IRRIGATION WELLS IN NORTHERN OMAN – A CASE STUDY

J.A. BALDWIN
Scott Wilson Kirkpatrick & Partners, Basingstoke, UK

Abstract
This paper provides an example of the effect of rehabilitation of irrigation wells on the potential for an increase in agricultural production in an arid zone. The practice of installing pumps in drilled wells without some form of development is shown to be uneconomic for the farmer and inefficient in use of the groundwater resource.
Keywords: Rehabilitation, Wells, Irrigation.

1 Introduction

There has been a rapid increase over the last 10 years in the drilling of water wells for agricultural use throughout all the main farming areas of the Middle East. An increase in saline intrusion along coastal farming strips and a decline of superficial aquifer water levels due to over pumping has lead to hand dug wells being neglected and replaced at a rapid rate by deeper drilled wells.
In many cases, the control of quality and maximisation of sustainable yield of a drilled well have been abandoned in favour of rapid completion of drilling and the installation of a pump. Often the pump is the only one available and if the well yield is insufficient for the farmer, a second well is the solution offered by the drilling company. This is good business for the drilling company but costly for the farmer.
This case study evidences the actions taken on a farm in northern Oman to improve the yield in a number of irrigation boreholes drilled four years previously without proper development or pump sizing.

2 The Location

The farm is located 3 kilometres inland from Barka on the Batinah Coast of Northern Oman (Figure 1). The coastal strip is a traditional farming area with extensive date gardens interplanted with alfalfa and new larger farms providing a variety of vegetable crops for the Capital Area, Muscat.

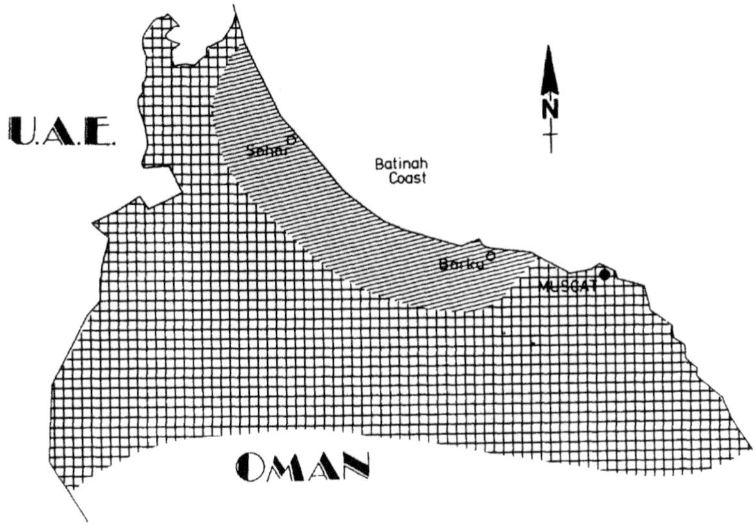

Figure 1 - Location

3 Hydrogeological Background

Recent test drilling has shown that much of the farm is underlain by two extensive aquifers. The existing wells on the farm and most of the wells in the surrounding area exploit the superficial aquifer. This aquifer contains good quality water and is subject to annual recharge during rainfall events in the Oman mountains.
The second deeper aquifer at Barka represents a much larger groundwater occurrence. The top of this aquifer lies between sea level and 10 metres below sea level. It reaches a maximum thickness of some 200 metres and is 100 to 120 metres thick under the northerly and westerly parts of the farm (Figure 2). Initial indications are that the deeper aquifer contains substantially more reserves than the shallow aquifer.

4 Existing Situation

4.1 Farming practices

The farm covers an area of 400 hectares and is located on the gravel plain of the Wadi Ma'awil catchment (Figure 3). Farming prior to rehabilitation consisted of growing water melon and tomatoes using flood irrigation on less than 1% of the farm area (2-3 hectares).

4.2 Existing wells

Three wells (W1, W2 and W3) had been drilled and equipped with Liste diesel engine driven line shaft turbine pumps to provide water for irrigation and potable use by the farm labour force. The basic well characteristics are shown in Table 4.1. A fourth well (W4) had been constructed 1.5 kilometers further inland within the farm boundary but had never produced a viable yield (Figure 4)

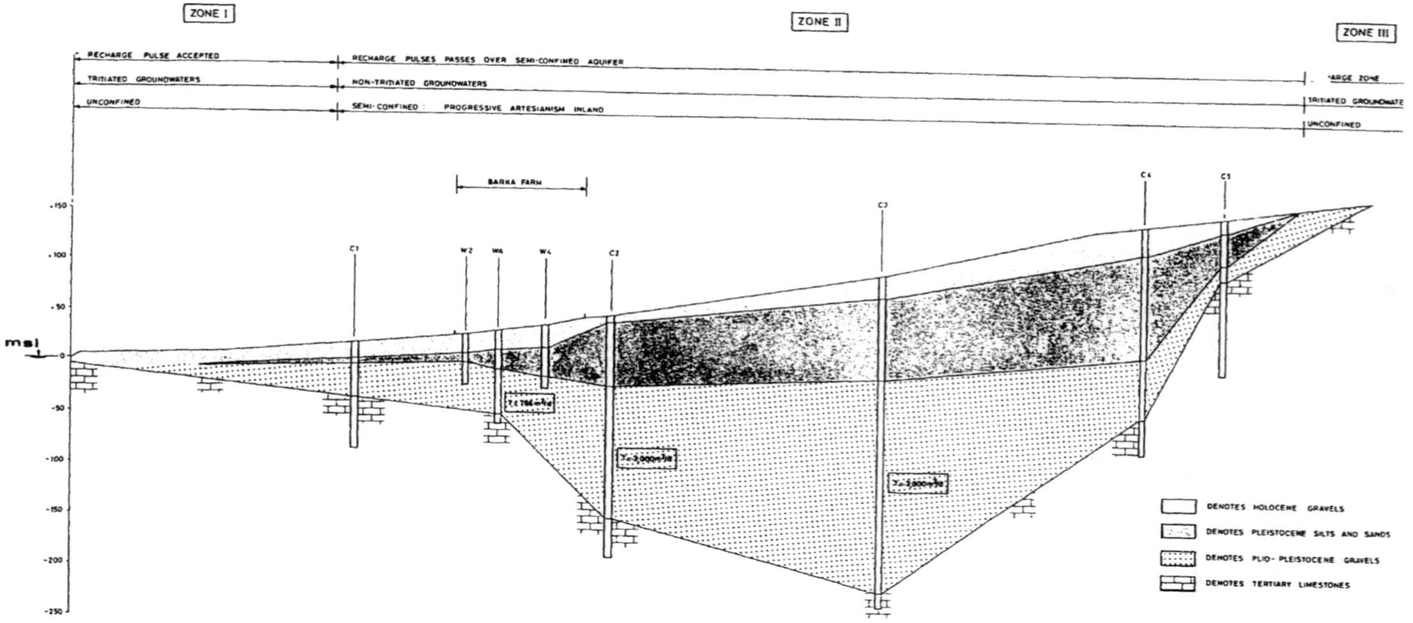

ZONE I
ZONE II
ZONE III
RECHARGE PULSE ACCEPTED
RECHARGE PULSES PASSES OVER SEMI-CONFINED AQUIFER
TRITIATED GROUNDWATERS
NON-TRITIATED GROUNDWATERS
UNCONFINED
SEMI-CONFINED : PROGRESSIVE ARTESIANISM INLAND
TRITIATED GROUNDWATE
UNCONFINED
BARKA FARM
msl
+150
+100
+50
0
-50
-100
-150
-200
-250
C1
W2
W4
C2
C3
C4
C5
DENOTES HOLOCENE GRAVELS
DENOTES PLEISTOCENE SILTS AND SANDS
DENOTES PLIO- PLEISTOCENE GRAVELS
DENOTES TERTIARY LIMESTONES

Wadi Ma'awil Aquifer

FIGURE 2

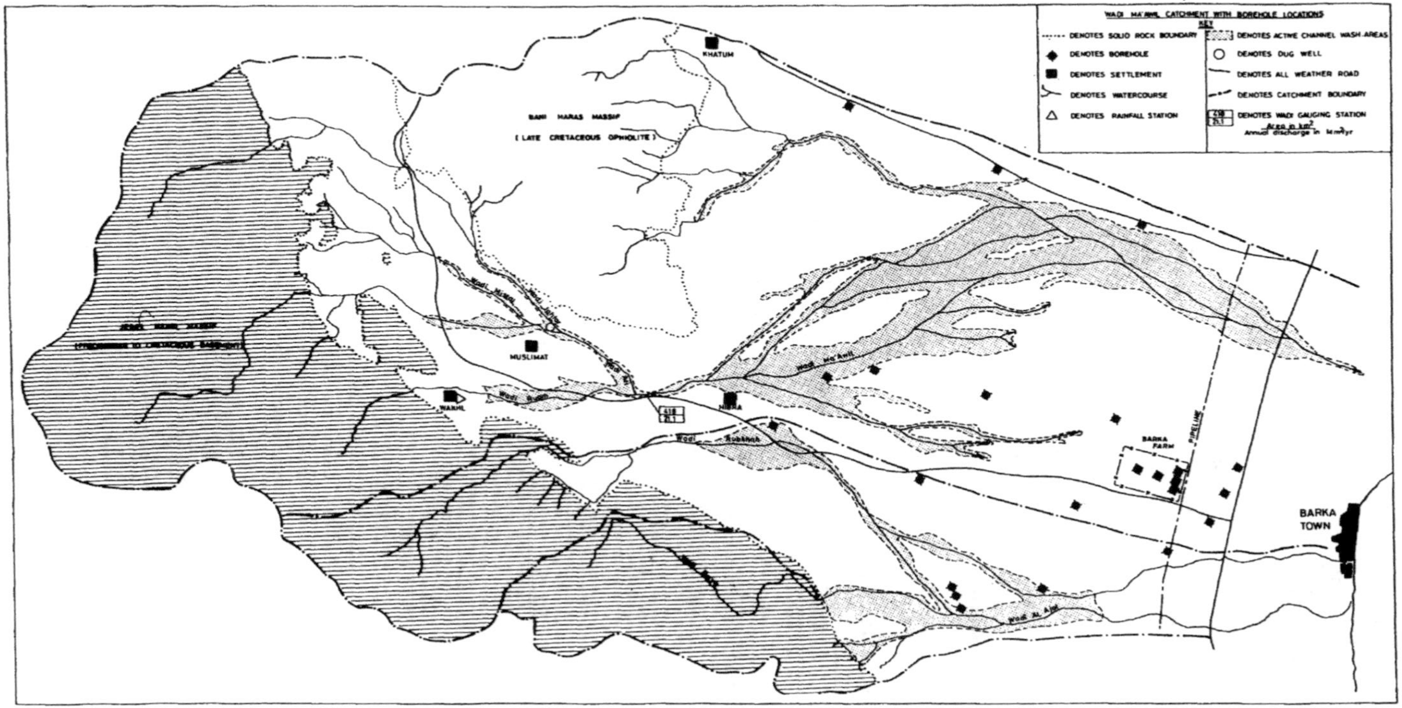

Wadi Ma'awil Catchment
FIGURE 3

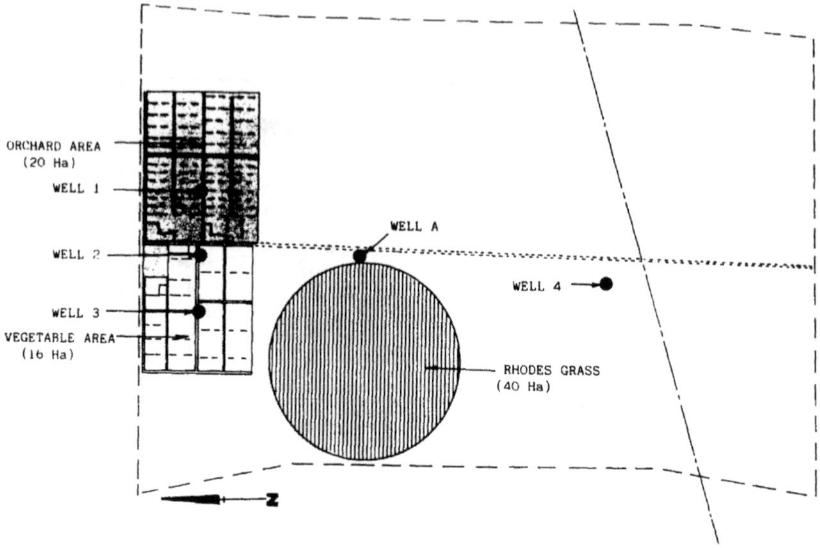

Figure 4 - Barka Farm

Table 4.1. Well Characteristics

Well	Casing dia (inches)	Well Depth (m.b.g.l)	SWL (m.b.g.l)	Pump Setting (m.b.g.l)	Engine HP	Pump Diam (mm)	Pump Stages
W1	10	39.5	26.9	33.8	16	210	8
W2	10	42.0	26.7	36.5	16	210	8
W3	10	38.9	26.7	33.8	16	210	8
W4	12	60.5	35.7	53.9	26	210	8

5 Rehabilitation

5.1 Preliminary inspection

After a preliminary inspection of the diesel engines and lifting of the line shaft turbine pumps it was decided to carry out a programme of rehabilitation followed by test pumping to improve the long term yield of the wells and extend the potential irrigated area. Decisions on re-use of pumps or replacement were held pending the results of the rehabilitation programme.

5.2 Rehabilitation programme

A standard programme of rehabilitation was established to improve well yield incorporating the following elements:-

1. Removal of existing pumps
2. Swabbing of the well with wire brushes and air lifting 24 hour period)
3. Injection of sodium tripolyphosphate, premi..ed in the proportion 2kg to 100 litres of water and left standing overnight. A total of 6kg used in each well.
4. Airlifting until clear water ejected
5. Test pump well in 4 steps of 100 minutes followed by a recovery period and then a continuous 24 hour pumping with monitoring of adjacent well standing water levels and recovery in the test well.

5.3 Test Pumping

The results of test pumping Well W2 were similar to those for wells W1 and W3 and are shown below. Well W4 was abandoned after Stage 4 as insufficient water was available for test pumping.

Pump test

A step draw down test consisting of 4 steps each of 100 minutes duration was performed on 26th June, 1986. An SP 70-7 model Grundfos electric submersible test pump was installed into the borehole at a setting depth of 39m. The borehole was pumped for 400 minutes increasing the discharge rate from 5 lps, to a maximum of 29 lps in 100 minute increments. The results are tabulated below:

Step No.	Duration minutes	D.W.L (M.)	Draw down (M.)	Cumulative drawdown (M)	Discharge lps.
1	100	26.97	0.12	0.12	5
2	100	27.16	0.31	0.43	10
3	100	27.66	0.81	1.24	20
4	100	28.32	1.47	2.71	29

After allowing the water level to recover, the borehole was then pumped continuously for 24 hours on 28th June, 1986. The discharge rate was set and maintained at 20 lps. The results are as follows:

S.W.L.	: 26.86 M.
D.W.L.	: 27.625M. after 24 Hrs.
Test discharge	: 20 lps.
Test duration	: 24 hrs.
Draw down	: 0.765 M.
E.C.	: 920 umhs-cm
Temperature	: 34 deg. C.

During shut down of the pump, the water level was observed for 2 hours. Water level rose to 26.87 M. The borehole recovered 95% of the draw down within 1.5 seconds. The water levels in the other boreholes were monitored during the constant discharge test and recovery period.

B.H.No.3

S.W.L. 26.67 M.
Water level after 24 Hrs. pumping 26.685 M.
Water level after 2 Hrs. recovery 26.68M.

B.H.No.4

S.W.L. 35.71 M.
Water level after 24 Hrs. pumping 35.72 M.
Water level after 2 Hrs. recovery 35.71 M.

B.H.No.1

S.W.L. 26.84 M.
Water level after 24 Hrs. pumping 28.86 M.
Water level after 2 Hrs. recovery 26.85 M.

5.4 Effect of rehabilitation

Following the rehabilitation programme it was calculated that wells W1 and W2 would have a long term safe yield of 20 litres/second and well W3 a long term safe yield of 15 litres per second. This showed a considerable improvement on the previous discharge rates and allowed irrigation from the same wells, using modern drip techniques, to be carried out on 16 hectares of vegetables together with bubbler irrigation of 6500 fruit trees over an area of 20 hectares. This was a dramatic increase in effective farming area. After two years of almost continuous daily use, no appreciable permanent drawdown of the SWL was evident.

6 Conclusion

It is evident from the work carried out at Barka Farm and from the general lack of use of development techniques in the region that there is potential for substantial sustainable increase in yield from a large number of undeveloped irrigation wells in the Batinah coast area. The same situation could apply in many other farming areas in the Middle East where similar drilling techniques are applied.

Efforts should therefore be made on an international front to publicise the situation and advise on appropriate rehabilitation methods for existing undeveloped wells and for recommended practice for development of newly drilled wells. In this way better control of aquifer development will be introduced and the economics of groundwater abstraction will be improved.

PART SIX

ECONOMIC AND MANAGEMENT FACTORS

31 BENEFITS OF PROPER MONITORING AND MAINTENANCE

C.G.E.M. VAN BEEK and F.A.M. HETTINGA
The Netherlands Waterworks Testing and Research Institute,
The Netherlands

Abstract
Proper monitoring and maintenance pay for themselves during periods of peak demand of drinking water. This paper describes the requisites for monitoring and maintenance for ground water abstraction. Special attention is given to measurement and interpretation of drawdown and ways to rehabilitate clogged wells.
Keywords: Clogging, Monitoring and maintenance, Rehabilitation, Water well.

1 Introduction

The clogging of wells is a widespread phenomenon. It is estimated that 2/3 of the well-fields exploited by the waterworks in The Netherlands have trouble with clogging. For drinking water supply companies a continuous and reliable supply of groundwater for treatment to drinking water is of prime importance. In order to achieve this, regular monitoring and maintenance is necessary. If not, problems with wells will manifest themselves just during critical periods, i.e. during periods of peak demands.

At present wells are equipped with submersible pumps. These pumps are, depending on their head-capacity curve, not very sensitive to clogging, that means their production can remain fairly constant over time while their drawdown is increasing. This drawdown may go so far that the pump starts sucking air, leading to a mass of iron deposits on the pump, and clogging of the pump, raw water lines etc. As an emergency measure the pump may be installed lower but this offers only a short term solution, but hopefully sufficient to overcome the critical period. However when the diameter of the casing at depth is less then the diameter of the pump this may not be possible.

From this it is clear that at present the production of a well remains fairly constant but the drawdown is variable. In former days the situation was just contrary, with the vacuum system where the drawdown was fixed and the production variable. Well clogging was then a much larger problem as the maximum possible production was determined by the degree of clogging. On the other hand a decrease in production was quickly noticed.

2 Definition

Clogging of a well is, for this paper, defined as a decrease of specific production, i.e. production over drawdown, or increase in specific drawdown, i.e. drawdown over production, in the course of time. The values obtained in this way are compared with the original value, i.e. the value at commissioning, after development, of the newly built well.

In this definition the actual value of capacity is not compared with a theoretical maximum capacity based upon the flow characteristics of the aquifer.

3 Monitoring

3.1 Drawdown

In order to be able to monitor the specific production of a production well, it is not necessary to know the exact value of the specific production. The values measured have to meet only one requirement i.e. the values should be comparable over time. This means that the procedure has to be executed each time in an identical manner.

In order to calculate the drawdown, the depth of groundwater during rest and during production should be available. As on a well field the groundwater level in the wells will be influenced mutually by the wells, it is of great importance to measure the levels in the course of time in an identical way, during identical operation, in the same sequence, the same period after stopping or starting the pumps etc. For example, if it is possible to divide the well field into two groups, it is advisable to measure the water levels in the producing and idle wells, to switch the wells to other way and to measure the water levels again.

The results of the monitoring should be recorded and processed immediately. In this way one can see immediately whether the measurements are in line with the foregoing results or not. If the results are not in line, something special has happened or the observer has made an error, for example a reading error of 1 (one) m.

Table 1 shows an example of how to report the data, measurements as well as calculations. Figure 1 shows the calculated data in a graph.

The frequency of measurement depends on the length of the rehabilitation interval and the history of the well field; is it a new or an old established one.

In order to accurately determine the need for rehabilitation it is advised to have 8 to 10 observations during the rehabilitation interval. As the rehabilitation interval may vary between 6 months to several years, this means that the interval for measuring drawdown varies between once a month to at least once a year.

These intervals are for an old established well field, where the clogging history is available. On a new well field this information is not available and it is advised to monitor initially more intensively.

Table 1a. Example of a sheet for noting the observations

Date.... Well in production							Well out of production				
Well number	pump started at .. [h]	depth to watertable [m]			volume flow [m^3/h]	Power consumption [Amp.]	pump stopped at ... [h]	depth to watertable [m]			production hours
		time	level inside well	level in gravelpack				time	level inside well	level in gravelpack	
					(constant value, depending on the pump)						

Table 1b. Example of processing the observations

Well nr.					
Date	Volume flow [m^3/h]	Drawdown [m]	Entrance resistance [m]	Production hours [h]	Remarks since last observations
					(for example change of pump, rehabilitation)

3.2 Static level

Sometimes the production of a well field is too high compared to the geohydrological conditions of the well field, i.e. the recharge of groundwater is too small, and/or the aquifer is too small or not sufficient permeable to replenish the abstracted groundwater. Under these conditions the static groundwater level will start to decline, but this phenomenon has nothing to do with clogging but with over-exploitation. The only remedy available is to decrease the production.

3.3 Additional observations

The behaviour of the well field can also be derived from additional observations.

During the measurement of the depth of the water level when the submersible pump is working, it is advisable to listen carefully to ascertain whether the pump is running properly. When such is not the case the pump may be worn-out or the water level has dropped so deep that the pump is sucking air.

Also inspection of the consumed power may give clues as to the functioning of the pumps. In the Netherlands at some water supply companies it is custom to run the pump without regular service. When malfunctioning is noticed then the pump is replaced; depending upon the damage the pump will be either repaired or discarded. Obviously during rehabilitation of a clogged well, the pump has to be removed; this offers a nice opportunity for inspection and cleaning of the pump.

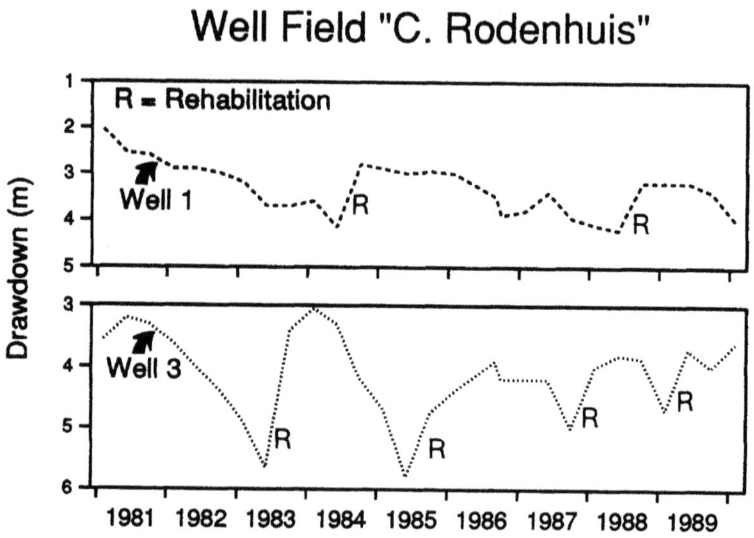

Fig.1. Characterization of well clogging by a continuous decrease in specific capacity. If the well is equipped with a submersible pump, the yield remains fairly constant and the clogging can be characterized by a continuous increase in drawdown.

4 Rehabilitation

4.1 Timing of rehabilitation

A right timing of the execution of rehabilitation is important. Rehabilitation of a well as soon as the specific production has been dropped to half its original value has been proven appropriate (van Beek, 1984). This is under conditions in the Netherlands where the cost of energy required to overcome the increase in drawdown (for example from 2 to 4 m) is negligible. This means that rehabilitation before this point is not worthwhile. Under conditions it may be required to rehabilitate such a well if one is not sure that the high production season will be covered.

The well field should be in peak condition before the season of high production.

It is not wise to permit that the specific production should decrease to below half its original value, as experience has shown that there is a greater chance that the specific production will not be brought back to its original value. Even a more intensive or repeated treatment will not reach that value.

Waiting too long will result in such high flow velocities through the slots of the screen that sand will be set in motion. The well will start delivering sand, and this may be considered as the final stage of clogging.

4.2 Causes of clogging

Several causes of well clogging can be distinguished (van Beek, 1984, 1989), see table 2.

When one is not sure about the cause of clogging inspection with a closed-circuit television (CCTV) is recommended (Howsam, 1988). CCTV is even more helpful when one is interested in the state of the well, for example when a well starts delivering sand.

Table 2. Summary of the various causes of well clogging (van Beek, 1984)

Location of the clogging	Are all the wells of the well field involved?	
	Yes	No
Slots of the screen	Mixing of incompatible groundwater chemistries	
Borehole wall/ Aquifer interface	Enhancement of already present processes	Mechanical causes

When all the wells of a well field experience clogging the cause must be attributed to the prevailing conditions (van Beek, 1984):

a. Mixing of incompatible water qualities. When there is a vertical stratification in the chemical composition of groundwater, for example oxygen near the top of the aquifer and iron near the bottom, abstraction will result in mixing of these water types. Through mixing, deposits of iron hydroxides, manganese oxides or aluminum oxides will develop and clog the slots of the screen. This type of clogging occurs in The Netherlands mainly in sandy soils with phreatic aquifers.

b. Enhancing microbiological processes. Due to the increased velocity of the groundwater flow, the intensity of microbiological processes already occurring in the aquifer, may be enhanced, resulting into the formation of deposits upon the well bore/aquifer interface. Clogging may occur by deposits of iron-sulfides by the action of sulfate reducing bacteria. This type of clogging is in The Netherlands often encountered in wells abstracting bank filtrate.

When not all wells of a well field experience clogging but only a few the cause must be attributed to accidental factors.

c. Mechanical clogging. When the diameter of the gravel of the gravel pack has not been correctly chosen for layers of (very) fine sand, when the gravel pack has not been placed properly or when the screen is not centered properly in the borehole, the well may start yielding sand and experience clogging. Until now this type of clogging has only been found in wells, abstracting deep anoxic groundwater. Recent research (Ryan and Gschwend, 1990) has shown that anoxic groundwater contains more colloidal particles than oxic groundwater.

Except the last cause, these wells can be rehabilitated successfully, when the right method is applied. As the clogging agents are often a mixture, it is advised not to apply only one identical method, but for example to vary the chemicals used.

4.3 Prevention

With the multitude of geological and hydrochemical conditions it will be clear that no general applicable method for prevention is available. Depending upon the situation very specific methods may be applied, such as separate abstraction of oxygen- and iron-containing water and removal of iron and manganese in the aquifer (van Beek, 1984).

However in order to prevent problems it will be clear that thorough development of the well will be of prime importance. Remnants of the (natural or artificial) cake on the borehole wall or in the aquifer will initiate processes leading to clogging. For that reason wells should be developed until the specific production does not increase anymore.

If it is known that the aquifer contains fine sand, for example despite exhaustive development the well keeps delivering fine sand, it is recommended to install well screens coated with pre-packed gravel packs. This type of screen has been available for a few years and seems to be suited to these conditions (Temmink, 1986).

Moreover continuously producing wells clog slower than intermittently producing ones (Moser, 1979). Each time a submersible pump is started a small amount of sand is produced by the well, indicating that sand in the gravel pack or aquifer has been set in motion. Moreover each time production stops, the cone of depression due to drawdown in phreatic aquifers will fill up again and give rise to mixing. Under conditions the well screen may act as a short circuit, in which water from the bottom part of the aquifer flows to the upper part.

5 Costs of monitoring and maintenance

The costs of monitoring form an intrinsic part of water production. They include costs of operation and supervision.

Costs of maintenance comprise amongst others the costs of rehabilitation. Rehabilitation usually takes 2 to 5 days:

a. removing the submersible pump;
b. cleaning the well and gravel pack;
c. pumping to waste, and
d. installing the submersible pump again.

The costs of rehabilitation comprise costs of labour and of chemicals and one time expenses for special devices and tools.

Nowadays specialized firms offer to rehabilitate clogged wells. Often they can only execute one method. It is the responsibility of the waterworks manager to judge whether this offered method is appropriate for his well field or not.

For some years a clogged well was often replaced by a new well. This new well was often sunk near to the clogged one, as in that case the costs for extending the raw water lines and power supply was minimal. From the foregoing it may be clear that if the clogging is caused by the local geological and/or hydrochemical conditions, the new well will also clog.

Often it takes more time before a new well needs its first rehabilitation then when it has already been rehabilitated once, i.e. the first rehabilitation interval is often longer than the consecutive ones. The cost of rehabilitation amounts on an average to 10 to 15% of the cost of sinking a new well. As the cost of rehabilitation remains fairly constant, the lower percentage refers to more expensive, deeper, wells. This means that only in the extraordinary case where the first rehabilitation interval is 7 to 10 time longer than the consecutive intervals, it is advantageous to sink a new well instead of rehabilitating the clogged one.

This relation is demonstrated in figure 2.

6 Acknowledgement

The research described in this paper has been carried out by KIWA (The Netherlands Waterworks' testing and research institute) as part of the research program of VEWIN (The Netherlands Waterworks Association) in close cooperation with interested waterworks.

The authors wish to express their thanks for the critical remarks by Ing. H.J. van Dijk, Water Supply Company "South Holland-East", Ir. J.G.H. Philips, Water Supply Company "East-Brabant" and Dr. P. Howsam, Silsoe College.

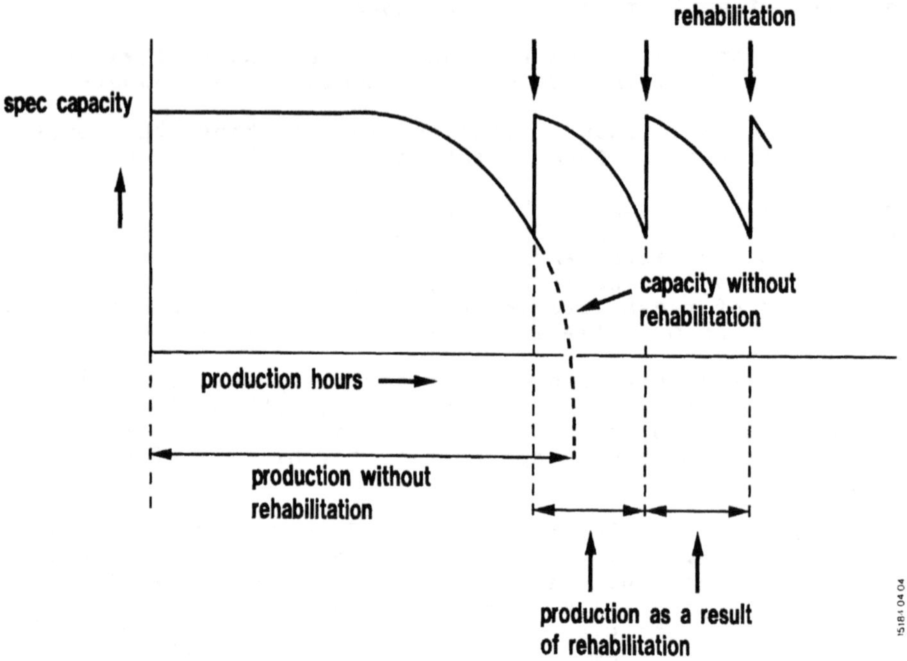

Fig.2. Alternatives when well clogging occurs: rehabilitation of the clogged well or replacement by a new one.

7 References

van Beek, C.G.E.M. (1984): Restoring well yield in The Netherlands, **J. Am. Water Works Ass.** 76 (10) 66-72

van Beek, C.G.E.M. (1989): Rehabilitation of clogged discharge wells in The Netherlands, **Quarterly J. Eng. Geology**, London, 22 (1) 75-80

Howsam, P. (1988): Biofouling in wells and aquifers, **J. Inst. Water Engineers Scientists**, 2 (2) 209-215

Moser, H. (1979): The clogging of vertical screen wells, the technical and economical aspects, explained by example of the well field of the Waterwork Mannheim-Käfertal, **DVGW Schiftenreihe 201**, 385-418 (in German).

Ryan, J.N. and P.M. Gschwend (1990): Colloid mobilization in two Atlantic coastal plain aquifers: field studies, **Water Resources Res** 26 (2) 307-322

Temmink, B.G. (1986): Personal communication

32 OPTIMISATION OF GROUNDWATER SOURCES

J.E. CAMPBELL
North West Water Limited, Warrington, UK

Abstract
A programme of testing has been undertaken in North West Water to promote the efficient use of water supply boreholes. Frequently energy saving schemes only address pump efficiency and are designed to identify any deterioration in the performance of the pump. However further savings are possible if the efficiency of the borehole and the supply system are examined. Step drawdown pumping tests were conducted and a yield/drawdown curve established for each borehole. The more efficient boreholes are those with a high yield and small drawdown, and maximum use should be made of them. Where a number of boreholes operate in a group to supply a single storage reservoir, a comparison between the boreholes in terms of their static head lift can be made. This is achieved by plotting the water levels and drawdown curves relative to the elevation of the reservoir. Water pressure measurements were taken in the delivery main and compared with calculated heads to identify high head losses or excessive system pressures. Results from two groups of boreholes are presented where savings of 10% in annual electricity costs were identified.
Keywords: Boreholes, Pumping Tests, Energy Savings, Source Optimisation.

1 Introduction

Some 240 boreholes are used by North West Water to provide an average of 11% of the water supplied to customers. A number of the boreholes were commissioned around the turn of the century when pumping test methods were crude and poorly recorded. Many of the rest were drilled in the 1960's and early 1970's when local water boards attempted to meet increased demand by exploiting the groundwater resources within their own boundaries. In some cases the increased abstraction led to water levels falling by 30 metres. As a result, although data are available from the commissioning trials on the boreholes, the results are no longer valid.

With the creation of North West Water in 1974, a regional approach to water supply was possible, and surface water sources were made available to areas previously supplied solely by boreholes. As a result demand on the groundwater sources has reduced and it is now possible to keep some boreholes on standby for emergencies while others remain in constant use. In deciding which boreholes to use, data on their yield and costs are required. To provide these a rolling programme of borehole testing has been undertaken.

The objectives of the testing programme were:

To provide a standard measurement of the efficiency of a borehole in terms of its yield and drawdown characteristics.
To provide data on the costs of pumping at different output rates.
To identify possible increases in peak output for emergencies.
To identify possible standardisation of pumping plant.
To provide data to optimise the usage of groups of boreholes.

2 Test Procedure and Analysis

Step drawdown tests, as described by Clarke (1977), were carried out on individual boreholes using the installed pump unit and flow meter. Water levels were measured at 1 minute intervals with either pressure transducers and data loggers or manual dippers. The standard format of the test involved the borehole pump being switched off a minimum of 12 hours before testing to achieve a rest water level. The valve on the delivery main was partially closed to reduce output to approximately 25% of normal yield. The pump was started and run for 60 minutes at this reduced rate. The valve was then opened to increase the flow to 50% of normal, and water levels were measured for a further 60 minutes. Further steps at 75% and 100% flow were monitored, followed by switch off of the pump and measurement of recovery levels for 60 minutes.

The drawdown results were analysed using the method introduced by Eden and Hazel (1973). The method uses a computer program to calculate the correlation between yield and drawdown,and to produce estimates of Transmissivity and Storativity. The correlation can then be used to predict the drawdown for different pumping rates at any point in time. A detailed description and listing of the program was published in a recent paper by Vines (1989).

3 Test Results

Results from two groups of boreholes are presented below to illustrate how the tests are used to assess the relative efficiency of individual boreholes.

3.1 Lightshaw boreholes

This group comprises 8 boreholes feeding water to Lightshaw Water Treatment Works(W.T.W.) which lies to the north of Warrington. Figure 1 shows the location of the boreholes, and Table 1 contains details of the boreholes.

All the boreholes are drilled into the Permo-Triassic sandstones which overlie the Carboniferous Coal Measures. In general the sandstones thin towards the north, but faulting complicates the structure. A major fault runs north-south through Golborne and Winwick and brings the Coal Measures to the surface. This creates a

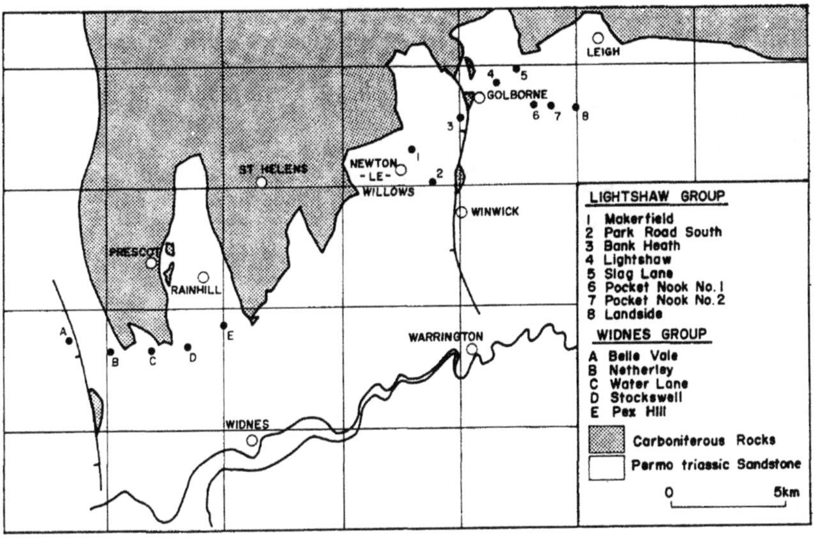

Fig.1. Borehole location map

Table 1. Details of the Lightshaw group of boreholes

Borehole	Ground level (m O.D.)	Rest water (m b ground)	level (m O.D.)	Specific Capacity (m^3/day/m)
Bank Heath	38.7	12.5	+26.2	75
Landside	18.2	28.4	-10.2	292
Lightshaw	34.0	52.7	-18.7	186
Makerfield	37.6	21.8	+15.8	248
Park Rd South	18.0	13.2	+4.8	329
Pocket Nook 1	31.0	40.0	-9.0	213
Pocket Nook 2	25.4	37.4	-12.0	357
Slag Lane	24.0	44.1	-20.1	380

barrier to groundwater flow in the sandstones, and effectively divides the aquifer into two separate blocks. Rest water levels at Bank Heath, Makerfield and Park Road South, which lie to the west of the fault, are all above Ordnance Datum (O.D.). In contrast all the boreholes to the east of the fault have water levels below O.D. (see Table 1).

Figure 2 shows a plot of measured drawdown at Pocket Nook 2 borehole to illustrate a typical test. Similar tests were carried out on the other boreholes in the group, and Figure 3 illustrates the

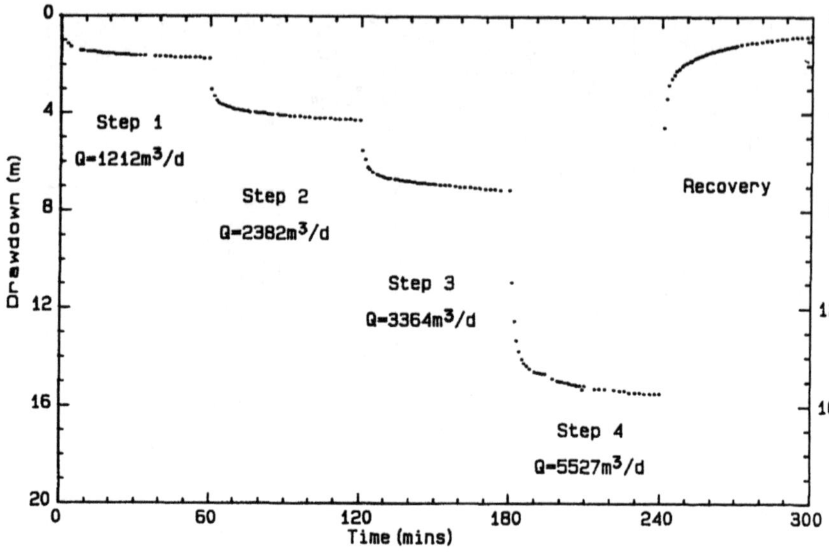

Fig.2. Step test at Pocket Nook No.2 borehole

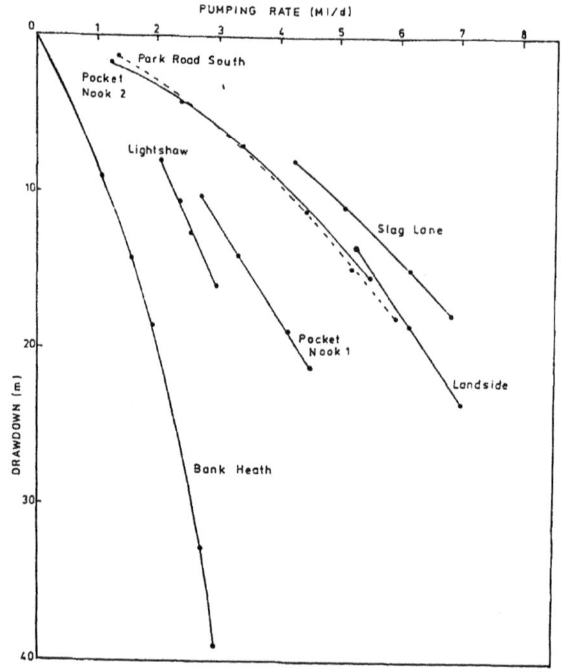

Fig.3. Yield/drawdown curves after 60 minute pumping steps

variation of yield/drawdown characteristics among these boreholes. For each borehole the total drawdown measured at the end of each step is plotted against the flow rate during the step. The most efficient boreholes are those to the top and right of the figure (maximum yield with minimum drawdown).

Bank Heath is the least efficient borehole, and its yield/drawdown curve demonstrates an important feature of borehole behaviour during pumping. At flow rates below 2 Ml/d there is a linear relationship between flow and drawdown. At rates above 2 Ml/d there is a disproportionate increase in drawdown and the curve steepens. The amount of drawdown is governed by two factors, the permeability of the aquifer and the well loss factor. At low flow the well loss factor is insignificant, and permeability controls the drawdown. As the flow rate increases, well turbulence generates extra drawdown and a head difference develops between the borehole water level and the water level in the aquifer adjacent to the well face. This is referred to as the seepage face.

The steepening yield/drawdown curve for Bank Heath above a flow rate of 2 Ml/d represents the development of a large seepage face. It is obviously less efficient to pump this borehole at a rate of 3 Ml/d, and to optimise pumping costs the creation of a large seepage face should be avoided.

During short duration step tests very few boreholes reach equilibrium conditions and it is important to consider what drawdown will result after several days or weeks of pumping. Using the correlation constants obtained by computer analysis of the step tests, predicted yield/drawdown curves can be drawn for each borehole at any point in time. Figure 4 shows predicted curves for Pocket Nook No.2 borehole after 7 days and 30 days continuous pumping.

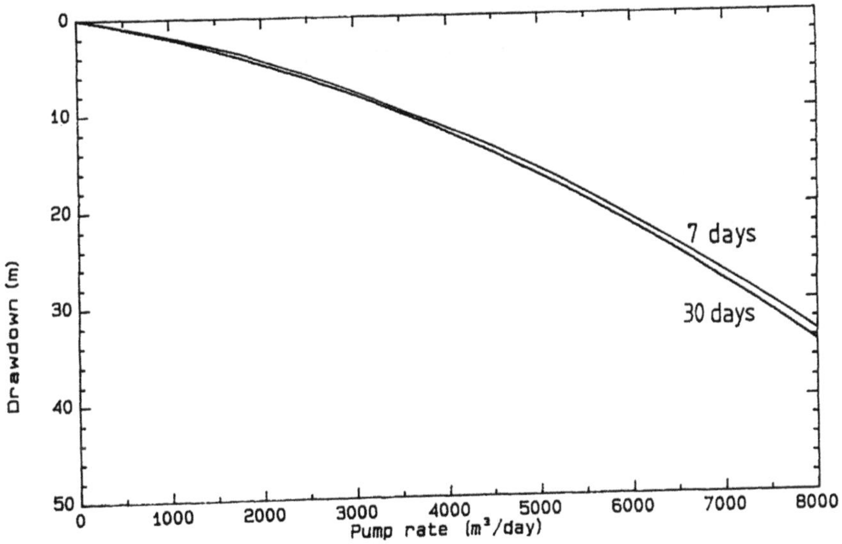

Fig.4. Predicted yield/drawdown curves for Pocket Nook 2 borehole

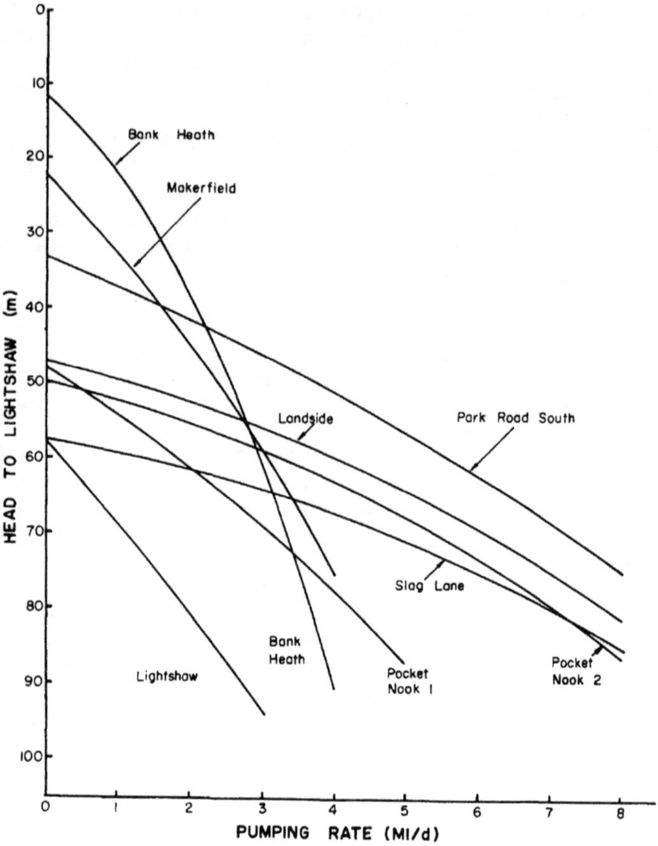

Fig.5. Predicted yield/drawdown curves after 7 days pumping

On Figure 5 the predicted curves after 7 days pumping are plotted for all 8 boreholes, and the drawdowns from actual rest levels are shown relative to the level of the discharge point at Lightshaw W.T.W. (38m above O.D.) to where the boreholes pump. This allows direct comparison between the boreholes in terms of their static head lift. The higher rest water level at Bank Heath and Makerfield (see Table 1) results in these boreholes having the lowest head lift at rates up to 2 Ml/d. However at higher pumping rates Park Road South is the most economic followed by Landside, Pocket Nook No.2 and Slag Lane.

3.2 Widnes Boreholes

This group comprises 5 pumping stations all with multiple boreholes. The boreholes feed to Pex Hill service reservoir(S.R.); Pex Hill and Stockswell boreholes pump along independent mains, but the other stations pump along a common main. The location of the boreholes is

Table 2. Details of the Widnes group of boreholes

Borehole	Ground level (m O.D.)	Rest water (m b ground)	level (m O.D.)	Specific Capacity (m^3/day/m)
Belle Vale	17	0	+17	82
Netherley	13	20	-7	1488
Pex Hill	64	77	-13	418
Stockswell	18	47	-29	2989
Water Lane	15	22	-7	482

shown on Figure 1, and their details are listed in Table 2. The rest water level at Belle Vale is high to the extent that groundwater discharges on the surface. Further to the east water levels at all the stations are below O.D., reflecting the effect that long term abstraction has had on the Permo-triassic sandstone aquifer.

Figure 6 illustrates the yield/drawdown curves that were obtained from testing the boreholes. Belle Vale is the least efficient borehole, and Stockswell is the most efficient with Netherley a close second.

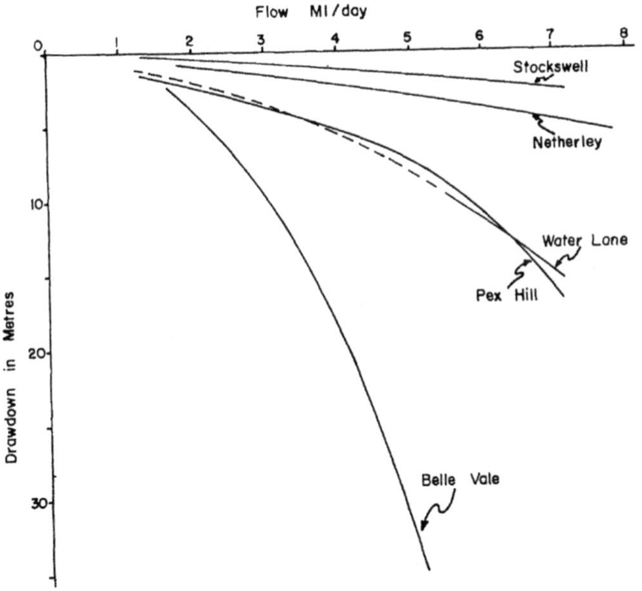

Fig.6. Yield/drawdown curves after 60 minute pumping steps

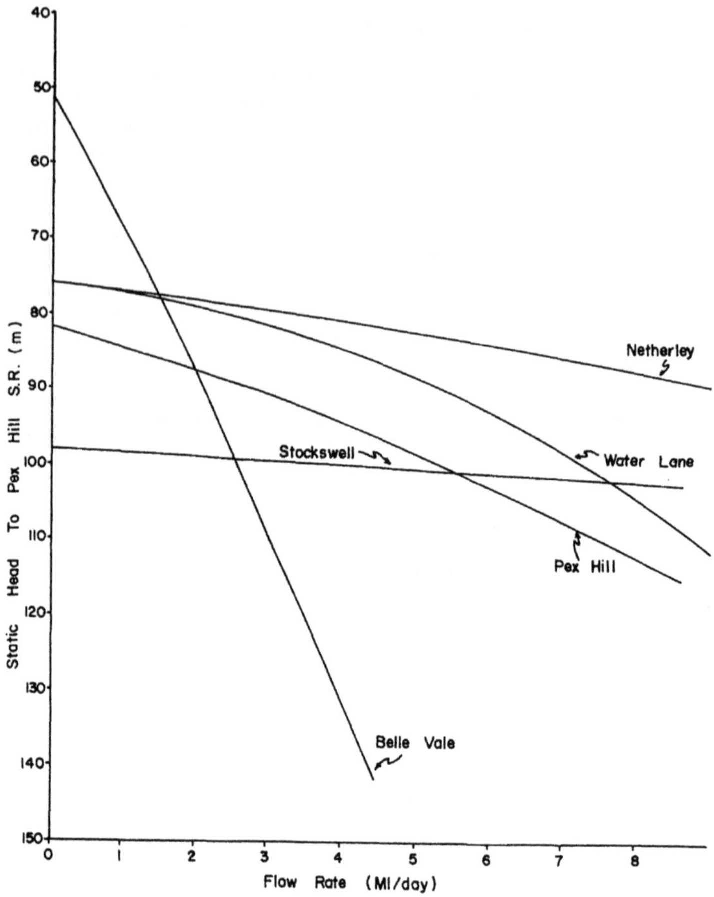

Fig.7. Predicted yield/drawdown curves after 7 days pumping

On Figure 7 predicted curves after 7 days pumping are plotted, and as before the drawdowns from actual rest levels are shown relative to the level of the discharge point at Pex Hill S.R. (69m above O.D.). The high rest level at Belle Vale results in this borehole having the lowest static head lift at flows up to 2 Ml/d. By contrast Stockswell is the most costly to operate at low flow rates based on its static head lift.

However in calculating pumping costs, the total head lift (static head plus pipework head loss) needs to be considered. Within the Widnes group pipework head losses are significant because of the length of pipework. The main from Belle Vale to Pex Hill S.R. is 9.5km long and is shared by the stations at Netherley and Water Lane. If more than 5 Ml/d is pumped into the shared main an in-line booster pump has to be run, adding to the operating costs.

Given that Belle Vale has to be pumped to prevent local flooding by groundwater, the optimum pumping pattern to satisfy the average

Table 3. Optimum pumping for average demand

Station	Flow rate (Ml/d)	Static head (m)	Pipe loss (m)	Total head (m)
Belle Vale	2.0	88	2	90
Stockswell	7.0	102	5	107
Pex Hill	5.0	99	1	100

daily demand of 14 Ml is as listed in Table 3. Netherley has a static head lift of 86m at a pumping rate of 7 Ml/d and pipe losses of some 15m, giving a total head of 101m. Although this is less than Stockswell, the additional cost of operating the booster outweighs any saving.

Demand up to 20 Ml/d can best be met by operating a second borehole at Stockswell. Above 20 Ml/d Netherley and finally Water Lane can be operated together with the booster pump.

4 Pump standardisation

Within the Lightshaw group, the total head was calculated for the boreholes at Park Road South, Landside, Pocket Nook 2 and Slag Lane for a pumping rate of 6.5 Ml/d.(See Table 4).

Table 4. Pump head at pumping rate of 6.5 Ml/d

Borehole	Static head (m)	Maximum total head (m)
Park Road South	65	77
Landside	72	83
Pocket Nook 2	76	86
Slag Lane	77	83

As the table shows, identical pumps capable of 6.5 Ml/d at a head of 85 metres could be installed in the boreholes. Similarly, pumps capable of 2.25 Ml/d at a head of 47 metres were installed in Bank Heath and Makerfield boreholes.

5 Projected savings

5.1 Lightshaw boreholes

Savings can be achieved by operating the cheapest sources (Bank Heath at 2.25Ml/d, and Park Road South, Landside and Slag Lane at 6.5 Ml/d) to meet the average demand of 20 Ml/d. Replacing the existing pumps with new more efficient ones will produce further savings. It is projected that the average cost per megalitre will reduce from £15.50 to £14.00, a saving of 10% on electricity costs. This is equivalent to an annual saving of £10,000.

5.2 Widnes boreholes

An annual saving of £12,000 can be made at Belle Vale by replacing the existing pump which is capable of 4.5 Ml/d with a 2 Ml/d unit. By reducing the abstraction rate, it is predicted that the pumping head will reduce by some 60m (Figure 7). Similarly, an estimated £3,000 can be saved at Pex Hill by reducing the pump capacity from 7 Ml/d to 5 Ml/d. The shortfall can be made up by increased pumping at Stockswell as outlined in Table 3. The total estimated saving of £15,000 is equivalent to 14% of the current annual electricity costs for the Widnes group of boreholes.

6 Further work

Pump scheduling techniques are being used to achieve further savings. Where it is feasible, time switches are installed to switch off pumps during high electricity tariff periods. Where there is sufficient storage in the supply system, daytime pumping is reduced to a minimum and maximimum use is made of cheap overnight electricity. Computer programs to optimise the operation of supply systems are being evaluated. Operating strategies to minimise costs can be developed by using computers to analyse data on source yields, electricity costs, pipe networks and demand requirements.

7 Acknowledgements

This paper is published with the permission of the Chief Scientist of North West Water Limited, although the views expressed are the author's own. Company staff in North Mersey Supply District and the Geological Services section are thanked for their assistance in carrying out the work.

8 References

Clark, L. (1977) The analysis and planning of step drawdown pumping tests. **Quart. J. Engng. Geol.**, 10, No.2, 125-143.

Eden, R.N. and Hazel, C.P. (1973) Computer and graphical analysis of variable discharge pumping tests of wells. **Civil Engng. Trans. Inst. Eng. Australia**, 5-10.

Vines, K.J. (1989) Ednhaz: A program for analyzing step drawdown tests. **Computers & Geosciences**, 15, No.6, 965-978.

33 REHABILITATE OR RENEW? AN EXAMPLE IN WESTERN ZAMBIA

S.E. SUTTON
SWL Consultants, London, UK
J.S. SUTTON
Sir Alexander Gibb and Partners, Reading, UK

Abstract
In the Western Province of Zambia, a programme of rural well construction has been running for ten years. Many early wells have declined in performance, and give rise to the dilemma of whether to rehabilitate or renew. The cost/benefits of renewal and rehabilitation need to be compared, with special regard to relative costs, but also to affordability of different solutions to those who will pay for them. Capital costs tend to be covered by outside funding agencies, recurrent costs by local government or by consumers.
In remote areas, transport costs become a major element in well construction, and in any well servicing activities. Thus, if rehabilitation will still lead to a need to re-visit more often than the construction of a new well with better design and materials, its cost advantage is very short lived. In this instance the solution therefore seems to be to increase the capital costs, borne by funding agencies and central government, thereby reducing recurrent costs and maintenance technology to a level which can be realistically sustained by the consumer.
Keywords: Rehabilitation, Renewal, Cost/Benefits, Sustainability, Consumer.

1 Introduction

Western Province, Zambia is located either side of the Zambesi flood plain upstream of the Zambia/Botswana border. The flood plain is generally some 20-30 km wide, rising gradually over several hundred kilometres towards the Angolan border, and with a more well-defined eastern boundary scarp, 10-30m high.

Over almost all of this area of approximately 100,000 square kilometres, the surface is directly underlain by fine-grained, wind-blown sands extending to depths of over 100m. The groundwater table is generally at, or close to the flood plain level, rising away from the plain at a shallow gradient of around 1-2:1000. Groundwater is often very acidic, and with almost no dissolved solids.

Traditionally the rural population, which numbers almost 500,000, has lived in small, scattered villages along the margins of the flood plain and around low-lying depressions (dambos). In each of these areas water is generally accessible within 1-3m of the ground surface for most of the year. Much of the development of the past 30-40 years, however, has led to the growth of villages in slightly more elevàted areas. Institutions such as schools and rural health centres have frequently been built on low hill-tops, ribbon development has occurred along transport routes, and the growth of the six District administrative centres has led to the expansion of peri-urban villages. Within these areas the depth to water may be up to 60m and averages 20-30m.

Only one tarmac road reaches the area, connecting three of the six district capitals with Lusaka. Most other roads do not receive any maintenance, and most villages are only served by rough tracks through very loose sand, for which the local, and most appropriate, form of transport is an ox-pulled sleigh. Imported materials have to travel 3000 km from Dar es Salaam by road, to central stores in the provincial capital, Mongu. From there, most distribution is on unpaved roads, requiring four-wheel drive vehicles, with 75% of people living more than 3 hours drive (which may only be 50 km or less) from Mongu. The combination of deep sand and areas of marshland are very hard on vehicles, leading to high maintenance costs and short life expectancy. For the West bank settlements boats are used to ferry materials some of the way.

2 Development of Rural Water Supply

Within Western Province, NORAD began funding extensive rural water supply works in the late 1970's (described further in Sutton 1986a,b). The principal targets within the Province were the schools, rural health centres (300-400 in number) and the larger villages, to achieve 50% coverage of the rural population, and complete 800 wells.

For several years the project constructed shallow hand-dug wells with concrete rings, and drilled boreholes with percussion drilling equipment which was not ideally suited to the ground conditions, and used well designs which were not the most appropriate. This resulted in the completion of several hundred wells which suffer from one or more of the following defects -:

- Insufficient, poorly designed, or poorly installed gravel pack
- Inappropriate construction materials leading to both corrosion and/or clogging
- Oversized screen slots
- Insufficient penetration of screen, and/or screen length below the water table
- Well casing (2", 50mm) serving also as pump riser

These defects in turn gave rise to poor reliability of supply and resultant low usage of facilities, which were recorded by annual monitoring programmes. In particular, excessive drawdown was suffered because of unnecessarily large well losses (sand clogging or iron precipitation), and pumps frequently broke down because of sanding wells.

In the mid-1980's, in parallel with a series of changes in approach and administration, the project was re-equipped with robust rotary rigs and a full range of ancillary equipment for direct circulation, mud-flush drilling and well development using jetting, surging and air-lifting techniques. Five-inch diameter, uPVC well screen and casing were adopted as standard, and hand pump models restricted to two proven models (Consallen and Blair), with final selection depending on pump lift.

Arising from these technical and administrative changes, the rate of well construction and the reliability of completed wells improved. However the dilemma of rehabilitation or renewal for wells completed earlier in the project's history required a policy decision, which would have major effect both on immediate budgets, and upon the long term responsibilities of the project, local government and consumers.

3 Rehabilitation

In the mid-1980's almost 80% of the completed wells were not operational, and the underlying cause of failure in almost all cases was sanding or iron precipitation. Airlift pumping provided a rapid means of sand removal and with pump repair, wells could then be quickly brought back into operation. In some wells, fabric-wrapped screen could be installed, but annular space was insufficient for gravel pack installation and continued movement of sand through the outer (older) screen, would have meant the build up of a very fine sand pack, whose permeability would be much less than that of the aquifer. Drawdown would then be increased in wells which anyway often barely penetrated enough of the aquifer when first constructed. In many wells, screen had been attached to a 2" (50mm) cylinder and casing which also acted as riser pipe, and thus new screen could not be inserted, although sand could be cleared out.

In the case of wells with high iron content, screen openings could be partially unblocked with buffered or weak acid, and sterilised to remove iron-fixing bacteria, where the screen was still strong enough to withstand such treatment.

In both cases, whilst wells could often be brought back into production, the underlying cause of poor performance could seldom be removed permanently. This meant that rehabilitation would provide short term benefits, but the life expectancy of wells so treated would not be as great as that for new wells, designed and constructed to different specifications.

4 Renewal

New wells, completed at larger diameter, with graded gravel pack, and plastic screen and casing, may seem an expensive and time consuming alternative to air-lift clearance and repair. They do, however, provide the opportunity to construct a sand-free well to sufficient depths to allow for water level decline, and long-term deterioration in well performance. 98% of new wells have not continued to sand after development, even in such a very fine, predominantly single-sized sand aquifer. Such wells may be expected to continue functioning without maintenance, for the lifetime of the plastic screen. Even for the few wells where sand ingress is a problem, (probably because of poor gravel pack insertion), construction of wells which are over capacity

(10 m deeper than theoretically necessary) means that well depths will remain sufficient for at least twenty years of sand accumulation, and screen intake area is generally far enough below pump intake for sand grains to fall to the bottom, rather than enter the pump and cause damage.

5 Comparative costs

It might be expected that the costs of rehabilitation would be considerably less than for renewal, since for the former there are no drilling/excavation costs, usually a lower expenditure on materials, less complicated equipment, and no siting costs. The costs of a single well repair/servicing are certainly less than those for a new well, but if that well then needs rehabilitation twice as often as a new well, it becomes the more expensive alternative, whatever the form of well construction (see Table 1)

Table 1. Breakdown of well costs (in £1000's/well)

Item	Shallow hand-dug well (7m)		Drilled borehole (average 40m)		Jetted wellpoint (up to 15m deep)	
	New	Rehab	New	Rehab	New	Rehab
Transport	1.8	1.3	2.4	1.7	0.9	0.7
Labour	0.3	0.2	0.35	0.2	0.2	0.2
Materials	0.35	0.25	0.7	0.4	0.35	0.1
Pump/wind-lass	0.2	0.1	0.35	0.35	0.2	0.2
Equipment/ Consumables	0.1	0.1	0.35	0.3	0.1	0.1
Siting/ CEP	0.4	0.2	0.4	0.1	0.4	0.1
TOTAL	3.15	2.15	4.55	3.05	2.15	1.4
Cost ratio Rehab/renewal	68%		67%		65%	

The above figures illustrate two main points. Firstly that transport makes up approximately one half to two-thirds of the total costs for both construction and rehabilitation works, and materials constitute only 10-20% of these costs. Secondly that rehabilitation is likely to cost approximately two- thirds as much as renewal in this situation.

The overall significance of transport to costs is similarly reflected in the project funding, where in 1990 rural water supply construction and maintenance receive only 65% of the funding which transport and workshops (which admittedly also serve eight small town piped water supply systems) receive. This is logical, since no activity can continue in the area if transport and rig maintenance are neglected, but underlines the fact that well construction or rehabilitation both need expensive infrastructure.

In terms of equipment needed, the cost of a service rig, set up for full rehabilitation is almost exactly half that for a new drilling rig. A Mercedes Benz 4x4 truck, with ancillary crane, compressor, pumps,tools, spares and riser pipes for well servicing was quoted at almost exactly £100,000 at the end of 1989, and a Knebel HY77 "Buggy Drill" at approximately £200,000, including all spares for two years' operation, at the same date.

6 Consumer interests

6.1 General

The chief primary objectives of rural water supply are to provide a source which is reliable, adequate in yield, and acceptable in quality and siting to consumers. Increasingly the consumer is then required to play a part in maintaining these aspects, particularly where central governments seldom have the infrastructure and finance to bear this responsibility in part or in full (Narayan-Parker 1989).

Where the source fails in any one or more of these attributes, the consumer soon becomes disillusioned, and efforts by the community to maintain or improve the supply soon decrease, and its benefits may be lost. There may even be a negative effect where health problems are exacerbated by return to traditional sources after some time with improved water. Also resistance to further involvement with fund-raising and maintenance activities may grow where supplies repeatedly break down. Temporary resuscitation of supplies may then be little better than re-adoption of traditional sources, if these are within 1-1.5 km of the settlement.

6.2 Reliability

Where rehabilitation can produce a well that is as reliable as a new one, it is obviously worth doing. Where, as seems the case in Western Province, this would seldom be so, it would have only very short term benefits. Well servicing needs to be done before the well reaches a state where pumps will be damaged, and long before it will stop producing water. A well that needs cleaning out every two years would be over five times more expensive than a new one, if new wells have a design life of fifteen years (which seems reasonable from present performance).

6.3 Water quality/ siting

Wells which suffer from iron clogging, or major corrosion will also tend to have water of a high iron content, which is generally unacceptable to consumers. It will only be used for drinking, if no better source exists within a kilometre or so. Rehabilitation of such sources may improve yields, but will not do anything for acceptability to consumers for most uses. In Western Province, it is often possible to re-locate sources, or re-design them to provide more acceptable quality, or accessibility.

6.4 Affordability/sustainability

An average rural water supply householder in Western Zambia will earn around £10 a month. On this he will support four or five other people.

The per capita cost of a new handpump/borehole supply will be some £30, or just over a year's income for the family. One rehabilitation of that well will cost the whole of approximately 9-10 months income. It is apparent that consumer financing of rehabilitation is as impossible to achieve as new construction, and both must rely on outside funding.

It is not in the consumer's long term interest to rehabilitate poor wells, if funding is available for a limited period, and consumers will shortly be back in the same situation again. In this case, the slower, but more long-lasting production of new wells, will provide a more reliable, and acceptable supply, which can then be locally maintained.

Consumers have proved able to support maintenance costs where these are limited to those of the pump itself, and the well surroundings, although transport costs are still to some extent (but decreasingly) subsidised. Such consumers require wells whose design life does not require frequent extension by rehabilitation. Designs therefore may need to be to higher specification than where recurrent costs can be more easily supported. To drill and screen an extra 10 metres of borehole may double the life of the well, but only increase capital costs by 5%

7 Conclusion

Rehabilitation in the developed world may be taken to be an economic way of extending the design life of a well. In many Third World instances, the poor performance of a well may be a symptom of bad design, or construction, which can be temporarily cured. What has to be judged, is whether such improvements of the source are sufficiently long term to offer the best alternative with funding available, and with consumer and local/central government interests in mind.
In this case, it appeared that few existing wells could be improved to provide the same reliability as replacement ones, and that in the long term new wells would work out to be cheaper, and more acceptable to all parties concerned. This was chiefly because of the high cost of transport in the area.

Acknowledgements. The authors wish to thank M.K. Samani, of the Department of Water Affairs, Western Province, Zambia for help in up-dating costs.

References.

Sutton S.E. (1986) Percussion vs. Rotary - Myths exploded. **World Water**, Nov.

Sutton J.S. and S.E (1986,1987) Diaries from Mongu 1,2,3. **World Water** Sept '86, Jan and April 1987.

Deepa Narayan-Parker (1989) Goals and Indicators for integrated water supply and sanitation projects.**UNDP/PROWESS** New York.

34 MONITORING AND DIAGNOSIS FOR PLANNING BOREHOLE REHABILITATION – THE EXPERIENCE FROM LIMA, PERU

S. PURI
Scott Wilson Kirkpatrick & Partners, Basingstoke, UK
C.V. FLORES
SEDAPAL, Direccion de Aguas Subterraneas, Chacarilla del Estaque, Lima, Peru

Abstract
In order to satisfy the medium term water demand for Lima and its coastal port of Callao, additional supplies are required from the aquifer underlying the city. Of the 300 boreholes providing the present demand many are considered to be exhausted. Nearby new holes do however produce water and therefore the need to investigate, ie diagnose, these exhausted holes arose. The scope of the investigation was expanded to obtain a wide ranging overview of the reasons why existing wells do not produce yields as high as they did when first drilled. Seventy boreholes were selected, based on predefined criteria and a comprehensive investigation was completed to diagnose the problems with the wells. This case history describes the approach adopted and sets out the factors which may be used as a guide elsewhere. The outcome of the diagnosis was production of a systematic rehabilitation programme for each of the holes and preparation of monitoring guidelines to help in advance warning of possible forthcoming yield declines and other well yield problems.
Keywords: Borehole diagnosis, Aquifer monitoring, Borehole Rehabilitation, Lima, Peru, Borehole instrumnetation, Pumping equipment.

1 Introduction

There is an increasing awareness in the water industry, that boreholes which at the time of construction had been designed for 30 or 40 year life, decline in efficiency and yield far sooner than expected. As this situation becomes obvious to the operators, some form of monitoring is initiated to get a clearer prognosis of the forthcoming problem. Normally, this evidence of falling yields and levels is used to justify either abandonment of the sources, or to plan the construction of more boreholes. Recently water supply professionals have questioned this philosophy and have turned to an alternative approach, ie to undertake rehabilitation of holes with the poor yields.

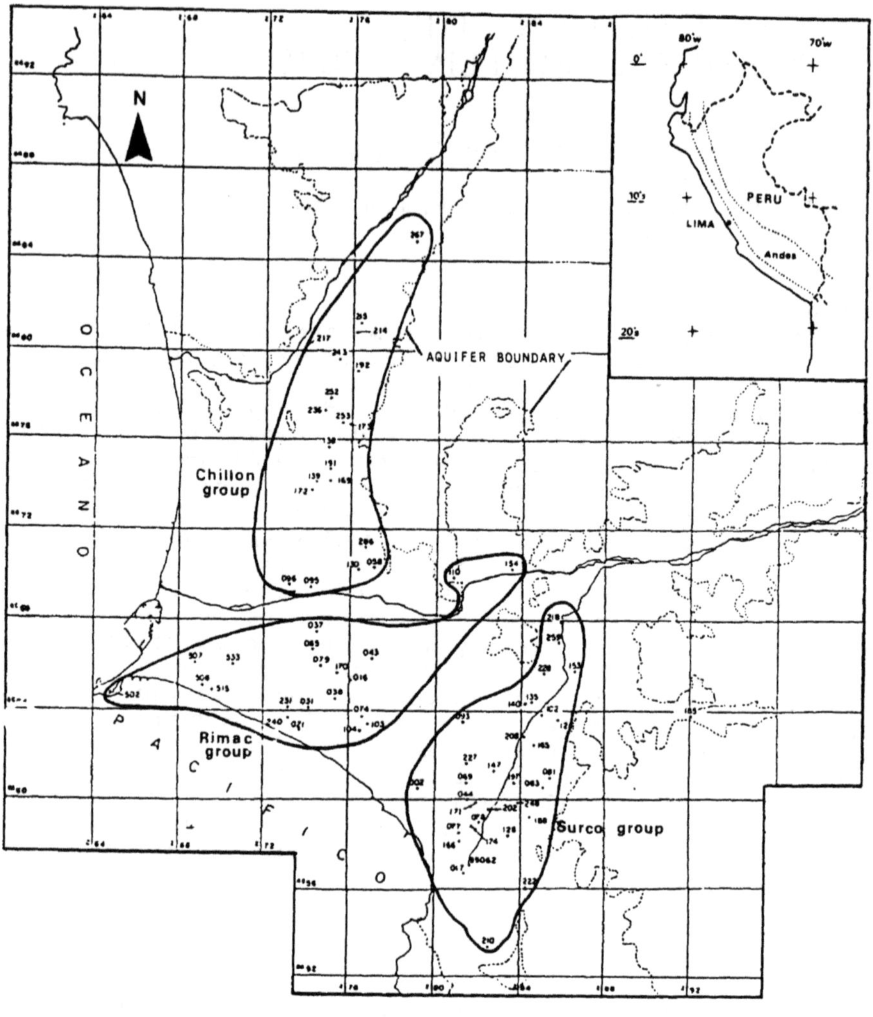

FIGURE 1 LOCATION MAP AND AQUIFER BOUNDARIES

Experience with rehabilitation has resulted in mixed degree of enthusiasm for it. When rehabilitation is successful, its justification is unquestioned but when the results of rehabilitation are marginal or minimal, then its value is scrutinised critically.

Professional hydrogeologists have long been aware that borehole rehabilitation is a science, rather than an art to be learnt by trial and error. For each aquifer and borehole construction, there are usually specific conditions and reasons for falling yields. To restore the yields and to operate boreholes more efficiently, a combination of appropriate rehabilitation techniques has to be utilised. The special skills that the hydrogeologist can offer to the water industry are to assist with selection of the right combination of the techniques that are appropriate to each aquifer-borehole construction situation. However, in order for the hydrogeologist to apply his scientific know how, he has to first diagnose the problem. The diagnosis that can be performed has to be based upon an investigation of the symptoms and upon consideration of the previous history of the ailing water source. Here the analagy with the medical professional ceases. The doctors patient is able o communicate his condition in the recent past and explain the nature of his ailment. The hydrogeologist in a similar situation is at a disadvantage, unless adequate previous monitoring had been undertaken in a methodical way.

The following case history from Lima, the capital city of Peru, is presented to illustrate the factors discussed above. In that project 70 town water supply boreholes were methodically investigated and all the previous monitoring data available was collated to systematically define the type of rehabilitation that would be effective (Binnie & Partners 1987). Prior to undertaking the actual rehabilitation, the objective of the diagnosis was to prognosticate the improvement in yield that would occur, in order to balance the costs of rehabiltation against the cost of new boreholes.

2 Background

2.1 Location & Geology

The cities of Lima and Callao, situated on the narrow coastal plain west of the Andes ranges, (Figure 1) lie at the foot of a large alluvial cone of an accumulated sequence of well rounded boulders, cobbles, coarse sands, lenses of fine silts underlain by granodioritic bedrock. The alluvial fan deposits developed as a consequence of Quaternary erosion of the Andes ranges, and reach a thickness of upto 500m below present ground surface. The rivers Chillon and Rimac flow through the city and their headwaters, in the high Andes, receive very high rainfall, while at the coastal area the annual rainfall is less than 50mm.

Lima and the port of Callao were modest towns upto the 1950's. From the mid 1960's there was a gradual increase in urbanisation and by the late 1970's it was clear that much of the alluvial cone underlying the city would be covered by urbanisation.

2.2 The Problem

During the Inca period most of Lima and Callao were given over to agriculture, sustained by irrigation from offtakes from the two rivers. Potable demand was met by direct abstraction from the rivers. A water treatment works was constructed upstream of the old city centres but with increasing demands boreholes were constructed. In the 1950's these boreholes had remarkably high yields which were exploited by high capacity shaft driven pumps. The borehole construction techniques were simple and as the aquifer appeared to be very potent, development proceeded at a rapid pace. By mid 1980's the boreholes that had yields of 90 l/s in the 1950's were producing 15 l/s or less. Monitoring of the aquifer conditions was perfunctory and what little there was, indicated that groundwater levels were dropping. However, it was interesting to note that new holes, drilled in the vicinity of older holes, still produced very high initial yields. As a consequence of this condition the assumption that older a hole, the less its yield, was made and often the solution that was adopted was to replace the source by a new hole and to abandon the 'exhausted' hole.

With increasing scientific awareness of the officials of the water supply authority, the idea that exhausted holes could be rejuvenated was born. At the same time, the need for diagnosing the reason for falling yield was also felt.

2.3 The Need for Diagnosis

About 40% of the current water supply demand of Lima is obtained from the aquifer underlying the city (Figure 2). The projected demand for the year 2000 is 34m3/s while the demand in 1986 was 21.9m3/s. There is increasing pressure for further urbanisation in the city and as a consequence water supplies are required for the numerous new residential districts that are being set up.

The aquifer receives most of its recharge from infiltration through the river beds, when high flows occur annually and from leakage, through the water supply network or due to excess watering of parks and gardens (Lerner 1986). The depth to water table is 60m in some parts of the city while generally elsewhere, it is 20m to 40m. There is no rainfall recharge. Abstraction exceeds recharge and as a result water levels have been falling in most of the city areas (Figure 3). Borehole yields have been falling,

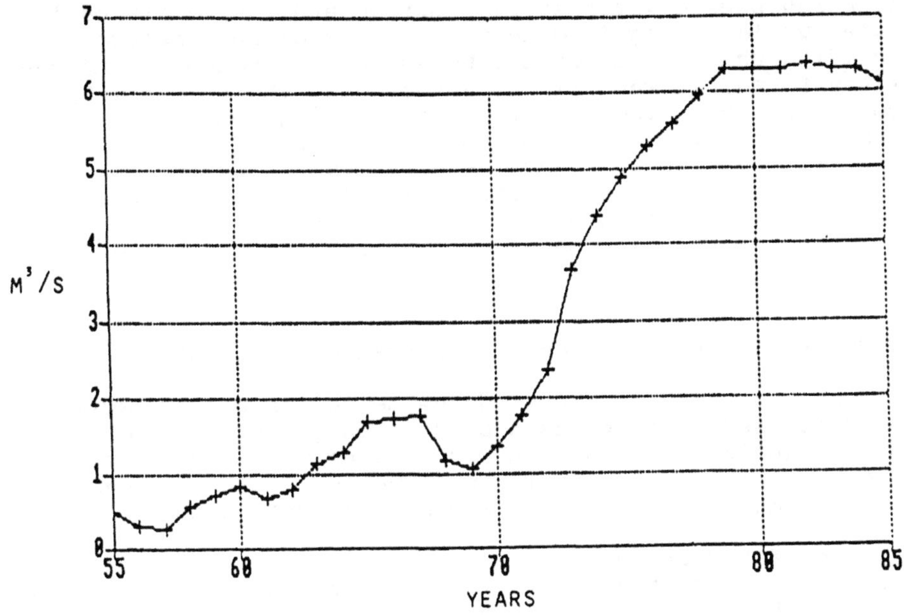

FIGURE 2 PRODUCTION RATE FROM BOREHOLES

but since the saturated thickness of the aquifer is thought to be up to 400m in places, deeper holes have been constructed to tap the aquifer. The cost for these deeper holes is increasing, as are the pumping lifts. There are over 300 public water supply sources and many additional holes would be required to provide the demand; before deciding that exhausted boreholes can be abandoned and new sites found, it was wise to investigate the possibility of rehabiltation. The potential cost savings would be particularly large in the urban conditions, where access is now limited, pipework and the distribution system is already in existence and the available infrastructure could be reincorporated into the supply system.

The need for diagnosis was therefore demonstrated and a programme of work initiated (Puri, et al 1989).

3 Criteria for Selection of Representative Sample

A number of factors dictated the selection of boreholes to be diagnosed. As it was necessary to gather a good understanding of these, it was decided that natural aquifer factors, man made borehole construction factors and the electro-mechanical factors would be investigated.

3.1 Aquifer Factors

Aquifer factors that would influence borehole yields in the case of Lima were grouped into:

Reduced annual recharge
Reduced aquifer storage
Hydrochemistry of the aquifer system

Over an extended period of time, boreholes will react to these regional aquifer factors in an indirect way. If recharge is decreasing because of changing land use (urbanisation) or river training works (to minimise flooding thus limiting the infiltration opportunity) then well yields will eventually decline. Similarly, reduced storage in the aquifer or even the changing hydrochemistry has an influence, which may be subtle in the timescale of most studies, but over longer periods these could be significant. In the case of Lima there was no data for the time scale in which these phenomenon becoming significant. However, knowing that in the previous two centuries the aquifer was overlain by agricultural land, irrigated by the annual flood flows, it is easy to conclude that total recharge may have reduced considerably. The need to investigate this as a cause of well depletion was therefore identified.

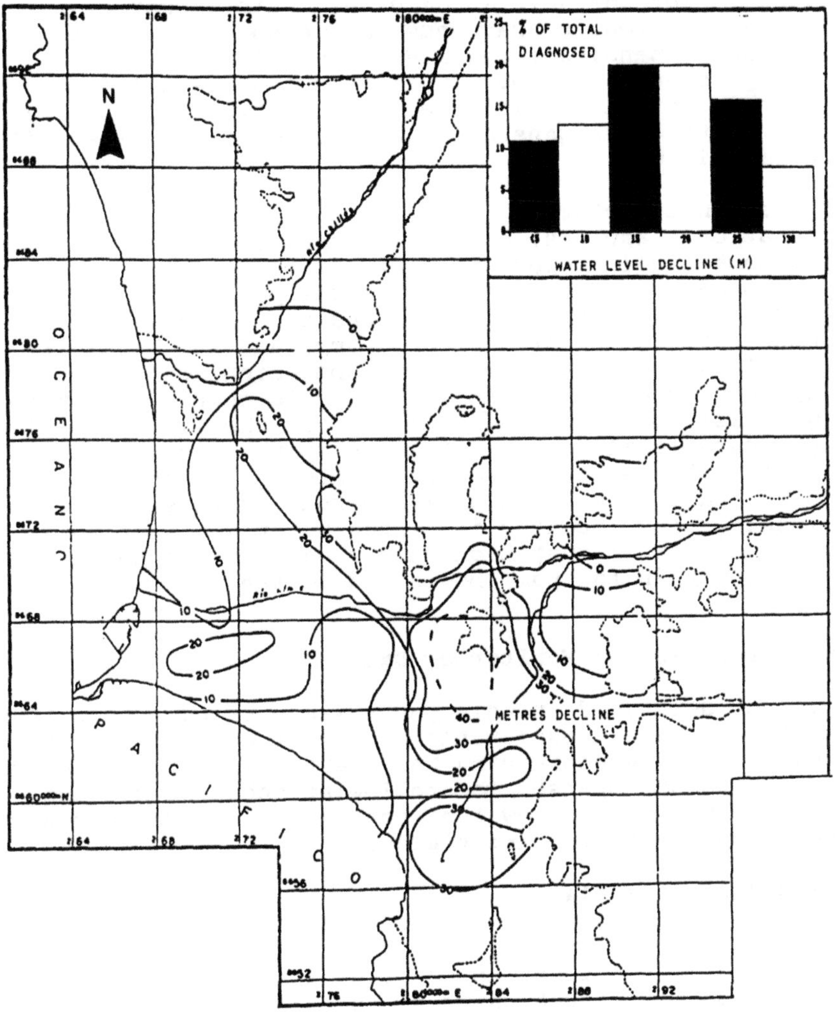

FIGURE 3 WATER LEVEL DECLINE 1969-1985

3.2 Borehole Construction Factors

The most obvious of the direct factors that would influence borehole yields is their construction. This is well understood by water supply professionals, and in the case of Lima, older wells were poorly completed and the construction methods were inadequate. With improved technology and greater awareness the methods used have improved and expensive screens are installed now. Previously screens were poorly slotted and installed but despite this, original yields were high, typically 100 l/s. Drilling depths were generally determined by the capability of the rigs in use.

3.3 Operational Factors

The operational aspects, especially the efficiency of the electro-mechanical components play a great role in the long term performance. The need to size the pumps appropriately in terms of their physical dimensions and yield characteristics is clear. Often, pumps and the pipework are installed soon after borehole commissioning for the then dynamic conditions. As water levels change, the pump will gradually become misdimensioned for the yield, leading to high operational costs. If the pumping rate is set too high, this will lead to frequent dewatering of the borehole. Apart from the negative effect of this on the pump itself, the cascading in of the water may lead to encouragement of encrustation on the wall of the borehole screen and within the pump bowls and intakes.

4 Interpretation of Monitoring Data

The project described in this paper was carried out in 1985 and at that time a considerable amount of monitoring was underway. Routine monitoring had started in 1978 but prior to that, data are sparse. The following sections discuss these data and explain how they contributed to definition of the reasons for reducing yields.

4.1 Groundwater Levels

Static and pumped water level data had been haphazardly collected previously, but it was the systematic monitoring from the mid 1970's that showed clear trends of water level declines albeit for some parts of the city. The data for pumped conditions was much less clear, however.

4.2 Abstractions

Records of public supply abstractions were based on twice yearly survey of pumping rate. The estimate of total abstractions was therefore based on these surveys and was thus subject to a deqree of inaccuracy. The private water supply sources were operated under licence but there was no data on the volume of annual private abstraction. Estimates were based on billings to water users.

4.3 Groundwater Quality

Very limited water quality data was available other than that required for environmental health reasons. Groundwater chemistry variations were therefore unknown as were temporal quality variations.

5 Design of the Diagnosis Programme

The programme of diagnosis was aimed at obtaining a wider understanding of the reasons why Lima borehole yields were reducing. To implement this, each of the selected boreholes were subjected to a series of tests and investigations. These can be grouped into three classes: those concerned with borehole performance, pumping equipment installation and chemical characteristics of the water in the aquifer. These are discussed below.

5.1 Borehole Performance & Construction

The performance of the borehole was checked by disconnecting the discharge from the supply network, and step pump testing the installed pump at five or more escalating rates. The pumped water was discharged to waste rather than to the pressurised supply network. The rates selected were based on records of previous performance and ranged from 30% to 150%. Records of time drawdown and pumping rate were made.

The pump and the rising main were then dismantled and a series of geophysical and CCTV logs were run. Verticality was measured in all cases and depth samples of water taken, if appropriate.

5.3 Abstraction System

During the pumping test, the electro-mechanical performance of the pump and its motor was monitored. This included a measure of the consumption of electrical energy for each pumping rate using independent voltmeter and amperimeter, pressure in the discharge line, pumping rate and pumping water level.

Following dismantling of the pump and motor, a comprehensive inspection of the parts was made, to assess the condition and to compile a detailed list of spares that would be required to reinstate the pump.

The pump house and all its installations, including the electrical control board, the cables and instruments were methodically checked and inventoried. The instruments included flowmeter, water level indicator, discharge and delivery manometers and the chlorination manometer.

5.4 Hydrochemistry

Parallel to the time drawdown measurements during the pumping test, the pH, Eh, EC, temperature and dissolved oxygen were monitored of the discharge water in sealed sampling pots. On-site titration measurements of alkalinity, CO2 and H2S were made. Water samples were taken for laboratory chemical analysis which included major anion-cations and Fe+2, Fe total and Mn.

Water samples were also taken for bacteriological analyses. In particular the considerable iron encrustation, which was thought to be due to biofouling, was to be evaluated on the basis of iron bacteria study. In the event this was an aspect that could not be completed as comprehensively as all the others were because of logistical reasons.

6 Rehabilitation Options Considered

A complete interpretation of the diagnoses of the seventy wells was carried out. Table 1 shows the nature of the evaluations completed and the aspects that were diagnosed. The selection of the rehabilitation options could now be made using the data obtained during diagnosis and from past historical monitoring.

Table 1. Diagnosed aspects to design rehabilitation

Investigation	Test Performed	Aspect Diagnosed
Borehole Hydraulics	Historical trends Step pumping tests Well Loss, Aquifer loss, efficiency	Regional & local aquifer influences
Physical Condition	Drilling & screen installation Verticality & alignment CCTV Geophysical logs	Borehole Construction
Well Head Chemistry & Bacteriology	On site & Laboratory analyses Iron & Carbonate Corrosion & encrustation Bacteriology	Aggressive, encrusting & eroding role of groundwater
Pump & Pipework Condition	Energy consumption Pump efficiency Inventory of pump Pipework & control valves Pump hose control equipment	Abstraction & delivery system

Rehabilitation options available for implementation in the situation of Lima, given the generally urban nature of the aquifer under consideration, were limited. These limits were imposed by the socio-economic and environmental constraints. From the plethora of rehabilitation techniques available, a systematic elimination process identified acceptable techniques that could broadly be classed into four groups. These are discussed below.

6.1 Basic & Minor Renovation
In nearly every borehole diagnosed, a modest amount of basic renovation work was required. This consisted of surging and bailing boreholes to clear them to their original drilled depth. Sand inflow was not generally a

problem. In addition to surging, every borehole had a high degree of encrustation on the casing and the screens, irrespective of the types of screen. This was to be removed by wire brushing.

Repair and replacement of worn or damaged parts of the pump and motor were required in most cases. The impellors and the bowls were found to be encrusted, confirming occurences of air pumping. The rising mains were also generally encrusted contributing to hydraulic head losses.

6.3 Replacement of Equipment

There were occasions when the borehole yield was found to be perfectly adequate, but the pump installed in the borehole was considerably misdimensioned. A simple replacement of the pumping equipment by appropriately sized pumps and motors was required.

6.4 Reconstruction & Redrilling

As a consequence of aquifer overdevelopment in some sections of the city, water levels has declined permanently by upto 40 to 50m. In these areas, boreholes constructed in the past were now simply tapping a much reduced saturated thickness of the aquifer. Partial penetration effects were overshadowing the aquifer drawdown and these, coupled with well losses, resulted in screens hanging above the water table. In situations of this type the only option was to drill through the existing borehole to tap the deeper sections of the aquifer. This would involve the installation of new, narrow screens, that would have to pass through the existing pipe and screen. In addition, the drilling rig to be mobilised would on occasion, have to manoeuvre through confined spaces between buildings and other structures. If there was a pump house, as there was in most cases, then the house would either have to have the roof plate removed, or in extreme cases the whole house would have to be demolished and reconstructed following the drilling operation. The decision to redrill was therefore based not only on the constraints imposed by the aquifer but also its urban infrastructure. Finally, the cost of every such precaution would have to be included in the rehabilitation works, leading to increase in the total cost.

7 Prognosis of Benefits

In order to define cost benefit of the rehabiltation programme that could be recommended, it was important to prognose the outcome of the work to be carried out. Some small scale rehabilitation work had been carried out in Lima in the past but there was not enough experience on the basis of which reliable estimates could be based.

Experience elsewhere suggested that rehabilitation has undeniable benefits. The detailed and comprehensive database obtained during the diagnosis work gave a reasonable basis on which to produce such a prognosis.

The primary and the essential factor that formed the basis of the prognosis was the yield characteristic of the rehabilitated borehole. The well drawdown parameters, obtained from the step test indicated the component of the drawdown that could be improved. If the well loss (rather than the aquifer loss) component could be reduced, then an estimate of the new yield could be made. If the hole was to be deepened then the aquifer yield for given drawdown added to the well loss could give an indication of the new yield. Mathematical modelling of the behaviour of the regional aquifer was performed in parallel to the diagnosis. The model was reformulated on the basis of the diagnosis data to predict individual borehole drawdown (Lerner 1989). The combined results of modelling and diagnosis recommendations provided a reasonable basis for anticipated yields. Given the anticipated aquifer yield and drawdown, appropriately dimensioned pumps, pipes, discharge valves and control systems could be costed.

It was estimated that a yield improvement of 56% could be achieved, from a current total production rate of the diagnosed boreholes of 917 l/s to 1628 l/s once they were rehabilitated.

Cost comparisons are not easy to make in the situation described above. The only basis of such a comparison could be the cost of construction of a new borehole. Generally the unit cost of rehabili- tation was estimated to be 0.5 to 0.2 of the cost of a new borehole. On average, the unit cost of rehabilitation of a well at 1986 prices was approximately ¯35000 not including purchase of new pumps.

The estimated costs of electro-mechanical repairs to pumps and motors that would restore them to an operating efficiency of 60%, by reducing current pumping costs, could be recovered within three years.

8 Conclusions

The case history given above describes a comprehensive diagnosis programme undertaken prior to rehabilitation. The scope of such an investigation could not have been reduced given the paucity of previous data on the basis of which a rehabilitation programme could be designed. Rehabilitation is a science and a 'try it and see' approach should never be adopted especially in situations where fragile socio-economic conditions prevail. In the situations of Lima, where the aquifer source is at threat, a detailed diagnosis was essential. All the indications are that an ongoing programme of rehabilitation of the 300 or so water supply holes will have to be carried out during the next 20 years, prior to possibly a new water source being provided. Funding agencies may care to note that the investment in conducting such an investigation pays off dividends in providing an excellent basis from which to plan, cost and execute rehabilitation.

Acknowledgement
The work described here was done by teams of field staff and could not have been completed without their considerable cooperation. Although the team members are unnamed here, their joint effort is nevertheless acknowledged.

The views expressed in this paper are those of the Authors'.

References

Binnie & Partners (1987) Management of the aquifer resources of Metropolitan Lima. **Final report to SEDAPAL**, Oficina de Proyectos Especiales. January 1989, Chacarilla del Estanque, Lima.

Lerner, D.N. (1986) Leaking pipes recharge groundwater. **Groundwater** V 24, No 5, pp 654-662

Lerner, D.N. (1989) Predicting pumping water levels in single and multiple wells using regional groundwater models. **Jnl. of Hydrology**, V 105, pp 39-55

Puri S, Petrie J.L., & Valenzuela Flores C (1989) The diagnosis of 70 municipal supply boreholes in Lima, Peru. **Jnl. of Hydrology**, V 60, pp 287-309

35 RURAL WELL REHABILITATION AND MAINTENANCE IN WELO, ETHIOPIA

M.J. JONES
Wimpey Environmental, Hayes, Middlesex, UK

Abstract
The Welo Well Rehabilitation Project inspected existing water supplies and equipment installed in the region and prepared an updated inventory of all boreholes and wells. The inventory was compiled into computer database files which contain records of 568 boreholes, 315 hand-dug wells and 564 pump installations. Based on these, rehabilitation and maintenance schedules were prepared and implemented. As the scedules were implemented, however, it became clear that some of the original borehole records were either incomplete or accurate and doubts arose over the design of certain boreholes in the region.

While the immediate concern of the project was to get hand and motorised pumps working, persistent problems were found with boreholes producing silt. Analysis of the database and field tests showed three main causes: i) The use in open hole completions in inappropriate ground conditions: ii) Inadequate developing and testing of boreholes and iii) Installation of too high a capacity pumps. In many cases the time and funds required to rehabilitate an existing borehole using a drilling or workover rig was estimated to considerably exceeded that required to drill a new borehole at the same site. Inflexible donor agency or client allocation of funds often, however, dictated rehabilitation.
Keywords: Ethiopia, Welo Region, Plateau Basalts, Rift Escarpment, Afar Depression, Well Database, Hand Pumps, Motorised Pumps, Borehole Siltation, Maintenance Programmes, Maintenance Routes, Satellite Maintenance Zones, Community Water Committees, Pump Attendants, Field Maintenance Teams.

1 Introduction

Successful development of groundwater to meet the increasing demand for safe, clean water has assumed a vital importance in the economic and social development of the rural areas of Welo Region. While traditional sources, springs and hand dug-wells still provide adequate water supplies for many communities, boreholes and deep well pumps can, and have, increased the available supplies to satisfy extra demands. These developments, however, rely on correct construction techniques, and require reliable and well maintained drilling rigs and pumping equipment. The region has received, mainly from Unicef, considerable assistance in acquiring the necessary techniques and facilities and by 1987, the well records held in the Welo regional offices at Kombolcha of the Water Supply and Sewerage Authority (WSSA) and the Ethiopian Water Works Construction Authority (EWWCA) listed approximately 400 boreholes and 250 hand dug wells. Of these, about 80 per cent were productive at the time of construction but during a preproject survey in 1986, only 50 per cent of all pumps installed were operating. Inoperation, at the time of the survey, was considered to be due in some cases to over pumping, and in others to a lack of water because of reduced water table levels following years of below average rainfall, and in others due to mechanical failure.

Up to 1988, EWWCA were responsible for all operation and maintenance arrangements for rural water supplies and WSSA only operated and maintained the large Urban water supplies in the region. In 1988, the rural water supply operation and maintenance responsibilities were transferred to WSSA. The two year Welo Project was largely directed to provide assistance in establishing WSSA's capability to undertake this work which was perceived to cover all aspects of rehabilitation and maintenance of hand wells and boreholes, pumping equipment and distribution systems. A computerised database established by the project and containing details of all rural water supplies formed the basis for planning the work and helped in meeting the four initial objectives of the project. These were:

- (i) Inspection and updating of the inventory of all wells and pumping equipment.
- (ii) Re-establishment and regular maintenance of existing water points.
- (iii) Assessment of drilled well performance, or potential performance.
- (iv) Implementation of a programme of development, or redevelopment of drilled wells.

Before initiating the programme to meet these objectives, it was necessary to consider it against a broad hydrogeological outline of the region.

2 Hydrogeological Background

Welo can be divided into a number of terrains each with its own development potential and problems. Fig. 1 is derived from the satellite imagery interpretation undertaken by the project and shows the main topographical, structural and hydrological features in the region. Geologically almost the entire area is underlain by extrusive igneous rocks. In the western highlands these are largely sheet flow basalts of Tertiary age. In the lowlands to the east of the main rift fault escarpment, effusive volcanic rocks predominate together with extensive tracts of fine grained lake deposits and Quaternary alluvial deposits.

Climatically the area ranges from humid to sub-humid in the western highlands and from semi-arid to arid in the eastern lowlands. Droughts occur frequently and there is a severe drought about every ninth year. There are no detailed regional hydrogeological studies of Welo. There are reports covering local surveys which give brief details on several of the hydrogeological terrains. From West to East these are:

The Plateau Surfaces - These are underlain by subhorizontal sheet basalt flows and have a variable cover of colluvial soil. The plateau evaluations are between 2000 and 4000 metres and have a seasonal rainfall of up to 2000mm. Run off

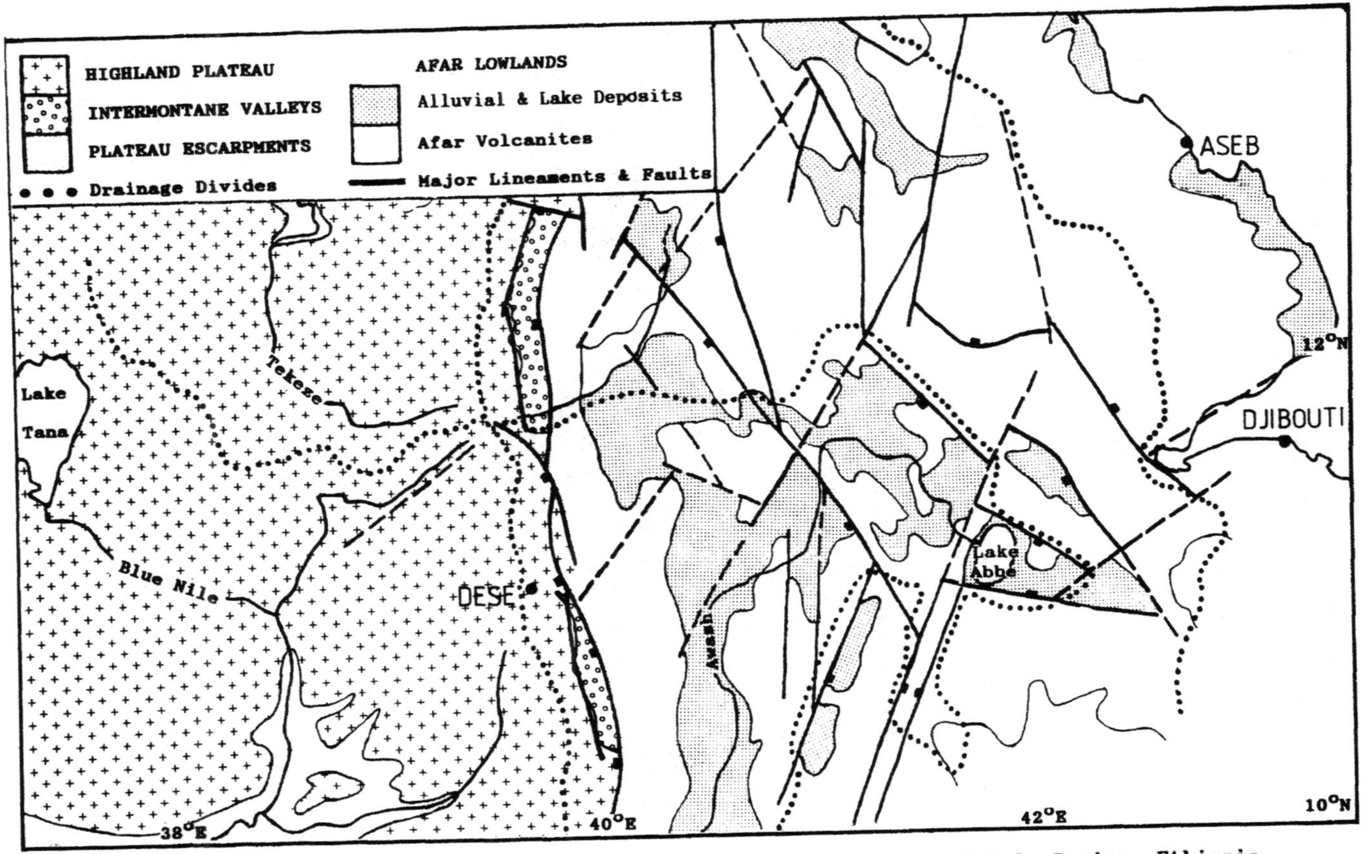

Fig. 1. Some Major Topographical, Structural and Hydrological Features of Welo Region, Ethiopia.

and erosion are high. The groundwater occurrences are associated with the fissures and inter-flow features within the basalts and the alluvial deposits along the stream beds.

The Plateau Escarpments (including the main rift fault) - The plateau surface has been deeply dissected and the upper parts of the escarpment are often precipitous. The lower flanks are covered with coalescing talus and outwash fans which receive substantial recharge from infiltrating streams.

The Intermontane Valley Floors - These are dominantly extensive graben valleys aligned approximately north-south parallel to the main rift faults. The grabens have a considerable thickness of alluvial fill which receives large quantities of recharge from the seasonal spate flows in the rivers. This alluvial fill forms the main aquifer bodies in Welo and these have been partially developed for town and rural water supplies. Local influxes of juvenile water associated with the rift faulting and recent volcanics are common.

The Afar Depression - Lying on the downthrow side of the main rift faults, this area is underlain by a complex series of effusive volcanic rocks dating from the Miocene to Present. The fissure flow basalts and stratiform basalts are punctuated by chains of volcanic vents and isolated caldera. Continental conglomerates and alluvial deposits occur along the foot of the rift escarpment and along the course of the Awash River. These deposits are superimposed on and interbedded with the volcanic rocks. There are several extensive areas of lake deposits, notably around Lakes Gamari and Abbe. The climate of the Afar depression becomes progressively drier to the east towards the Red Sea. The known groundwater occurrences appear to be limited and most developments rely on alluvial aquifers associated with the main water courses. Hot groundwater has been encountered at a number of locations and saline and gypsiferous soils are found associated with the lacustrine deposits. The amount of hydrochemical data that is available for the region, however, is limited and of uncertain quality.

3 Inspection and Inventory of Wells and Pumping Equipment

Following an exhaustive search of existing records and field collection of data, details of 568 boreholes and 318 hand dug wells were compiled and entered into the EWWCA database during the project. The WSSA database contains records of 564 pumps installed in the region, of these, 310 are borehole installations and 254 are hand-dug well installations. Full details of 86 of the known 97 mechanical pumps installed were recorded. There are details of 394 hand pumps (mostly India Mk2 or Mono design) out of the 467 hand pumps known to be installed. Inspection teams visited 95 per cent of the motorised pumping installations and 65 per cent of the hand pumps in the region. It was not always possible to record all details of the motorised pumps as this would have required removing the pump from the borehole. The inspection reports identified which water supplies had problems with the mechanical pumping equipment and distribution system and which had problems with the borehole itself. On the mechanical side virtually all the motorised pumping installations and 75 per cent of the hand pumps in the region were found to require some rehabilitation involving maintenance, repair or replacement. As maintenance work progressed, approximately 70 per cent of the problems encountered with pumps installed in boreholes could be traced to incorrect installation or selection of equipment.

Prior to 1988 all the India Mk 2 hand pump heads installed in the region were of the chain link type. When these pumps were set at shallow depths, persistent problems were found with the chain breaking as there was insufficient weight of rods to return the piston without an upward pressure on the handle. To overcome this, users had adopted a rapid, short-stoke, pumping action which contributed to the early chain failure. The project began installing solid link India Mk 2 heads in 1988 and no problems were reported within the first year. Other persistent problems with hand pumps related to poor quality pipe materials and treading and poor well top designs. Often pump heads worked loose when they were cut-off short and welded directly to the borehole casing.

Direct motor driven and submersible pumps were found to suffer from a variety of problems but the most persistent were the oversizing of the pump capacity compared to the borehole yield and mismatching of generators and electric submersible pumps.

The benefit of careful installation was seen in a few cases where hand pumps had been in continuous service for over 7 years and motorised pumps over 10 years and in one case over 14 years without major repair.

Problems within the boreholes became evident when the project test pumping and equipping programme showed that 8 of the 24 new boreholes tested produced significant quantities of silt in the water, even when the pumping was at rates considerably lower than the maximum yield of the hole. In addition when removing motorised pumps from existing water supply boreholes, 12 pumps were found to be blocked or damaged by silt. Minor iron encrustation was found on a number of pumps removed from boreholes located in the intermontane valleys. In the Afar lowlands, two pumps were found damaged by aggressive groundwater. In one case, the size of the rock debris lodged in the pump intake indicated that the borehole casing must have been seriously holed by corrosion.

While a few of the hand pump cylinders and leathers showed wear associated with sand and silt in the borehole, the use and yield of the majority of the hand pumps installed was insufficient for a siltation problem , if it existed, to become noticeable.

4 Well Database Analysis

Initially it was considered that the paucity of useful groundwater and construction data in the EWWCA database would frustrate any

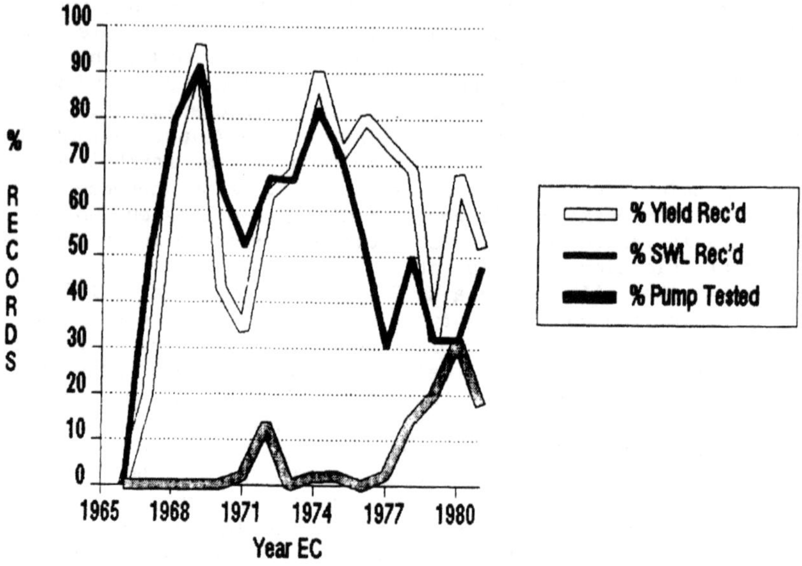

Fig.2. EWWCA Borehole Records, % Recorded Groundwater Data (EC - Ethiopia Calander, 1967 EC = 1975/76 AD)

objective assessment of the existing boreholes in the region. The very limited amount of groundwater data can be seen in Fig. 2 which shows the percentage of static water level (SWL), water struck levels, yield and pumping test data recorded anually between 1975 and 1988. Prior to 1986, pumping tests had been carried out in only 18 of the 540 boreholes drilled in the region. In addition, where borehole yields are recorded there is no indication as to whether these are the maximum yield of the borehole or of the pump used for the test.

Examination of the limited records in terms of borehole depths, static water levels and yields, however, provided information on potential pumping depths and yields. Figs. 3 and 4 show the resulting histograms, Fig 3 provides information on potential pumping heads and Fig 4 shows that 50 per cent of the reported static water levels are less than 30m below ground and the borehole yields are clustered around 2 to 3 l/sec. This information is somewhat limited but it provided sufficient basis for subjective conclusions to be made concerning pump operation and maintenance requirements in the region. These conclusions were:

Hand Pumps - about 50 per cent of the India Mk 2 hand pumps should be of the solid link type and that the average length of rising main and rods ordered for each pump should be 18m and not the customary 30m. (There were some 7 km of surplus India Mk 2 rods stored in Kombolcha from previous orders.)

Motorised Pumps - the percentage requirements for motorised pumps in terms of yield and pumping head within the borehole are given in Table 1.

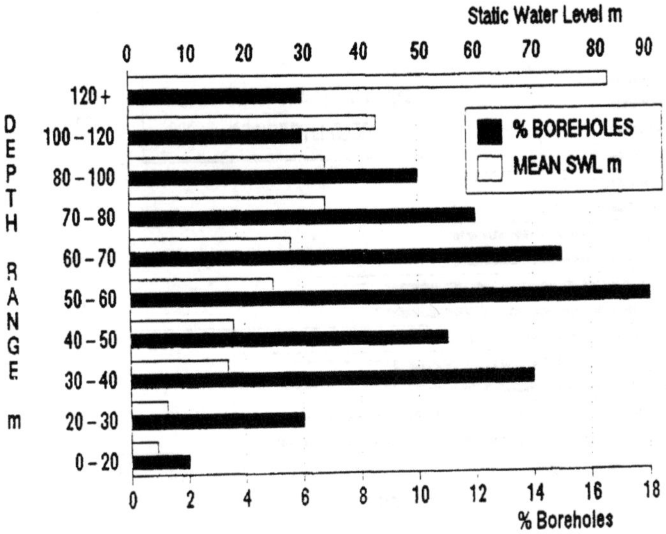

Fig. 3. EWWCA Borehole Records, Depths and Mean Static Water Levels.

Table 1. Welo Region, Borehole Pump Specifications

Yield Range (l/sec)	Percentage	Pumping Head (m)	Percentage
0.5 to 1.0	10	> 30	15
1.0 to 1.5	15	30 - 50	30
1.5 to 2.0	15	50 - 70	30
2.0 to 2.5	20	70 - 100	15
2.5 to 3.0	10	< 100	10
3.0 to 4.0	10		
4.0 to 5.0	10		
5.0 to 6.0	10		

Extracting the main statistics available from the EWWCA database (Table 2) enabled the problems with boreholes producing silt to be reviewed. These statistics showed open hole completions used in 202 holes out of the 411 successful boreholes constructed in the region between 1976 and 1988. Of the 202 open hole completions, 51 boreholes were equipped with motorised pumps, 93 with hand pumps and 57 were capped.

All 8 new boreholes which produced silt damage during the test pumping programme and 8 of the existing supply boreholes with silt-damaged pumps were found to be open hole completions. The remaining 4 silt damaged pumps were found to have capacities far in excess of the yield of borehole from which they were removed.

Unlike the lack of construction and groundwater data for the boreholes in the region, a considerable number of borehole geological logs were available. Examination of these showed that characteristic geological logs could be associated with each hydrogeological terrain.

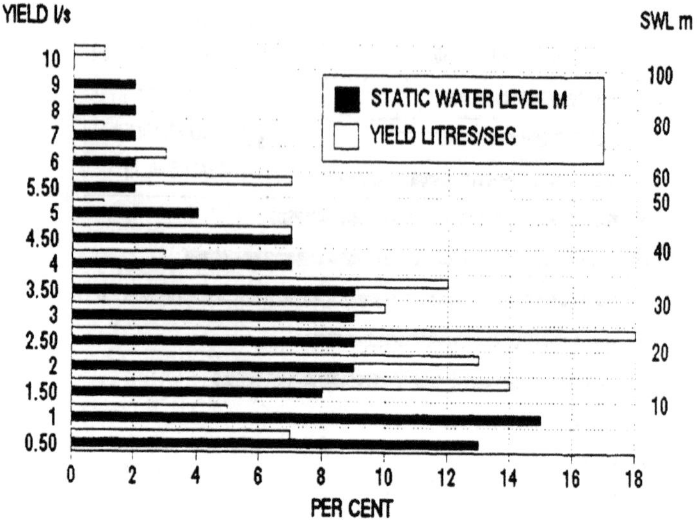

Fig. 4. EWWCA Borehole Records, Borehole Yields and Static Water Levels.

Table 2. Welo Region, Borehole Construction Statistics

Year EC.*	Total Drilled	Successful Boreholes	Open Hole Completion	Casing Shoe Depth <SWL	Casing Shoe Depth >SWL	Borehole Capped
1967	24	20	3	2	1	2
1968	20	6	5		5	
1969	24	22	9		9	
1970	37	26	10	1	9	1
1971	50	31	13	5	8	6
1972	52	38	22	3	19	1
1973	51	35	20	9	11	3
1974	51	45	14	6	8	7
1975	50	40	15	2	13	3
1976	51	45	32	15	17	5
1977	43	37	23	6	17	7
1978	42	30	16	8	8	6
1979	25	15	10	2	8	7
1980	25	21	10	9	1	9

(* Ethiopian Calendar, 1967 EC = 1975/76 AD)

On the plateau surfaces, the boreholes penetrated a thin soil or alluvial horizon which was underlain by a sequence of massive, fractured basalt lava flows which were separated with frequent palaeosoil and weathered horizons.

Few boreholes have been drilled in the plateau escarpments but the available geological logs show between 5 and 50 metres of soils and alluvium or colluvium overlying the same massive bedded basalts and palaeohorizons as on the plateau surfaces.

In the intermontane valleys floors, most boreholes are constructed in the alluvial and colluvial valley fill and rarely reach the bedrock.

In the Afar lowlands three main geological successions are recorded. Firstly there are thick sequences of Recent bedded basalts with scoriaceous interflow zones, secondly there are unconsolidated and semi-consolidated porous ash and clayey weathered ash sequences associated with late volcanic activity and thirdly, thicknesses of alluvial and lacustrine sediments associated with the course of the River Awash.

Of the 8 new and 8 existing boreholes with silt problems, 12 were located on the plateau surfaces, 1 in the intermontane valleys and 3 in the Afar lowlands. All 12 problem boreholes on the plateau surfaces were found to have very short lengths of surface casing set well above the static water level. The geological logs showed multiple soil and weathered horizons that had not been cased off. These would appear to be the source of the silt as no borehole which was cased to below the water table was found to have problems with silt. The record for the intermontane valley borehole showed that there was an unlined section at the bottom of the hole and that the pump suction was set in this unlined section. The 3 boreholes in the Afar lowlands suffered by being unlined through unstable formations which were basically only poorly permeable. Pumping, therefore, was producing excessive drawdowns which caused silt to enter the boreholes.

5 Borehole Rehabilitation Requirements

When reviewing borehole rehabilitation requirements and limiting work to holes producing silt, the project were aware that the field testing and surveys had covered only 40 per cent of potential problem sites. The options available for the 8 new and 8 existing boreholes demonstrated to have silt problems were limited either to firstly relining the boreholes with suitable casing and screening; or secondly lowering the capacity of the pump installed to a level where no movement of fines took place in the boreholes or finally redrilling and constructing an abstraction borehole appropriately designed for the geological conditions found at the site.

Adequate drilling equipment, casing and screening materials were available to undertake each of these options but the central allocation of funds was often initially at conflict with the most economic solution. Donor money budgeted for drilling of new boreholes at a given site was not necessarily readily transferable to the more economic rehabilitation of a borehole at the same site. Similarly, even though construction of a new borehole may have been cheaper than the cost of rehabilitating an existing borehole, inflexible funding allocation dictated that rehabilitation had to be attempted.

The project opted to fund the rehabilitation of these boreholes to prove the basic solutions involved. A typical site was at Wegel Tena, an administrative centre on the highland plateau. A water supply system had been implemented here in 1987 based on one production borehole equipped with an electric submersible pump. Persistent problems with the pump becoming blocked and damaged by silt had made the system nonoperational within 6 months of commissioning.

The construction records for the boreholes showed the holes to be uncased below short lengths (3m to 6m) of 10" steel surface casing. The static water levels were some 25 below ground and total depths were 74m, 118m and 123m. The latter hole was equipped with a 2.5 l/sec submersible pump after a reported pumping test. The pumping tests results however were not traceable and doubts existed as to whether the test had been carried out. The borehole geological logs showed the formations penetrated to be massive and jointed basalt lava flows separated by well-defined palaeosoil horizons. Three soil horizons were identified between the bottom of the casing and the static water level and these were considered to be the main source of the silt entering the hole. Additional information contributing to this conclusion was that a large number of holes constructed in similar geological sequences but plain cased to several metres below the static water level were not found to have siltation problems despite being pumped at higher rates. The project design for rehabilitating the well used 150mm plastic plain casing and 150mm diameter biwalled screening, prepacked with a graded sand filter medium. The internal diameter of this screening was less than 80mm and its use had been previously resisted as this was considered to be too small for a production well. The modified borehole was successfully pumped at 2.5 l/sec with less than 100 mgl of silt been produced with the water.

An example of inflexible planning and funding was the attempted rehabilitation of a borehole in the intermontane valley at Tisabalima. The borehole was drilled in 1987 and was 18.5 m deep. It was fitted with 6" diameter plain casing to 3.7m and slotted plastic screening to 13m. The hole was uncased from 13m to 18.5m through unconsolidated alluvial fill. When the hole was to be tested and pumping equipment installed in 1989, plumbing found it caved to 14m and to be extremely crooked from that depth to the surface. The EWWCA workover rig sent to undertake the work spent 18 days in pulling the casing and trying to straighten the hole. When consulted, the project considered rehabilitation of this hole impossible and stepped in with funds for the two days work and materials required to construct a new borehole at the site.

6 Pump Maintenance Programme

Prior to 1989, the majority of the rural pump maintenance in the Welo region was carried out on an emergency basis. Arrival of equipment and vehicles and the training of technicians enabled planned maintenance to begin along four defined routes totalling some 1,500 km.

WSSA's future planning for systematic and emergency maintenance of rural water supplies will require a decision on how the operation should be organised. Essentially there are two options:

(i) To establish a rural water supply maintenance section based in Kombolcha for the whole region that is clearly separating the urban and rural water supplies or,

(ii) Zoning the region based on the larger urban water supply offices where each urban water supply would have a number of satellite rural water supplies to look after.

The map, Fig 5 shows the existing maintenance routes in the region and the potential zoning option and satellite rural supplies. The map also indicates the distribution of motorised and hand pumps which reflects not only the population distribution but also road access and the areas where there are difficulties in getting adequate surface and shallow ground water supplies.

At present, WSSA's systematic maintenance work is developing along the lines of a separate section operating on all of the defined maintenance routes. Current practice for emergency maintenance is for a team to be specially mobilised from Kombolcha to do the repairs. In a short time however even working along defined routes there will be a need to open sub-regional rural maintenance facilities. Therefore given the shortage of trained manpower and the extreme logistical problems due to the size of the region, it is felt that the urban satellite system would prove the most efficient organisation for the future. The shorter lines of communication between the satellite rural areas and the urban centre would enable maintenance teams to include emergency repairs in their immediate programme. The urban water supply depots would keep stocks of spares and replacement units appropriate to their satellite zone and provide workshop facilities and technical staff as required.

WSSA's organisation plan for rural maintenance plan in the region envisaged 4 levels of input:

Level 1: Community Water Committee who are responsible for setting, collecting and controlling community water charges. Typically these are US$ 1.50 per household per month. They are also responsible for ensuring all water sources and points are protected from damage, are kept clean and well drained and in good order. WSSA employs community liaison officers who advise community water committees over the water charges and on the selection of pump attendants, as well as providing instruction on the general care of the water supply system and water related health matters.

Level 2: Pump Attendants who receive training in the day to day operation and maintenance of the equipment they are responsible for. Where responsible for motorised pumps they should carry out standard maintenance procedures and record fuel and oil consumption and hours of operation. Ideally they should also record water produced and water levels in boreholes. They should also be able to undertake simple maintenance tasks like replacing Vee belts and bleeding fuel lines. Where there are hand pumps they may be trained to undertake minor repairs depending on the type of pump installed. They should receive regular and reasonable payment for their services and should report all faults to the water committee who in turn should report to WSSA.

Level 3: Field Maintenance Teams responsible for preventative maintenance of equipment, correcting minor faults before they became major faults. The teams were instructed in carrying out frequent periodic inspections, to service engines, changing filters and oil

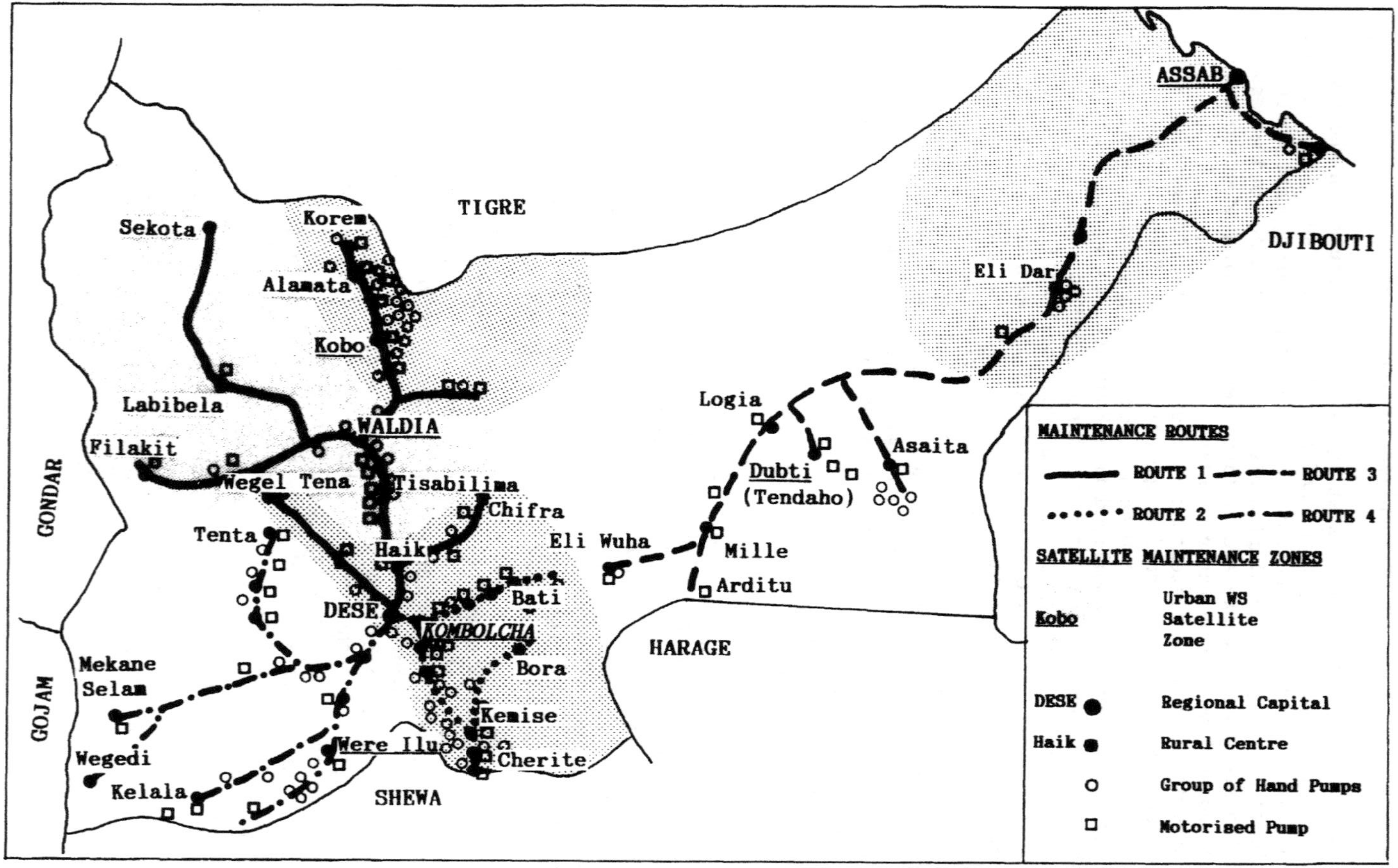

Fig. 5. Welo Region - Location, Maintenance Route and Satellite Zone Map

and to make any adjustments required. Where generating sets were installed the teams were only responsible for servicing and maintaining the engine. Generators were only checked for correct voltage, amperage and cycles. Engine revolutions were adjusted to those specified by the manufacturers. All electrical work was carried out by designated electricians. The teams completed minor running repairs in the field, but when major repairs were needed the defective equipment were removed for repair in a central workshop in Kombolcha and replacement units were fitted. To undertake this work three identical pump maintenance vehicles were fitted out. The main equipment provided included; a 3 tonne capacity, truck-mounted crane with a full 7.4m hook clearance for pulling and installing pumps and motors; 110 volt hammer drills, cutters and grinders for cutting casing, modifying well heads and concrete motor plinths and drilling new mounting bolt holes to secure replaced motors and pumps, pipe wrenches and threaders, a welding set and tool kits. The teams were expected to undertake full repair and replacement of all types of hand pump. Incidental tasks include care of well and borehole surrounds and aprons, fixing minor leaks and inspecting water fountains for problems. Full scale replacement of storage tanks and distribution systems is outside the role of the teams.

Level 4: The need was established for a permanent pump repair workshop in Kombolcha for the maintenance and repair of pumps, pump heads, power drives, generators nd electrical control panels. Construction of a 540 m^2 portal frame building in Kombolcha was started in 1989 and facilities included adequate secure and covered storage space and a pump testing tank with appropriate pipework.

The role of the project in these areas was largely directed at levels 3 and 4 and involved training workshop and field based mechanics and technicians in maintenance and repair of equipment; the fitting out of stores and instituting stores procedures; specifying, purchasing and equipping field mainentance vehicles; ensuring full inspection reports and details of repair work were recorded in the database and the use of this information in planning future maintenance programmes.

7 Summary

The Welo Well Rehabilitation Project was funded by a UK Overseas Development Administration grant of £1.5 million and was implemented by a four man techical cooperation team who were in Ethiopia for two years. This paper describing the teams work in rural water supply maintenance which covered about 50% of the total project activities. Other work involved the training of hydrogeologists, vehicle mechanics, drillers and supporting field technicians, the purchase and importation of £0.8 million of equipment and spares and the construction of the WSSA pump repair workshop in Kombolcha. While there are strong doubts over the long term impact of this particular project, there are no doubts over the need for considerably more donor attention and funds to be directed away from new development programmes to the rehabilitation and maintenance of existing water supplies.

36 THE MANAGEMENT OF IRRIGATION TUBEWELLS

I. SMOUT
Water, Engineering and Development Centre (WEDC),
Loughborough University of Technology, UK

Abstract
There is currently a great deal of interest world-wide in the scope for greater farmer involvement in irrigation schemes. The paper considers how this approach applies to irrigation tubewells, drawing particularly on the author's experience over several years in Indonesia, but also including material from Bangladesh and elsewhere. Following the general theme of the conference, but taking an irrigation perspective, observations are presented on the operation and maintenance of the pumpset and irrigation system, on government-initiated developments, and the possible turnover of these to water-user associations.

Compared with government canal schemes, irrigation tubewells are relatively small units which can be managed independently. However, problems can arise from the unfamiliarity of the technology and the complex problems of tubewell management. These include the need to raise considerable amounts of money to meet the costs of operation and maintenance, ensuring the availability of spare parts and trained mechanics, and the degree of cooperation and skill needed to avoid inefficiency, inequity, and breakdown.

These issues are examined, together with approaches to tackling them, such as training the officers of water-user associations, the collection of irrigation fees, and the scheduling of irrigation supplies. The discussion covers problems that can arise in practice, including some which arise from decisions made at the planning and design stages which set the framework for the eventual management and operation of the tubewell.
Keywords: Irrigation, Water-User Associations, Operation, Maintenance, Management

1 Introduction

One of the objectives of this paper is to provide an irrigation perspective at the Conference and focus on the pumpset and irrigation system, rather than the tubewell. This is particularly important when considering the rehabilitation of problem tubewells. The tubewell can be seen as the beginning of a process which ends with the production of irrigated crops. The efficiency of extraction of groundwater is important, and on some tubewell systems it is a critical constraint, but its benefits depend on the other parts of the process working in a complementary manner. In some situations, performance of the tubewell itself may not be a critical factor: for example, a survey of over 1000 deep tubewells in Bangladesh (Morton, 1989) showed that the tubewell test discharge had almost no influence on the area irrigated. Therefore in considering the rehabilitation needs of problem tubewells, it is necessary to look at the whole system and identify all the problems and the critical constraints, before focussing on a particular problem.

From an irrigation engineer's point of view, the major advantages of tubewell irrigation compared to gravity canal schemes are its small unit size, its flexibility in operation and its low capital cost. The major problems are the skills and equipment needed for installation, the difficulty of providing effective maintenance and avoiding breakdowns, the high recurrent costs, and the complexity of managing the groundwater resource.

Farmers see tubewells in a similar way, and many have invested in private tubewells where small units can be installed relatively easily and cheaply (open wells or shallow tubewells with centrifugal pumps or handpumps). Some governments have subsidised these (both through their nominal price, and through the credit system), but most direct government activity has been in the installation of larger systems with deep tubewells and turbine pumps. The operation and maintenance of these larger systems are the focus of this paper.

Operation of a system represents the direct link between the tubewell and the irrigated area with its crop production benefits. Maintenance is crucial for sustained utilisation of the tubewell, and requires effective management. Increasingly, governments are handing over these responsibilities to groups of farmers, often organised as water user associations (WUAs).

Tubewell irrigation command areas normally are less than about 50 ha which falls within the range of small-scale irrigation, and is an appropriate size for management by farmers. However some features of tubewell irrigation make it more demanding of efficient management than irrigation schemes from surface water: tubewell operation normally requires a trained operator and fuel, maintenance requires mechanical skills and spare parts, and timely provision of all of these requires management skills and funds. For small private tubewells these requirements are met by the tubewell owner and the local market. For larger tubewells installed by government, approaches vary. In some countries (Pakistan for example where the water discharges into canals which are part of the surface water irrigation system) the government is responsible for management of the larger tubewells. In others (for example Bangladesh and Indonesia), a WUA or cooperative is responsible for management of each tubewell and its irrigation scheme.

The key issues are discussed in turn: pumpset operation and maintenance (section 2), irrigation distribution (section 3), finance (section 4), public and private management (section 5) and water user associations (section 6).

2 Pumpset operation and maintenance

2.1 The importance of correct installation

It is widely understood that poor selection and installation of a pumpset causes problems and breakdown later, but many faults still go unnoticed in practice, and are often regarded as maintenance problems. Common examples are:

(a) Inefficiencies because of poor match of pump and engine power or speed.
(b) Pump running dry because set too high.
(c) Wear of impellers because of poor design of tubewell screen and gravel pack.
(d) Line shaft failures caused by poor, uneven connections.
(e) Excessive vibration caused by misalignment of pump drive and engine, resulting in operation at reduced speed or bearing failure.
(f) Pump damage caused by omitting to install a downward bend at the exit of the discharge pipe, with the result that children can throw stones into the pipe and down the tubewell, damaging the pump.
(g) Engine overheating caused by poor ventilation of the pump house.

2.2 The role of the operator

The operator normally works full-time on a tubewell scheme whereas other WUA officers are much less active. The operator can therefore have a strong influence on the use of the system, extending much further than basic mechanical operation and maintenance. This should be recognised in the training programmes for operators, which are very important. It should be remembered that if an operator is changed for any reason, it will be necessary to provide training for the replacement.

Each operator needs to be supplied with a clear operation and maintenance manual specific to the operator's tasks, and basic training in these. The importance of keeping an accurate tubewell log should be stressed, to provide a record of running hours, consumption of fuel and lubricants, and servicing.

The operator of a diesel engine may need to be guided on the optimum engine speed for maximum efficiency (assessed as fuel consumption per unit volume of water pumped). It is necessary to ensure that this speed does not cause excessive vibration, that the canals can convey the resulting discharge and that it is appropriate for the command area and irrigation interval. Other practical points are that diesel should be filtered before use, and heavy duty engine oil should be used for lubrication, rather than standard engine oil.

It is very important that operators of electrically-powered pumpsets pay attention to safety and to correct use of fuses and the voltage protection circuits.

Electric motors usually have lower running costs than diesel engines, and less maintenance requirements as long as they are operated at the correct voltage. Unfortunately, electricity supplies can be variable in many less developed countries, and protection circuits may be bypassed by operators who are frustrated at not being able to run the pumpset because of low voltage. However this frequently results in breakdown of the electrical equipment (Felton and Smout, 1989).

Diesel engines have the advantage of flexibility, which may be important to the farmers, for example by enabling the irrigation time to be reduced at peak periods by operating the engine at a higher speed.

2.3 Supervision of the operator

Scattered tubewells are difficult for government bureaucracies to manage effectively, and tubewell operators who are government employees can wield considerable local power, which may restrict the efficiency of operation. Johnson (1984) describes the problem vividly, referring to the Salinity Control and Reclamation Projects (SCARPs) in Pakistan:

> 'Tubewell operators comprise the largest number of staff working in SCARP circles. SCARP rules stipulate that the operator must be a local person, but he cannot come from the village or villages served by the well. In order to reduce the danger of misallocation, the rule restricts tubewell operators from working within a radius of 24 kilometres from their place of origin. Yet as the operator is always supposed to be present when the tubewell is in operation, normally 20 hours per day, this rule is clearly counterproductive. It forces the operator to cheat by leaving the system jammed open, thereby circumventing safey devices, or to turn off the tubewell even if water is scheduled. Usually farmers and operators work out some type of compromise which invariably costs farmers money...
>
> An additional problem is that tubewell operators are highly unionized and therefore difficult to punish or dismiss. It has also been suggested that another reason why officials find it difficult to control their subordinates is that substantial portion of the operators exactions are passed up the line. Whatever the truth of these allegations, it is clear that day-to-day operations of tubewells are very loosely, if at all, supervised. An absence of effective control over activities of field staff either by senior officials, by standardized cross-checking procedures, or by the farmers through some ability to reward or punish, all demonstrate that planners did not think out how tubewells were to be operated and maintained in practice.'

Similar problems in India have encouraged the spread of automatically controlled pumps with buried pipe distribution systems. According to Campbell (1984):

> 'In redesign of the tubewell system [in Uttar Pradesh] an effort has consequently been made to reduce the duties of the tubewell operator largely to recording daily performance of the well, and water use by individual cultivators. Starting and stopping of the tubewell has been made entirely automatic, actuated by operation of the outlet valves on the distribution system.

Such operation, virtually on demand, is possible only with a pipe system in which opening or closing of an outlet valve is automatically signalled back to the tubewell by a change in water-level in the regulating chamber, and the tubewell is started or stopped accordingly.'

This is an interesting example of technical innovation and investment to overcome the institutional problem of managing the operator effectively. Turning the system over to the farmers would be an alternative approach, so that the operator would be answerable to the WUA. Some external guidance may be needed to ensure that a reasonable payment system is developed for the operator, such as a seasonal honorarium plus an hourly payment at a suitable rate. Some of these issues are discussed further in sections 5 and 6 below.

2.4 Servicing and repair facilities

Provision must be made for the servicing and repair of the pumpset, or it will quickly break down and go out of operation. On government schemes particular attention must be paid to budgetary provision for servicing and repair, as the requirements are considerably higher per hectare than for maintenance of gravity irrigation schemes.

In many situations pumpset servicing and repair will have to be done through government, at least until the private sector gains familiarity with the equipment and access to supplies of spare parts. Private workshops may have advantages of greater incentive to repair pumpsets quickly, and pay flexibility to attract scarce skilled mechanics but quality control may be poor.

A systematic maintenance system needs to be established, setting out procedures for rectifying each type of fault, together with manufacturers' manuals to be followed for the detailed work. It will be important to differentiate between faults which should be corrected on site, and those which require the pumpset or faulty item to be removed to a properly eqipped workshop.

Stock control and reordering procedures are important for the smooth running of a workshop, whether government or private. If the tubewells are turned over to be managed by the farmers, it may also be necessary to set procedures for costing repairs and charging for these.

2.5 The role of the mechanic

To the farmers with a broken-down pump, it is crucial that the mechanic responds urgently and repairs the pumpset as quickly as possible. This requires well organised and motivated mechanics, with appropriate transport.

As well as the necessary manual skills, a basic requirement of a mechanic is skill in diagnosis. In particular, the mechanic needs to be able to identify and correct faults on site and without dismantling the pumpset. This saves resources and time, and also keeps the equipment in its factory-tuned condition, which is very difficult to re-create.

This skill and its importance may be neglected in training provided by manufacturers, which is often based on dismantling and reassembling an engine to familiarise the mechanics with the various components. If a conscious effort is not made to correct this, mechanics may then regard dismantling the engine as the usual procedure for dealing with problems, even though considerable resources are needed to remove an engine, take it apart, reassemble it and reinstall it on site, and many engines do not work as efficiently afterwards.

3 Irrigation distribution

The distribution of irrigation water depends both on the physical infrastructure of canalisation, and on the methods of scheduling and delivering water each day.

3.1 Canalisation

A key factor in the economic performance of an irrigation tubewell is the area which is actually irrigated. In many situations this partly depends on the extent and intensity of canalisation, and there is considerable debate in tubewell irrigation about appropriate distribution systems.

Distribution systems may be constructed as part of the project or left to the farmers, and they may comprise earth canal, lined canal, or buried pipe systems. The examples below illustrate different approaches and levels of investment adopted on different projects.

In Bangladesh, Mandal et al (1988) found that lined canals were generally not cost-effective compared with properly constructed unlined canals. This conclusion is followed on the Deep Tubewell Project II, where an earth canal system is designed for each tubewell, to deliver water to standard size outlets. This is intended to be constructed by the farmers, but few seem to have been completed to date. Many deep tubewells have poorly constructed earth canals which do not extend over the design command area.

On the Madura Groundwater Irrigation Project (MGIP) in Indonesia, a contractor builds a concrete-lined "tertiary" canal from each tubewell to supply seven 6 ha rotation blocks. Initially the project attempted to persuade farmers to construct earthen quaternary canals within each block to distribute the water, but their reluctance to do this led to the construction by contract of partially dry-lined quaternary canals. The lining provides a permanent canal system, serving the whole command area, but it may not improve irrigation efficiencies: Hyder and Ward (1989) found that about 50% of the water pumped from the tubewell was lost as seepage from these lined canals, mostly through poor joints between the concrete slabs and the bed.

A different approach has been adopted on various World Bank funded projects in India, where buried pvc pipe is used, delivering water to outlets which each serve 3 to 5 ha (Campbell, 1984). This system seems to offer considerable benefits of leaving the natural drainage unchanged, together with reduced land-take, more convenient operation, lower losses and increased command areas compared to canal systems. However it requires a higher investment per tubewell which may not be acceptable in some countries.

3.2 Schedules

A schedule is a timetable for the operation of the tubewell and the delivery of water to the outlets or to the fields of individual farmers. The schedule is intended to provide an efficient and equitable distribution of the water, in accordance with crop water requirements.

The day block rotation system is commonly the basis of design, with irrigation scheduled to be carried out on a particular block one day, and shifted to another block on the following day. In practice this does not seem to be followed for various reasons:

(a) irrigation often cannot be completed on the block in one day; this may be because actual operating hours are lower than designed or due to low efficiencies or high water requirements (eg for land preparation or because of long irrigation intervals); therefore irrigation is continued for "as long as it takes";
(b) the tubewell discharge may be divided between two blocks, to give lower discharges in each, which are easier for farmers to manage and require smaller distribution channels (less land-take);
(c) farmers may prefer to use a demand or indent based system, to obtain water at the times they require it - though this may be inefficient, especially if transit losses are high;
(d) farmers may develop their own schedule which takes account of local variations in cropping, soils etc.

Whatever method is used, it will be more efficient if adjacent fields are irrigated at the same time, to reduce transit losses etc in the canal, and this requires coordination of farmers and organisation of water distribution during the day. This may be done by a full-time water bailiff, preferably with assistance from block leaders (see section 6 below).

3.3 Maintenance

Canals and structures need to be maintained to prevent their condition deteriorating, leading to losses by seepage, overtopping of low banks or leakage at structures. Government however may not have funds available for this, and farmers are reluctant to do it if they think it is government's responsibility. A clear system is needed, to assign responsibilities and ensure that the necessary maintenance is done. The costs also need to be included in the charging system for water.

4 Finance

Tubewell irrigation cannot function without funds. Cash is needed to purchase fuel, lubricants and spares, and to pay the operator and servicing mechanic. Therefore collection of funds is a crucial part of tubewell management. This is commonly through an irrigation service fee, based on the area irrigated, or a water charge related to the volume of water supplied.

The costs listed above are largely dependent on the operating hours of the pump, so it is logical to set a water charge per hour of pumping. This is an easy system for budgetting, and it allows flexibility for new crops and crop mixes, so it may be particularly useful in the early years of tubewell operation, before experience has been developed. However, the system has high administration requirements, for example for accurate record keeping of the times that each cultivator receives water, or collection of the charge at each irrigation. It is easier to collect a standard irrigation service fee per hectare each season from each cultivator (perhaps with different rates for the various crops). This requires the calculation of the costs of tubewell operation and maintenance over the season, which is relatively easy in retrospect, but more difficult to predict in advance.

An alternative system is to levy both an irrigation service fee per season, and a pumping charge to be paid to the operator (for example for fuel and operator's services) at the time of irrigation. In some places the farmer supplies the fuel, but this is usually unsatisfactory because diesel available in a local market is likely to be poor quality and may damage the engine; it is also an expensive arrangement for the farmer, both because of the local price of small quantities of fuel, and because the resulting special arrangements to irrigate the individual farmer's holding lead to higher losses of water from the canal than rotational irrigation of a number of adjacent plots. It is important to ensure that the fee element includes sufficient funds for maintenance and spare parts.

Palmer-Jones and Mandal (1987) describe the irrigation service fee and "farmer's fuel" systems in use in Bangladesh, and also a share system, where the cultivator paid 25% of the gross output of the irrigated fields. This latter system appears to be based on the value of water to the cultivator rather than the cost of supplying it, and the authors show that tubewell management can make very high returns from this (for example a payback period of two years on the subsidised capital cost). They also show that agricultural production from the tubewell command area was significantly higher on tubewell schemes where the irrigation service fee was used, than on those with the farmers fuel or share systems. However the type of formal institution managing the tubewell made little difference.

Experience elsewhere suggests that the general type of system used may often be less important than the detailed methods of administering it: the setting of rates, collection of charges, storing of revenues etc. Weak WUAs may need support to develop effective procedures for these.

5 Comparison of public and private management

The comparative advantages and disadvantages of public and private development have been discussed by Carruthers and Stoner (1981). They list the following as favourable areas for considering public initiatives:

(a) for pioneer development, to demonstrate the potential of tubewells

(b) for poverty alleviation, though the authors argue that this is unlikely to be successful; this is consistent with recent reports

that in eastern India, public tubewells have been much less efficient than private wells, and though they are officially justified on equity grounds, small and marginal farmers rarely seem to have been the principal beneficiaries (Bottrall, 1989)
(c) for technological advantage, such as economies of scale
(d) for management system efficiency, for instance integrated management of surface water and groundwater
(e) to realise aid opportunities, which, the authors argue, favour public investment and capital intensive designs

It may be noted that these arguments mostly apply to the planning and installation phases of development, and management system efficiency is the only one which would require public management once the tubewell is commissioned. For example in Pakistan, O'Mara and Duloy (1984) estimated that conjunctive use had the potential to increase agricultural production by 17 to 20%, by diverting surface water to irrigate areas where the groundwater is saline. It appears however that little effort has been made on conjunctive use in Pakistan, and gravity supplies and tubewell supplies are managed independently by separate government "Circles", even though the water is distributed through the same tertiary canals. In general, conjunctive use faces the problem of how to set water charges when the sources have widely differing costs.

On private development, Carruthers and Stoner (1981) see the main advantages as

(a) high rate of development
(b) high operating efficiency achieved in practice
(c) use of low-cost local materials, enabling close linkages with local industry, and cheap innovations

In addition private tubewells may help to overcome the failures of public canal systems, for example the high water tables and low discharges at the tail of some canals.

Disadvantages are:

(a) possible over-investment (and Morton, 1989 argues that this has occurred in Bangladesh, encouraged by subsidies on the capital cost of tubewells)
(b) uncontrolled extraction of groundwater, leading to falling water tables and resource constraints.

Carruthers and Stoner make the following conclusion, which is directly related to management after commissioning:

'Farmers operate wells efficiently but cannot even conceive the problem of managing an aquifer. Public authorities can manage an aquifer but cannot operate wells efficiently for their best agricultural use...Clearly in the future, all wells will have to be licensed and the number and discharge restricted to avoid over-exploiting the resource. If this can be enforced, then there are clear advantages in allowing the farmer or farmer group to operate and integrate groundwater into the existing surface water system.'

Financial management of the tubewells is one area where government agencies often seem to have difficulty, and encounter particular problems both in setting a realistic water charge which covers their costs, and in achieving a reasonable rate of collection; receipts are usually considerably less than actual operation and maintenance costs. For example, the official fee for tubewell operation on a project in Pakistan was Rs 25 per ha per year when the estimated operation and maintenance costs were Rs 640 per ha per year (Lee et al, 1987).

Governments now seem to be under pressure, not least from aid agencies, to reduce this subsidy; given the political and administrative problems of increasing the revenue from charges, an attractive method is to turn the management of the tubewell over to a group of villagers. This group is usually a water-user association of all the farmers in the command area, but might otherwise be a restricted group of farmers (for example the KSS cooperative in Bangladesh) or a group of landless people who can then sell water to farmers who have land (for example the Proshika programme in Bangladesh).

Turnover usually takes place after the tubewell has been commissioned, and is intended to free government from the troublesome tasks of operating and maintaining scattered pumpsets and tubewells, and from the need to subsidise these due to the political and administrative difficulties of raising funds from the farmers. The government may retain ownership of the tubewell, which enables it to exert some influence over the farmers and in the last resort to remove the pumpset. Alternatively the government may sell the tubewell to the group through a credit system, (as in Bangladesh, where the KSS buys the tubewell from the government at a heavily subsidised price).

6 Water-user associations

6.1 WUA structure

A typical WUA structure (from Madura Groundwater Irrigation Project, Indonesia) is shown in Figure 1. The WUA is headed by a WUA leader (chairman), who supervises the activities of the other officers - secretary, treasurer, water bailiff and operator. The operator is employed by the project for the first two years, but thereafter the WUA is responsible for paying the operator's salary.

The WUA comprises all the households who farm in the command area. The area is divided into seven equal-sized rotation blocks, according to the position of the offtakes from the tertiary canal, and the farmers in each block elect a block leader from among themselves. The WUA leader (chairman), secretary, treasurer and water bailiff are elected by all the WUA members.

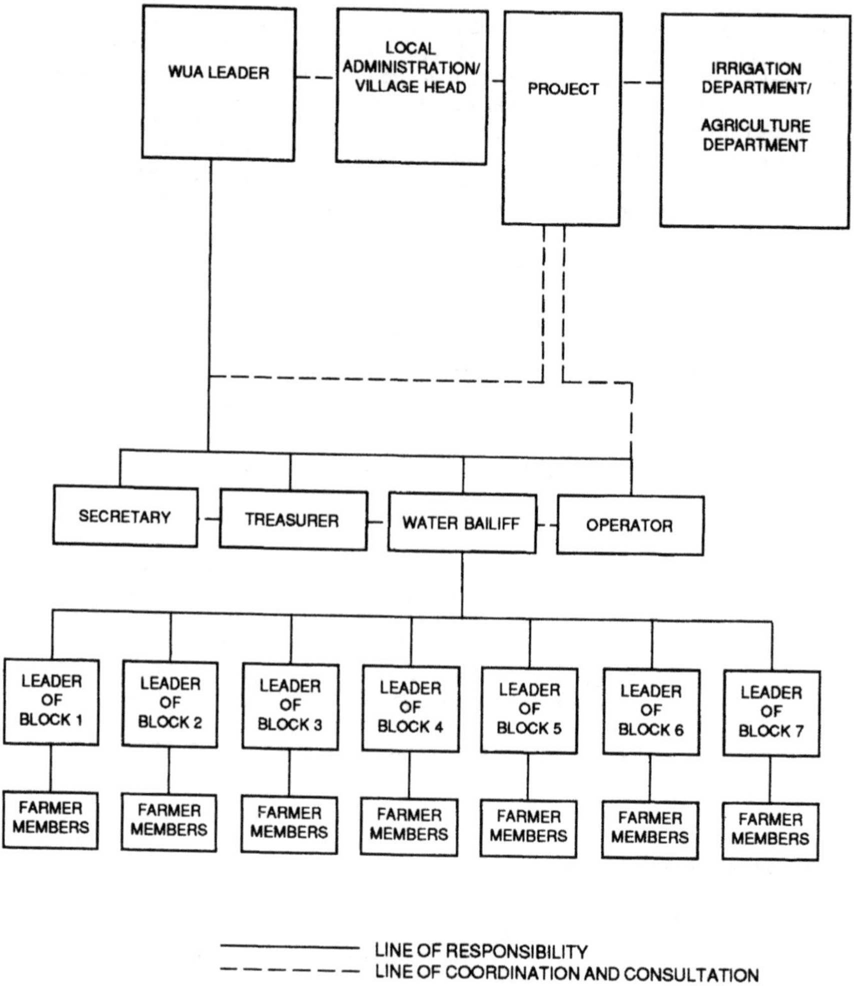

FIGURE 1: ORGANISATIONAL STRUCTURE OF A TYPICAL WATER USER ASSOCIATION (P2AT, 1987)

6.2 Responsibilities

Typical responsibilities of the WUA officers are listed below, based on MGIP (P2AT, 1987). The WUA meeting may decide to pay the officers for performing these duties.

The WUA leader (chairman) is responsible for the running of the WUA, and together with the secretary, liaises with the village head, the project, the government irrigation service and the government agricultural service.

The secretary is mainly responsible for administration, including purchase of fuel.

The treasurer keeps the cash book and operates the bank account, collects charges from the farmers with the help of the water bailiff and block leaders, and makes payments.

The water bailiff is responsible for coordinating requests and distribution of water, keeping necessary records of irrigation, and reporting any problems (including maintenance requirements) to the WUA leader.

Within each block, the block leader acts as a channel of communication between the WUA officers and the farmers, on matters such as timing of irrigation, charges and maintenance.

6.3 Finance

Tubewell management requires cash to purchase fuel, lubricants and spares, and to pay the operator and servicing mechanic.

On MGIP in Indonesia, the officers of each water-user association were trained to understand these costs, and to calculate hourly water charges or seasonal irrigation fees. Different approaches were adopted on different tubewells, according to local preference, but a common pattern was a seasonal fee system for tobacco which was irrigated daily, and an hourly charge for other crops.

The WUA treasurer usually collects the irrigation service fees or water charges in instalments during the season, according to the farmers' willingness to pay. The funds may be kept in a special bank account.

The treasurer is also responsible for keeping financial records and accounting for the funds. The methods of doing this need careful attention. They must be sufficiently simple to suit the villagers' skills and to be checked easily. The following procedures have been recommended on the Deep Tubewell Project in Bangladesh (MacDonald and Hunting, 1987):

(a) All transactions are to be recorded in a cash analysis book to separate items under different headings.
(b) Details of all receipts and expenditures are to be read out by the treasurer to a meeting of the irrigation cooperative (KSS) for approval.

6.4 Training of WUA officers

Villagers cannot be expected to take on the duties of WUA officers without training, because the roles involve formal responsibilities and procedures which will be new to them. MGIP provides an example of a training programme, using a mobile team of trainers with standard audio-visual material. Three days of training are held at each tubewell, for the 12 officers and the village head. The first is carried out at the time of commissioning, the second after one season's operation, and the third before the tubewell is handed over from the government to the WUA.

The basic organisation and the first two days' programme have been described by Smout (1986). These cover:

(a) WUA organisation and the officers' responsibilities
(b) financial aspects of managing the tubewell
(c) improving water management
(d) agriculture improvement, including increasing the area irrigated and the cropping intensity

The third day is described as a follow-up meeting, and is concerned with the following (GDC, 1989):

(a) WUA finances, including the level of charges
(b) physical changes in the system or its rehabilitation
(c) bolstering the WUA, including reforming the WUA and reappointing officers if necessary, and general preparation for handover.

In Bangladesh, similar training is carried out under the Irrigation Management Programme (IMP).

6.5 Shortcomings of WUAs

The following problems have been reported on some tubewells:

(a) officers using the WUA to exploit other farmers, for example by excessive charges which go into their pockets
(b) officers restricting the command area and refusing to supply some farmers
(c) passive officers allowing the tubewell to fall into disuse and disrepair

Government support in setting up the WUA and monitoring its performance can help to minimise these problems. To some extent they arise from local social factors, but these may be regulated by a formal framework for the WUA, including legal status.

As an example from India, Bottrall (1989) reports that evidence presented at a workshop on groundwater management suggested that wherever pump capacities and commands were relatively small and subject to competition or encroachment by others, WUAs were often subject to erosion of support and takeover by individuals or a few family members. This tendency might be reduced if the WUA received external support, especially at the start.

It would be advantageous to establish a system of monitoring WUAs from commissioning, using appropriate performance indicators, such as irrigated area, tubewell running hours, fuel use, water charge rates and financial balances. This needs some initial effort, to encourage the WUAs to maintain accurate records, and ensure that government staff do not have to spend a lot of time chasing these up.

7 Conclusions

By handing over a tubewell for management by a water-users association, a government can stop subsidising the operation of the tubewell, and overcome the difficulties of supervision. However the WUA will require training and support if it is to perform effectively, and in many circumstances the government will probably have to continue to assist with mechanical maintenance.

8 Acknowledgements

This paper draws on consultancy work carried out by the Mott MacDonald Group and funded by the Overseas Development Administration, UK. However the author's comments and conclusions are derived from consideration of a wide range of projects and publications.

9 References

Bottrall, A reported in Newsletter of **ODI/IIMI Irrigation Management Network**, Paper 89/1a

Campbell, D E (1984) Pipe distribution systems in groundwater development. **Agricultural Development**, vol 16, pp 209-227

Carruthers, I and Stoner, R (1981) **Economic aspects and policy issues in groundwater development**, World Bank Staff Working Paper No 496

Felton, M and Smout, I (1989) **Evaluation of the Tubewells Project, Pakistan**, ODA Evaluation Report EV 474

Groundwater Development Consultants (International) Ltd (1989) **Sustaining tubewell effectiveness**, Madura Groundwater Irrigation Project, Government of the Republic of Indonesia

Hyder, C and Ward, C (1989) Operation monitoring for water management, in **Irrigation Theory and Practice** (eds J R Rydzewski and C F Ward), Pentech Press, pp778-784

Johnson, S H III (1984) Large scale irrigation and drainage schemes in Pakistan: a study of rigidities in public decision making **Issues in the efficient use of surface and groundwater in irrigation** (Ed Gerald T O'Mara), World Bank Staff Working Paper No 707

Lee, P S Shaikh, A R and Youssef A N (1987) Left Bank Outfall Drain: integrated irrigation and drainage in Pakistan. **International Conference on Irrigation and Drainage** 13th Congress, Casablanca

MacDonald, Sir M & Partners Ltd and Hunting Technical Services Ltd (1987) **IMP administration and cooperative development,** Working Paper Nr 24, IDA Deep Tubewell Project II, Bangladesh Agricultural Development Corporation

MacDonald, Sir M & Partners Ltd and Pakistan Architects and Consulting Engineers (1987) **Operation, maintenance and monitoring review report,** South Rohri Fresh Groundwater Project, Pakistan Water and Power Development Authority

Mandal, M A S Dutta, S C Khair A and Biswas M R (1988) Feasibility of irrigation canal linings in Bangladesh. **Water Resources Development**, vol 4, no 3

Morton, J (1989) Tubewell irrigation in Bangladesh. **ODI/IIMI Irrigation Management Network,** Paper 89/2d

O'Mara, G T, and Duloy, J H (1984) Modelling efficient water allocation in a conjunctive use regime: the Indus basin of Pakistan. **Water Resources Research,** vol 20 no 11

Palmer-Jones, R W and Mandal, M A S (1987) Irrigation groups in Bangladesh. **ODI/IIMI Irrigation Management Network,** Paper 87/2c

P2AT (1987) **Tubewell irrigation manual, TW71, Pandanan,** Madura Groundwater Irrigation Project, Government of the Republic of Indonesia

Smout, I (1986) Training programmes for irrigation farmers. **ODI/IIMI Irrigation Management Network,** Paper 86/1e

INDEX

This index has been compiled using the keywords provided by the authors of the individual papers, with some editing and additions to ensure consistency. The numbers refer to the first page of the relevant paper.